# Cognitive Technologies

**Editor-in-Chief**

Daniel Sonntag, German Research Center for AI, DFKI, Saarbrücken, Germany

Titles in this series now included in the Thomson Reuters Book Citation Index and Scopus!

The Cognitive Technologies (CT) series is committed to the timely publishing of high-quality manuscripts that promote the development of cognitive technologies and systems on the basis of artificial intelligence, image processing and understanding, natural language processing, machine learning and human-computer interaction.

It brings together the latest developments in all areas of this multidisciplinary topic, ranging from theories and algorithms to various important applications. The intended readership includes research students and researchers in computer science, computer engineering, cognitive science, electrical engineering, data science and related fields seeking a convenient way to track the latest findings on the foundations, methodologies and key applications of cognitive technologies.

The series provides a publishing and communication platform for all cognitive technologies topics, including but not limited to these most recent examples:

- Interactive machine learning, interactive deep learning, machine teaching
- Explainability (XAI), transparency, robustness of AI and trustworthy AI
- Knowledge representation, automated reasoning, multiagent systems
- Common sense modelling, context-based interpretation, hybrid cognitive technologies
- Human-centered design, socio-technical systems, human-robot interaction, cognitive robotics
- Learning with small datasets, never-ending learning, metacognition and introspection
- Intelligent decision support systems, prediction systems and warning systems
- Special transfer topics such as CT for computational sustainability, CT in business applications and CT in mobile robotic systems

The series includes monographs, introductory and advanced textbooks, state-of-the-art collections, and handbooks. In addition, it supports publishing in Open Access mode.

Daniel Schulz • Christian Bauckhage
Editors

# Informed Machine Learning

 Springer

*Editors*

Daniel Schulz
Research Center Machine Learning
Fraunhofer Institute for Intelligent Analysis
and Information Systems IAIS
Sankt Augustin, Nordrhein-Westfalen,
Germany

Christian Bauckhage
Fraunhofer Institute for Intelligent Analysis
and Information Systems
Sankt Augustin, Nordrhein-Westfalen,
Germany

ISSN 1611-2482                         ISSN 2197-6635    (electronic)
Cognitive Technologies
ISBN 978-3-031-83096-9            ISBN 978-3-031-83097-6    (eBook)
https://doi.org/10.1007/978-3-031-83097-6

This work was supported by Fraunhofer "Center for Machine Learning" within the Fraunhofer "Cluster of Excellence Cognitive Internet Technologies".

This Springer imprint is published by the registered company Springer Nature Switzerland AG
The registered company address is: Gewerbestrasse 11, 6330 Cham, Switzerland

If disposing of this product, please recycle the paper.

# Preface

The past decade has seen substantial progress in the field of Artificial Intelligence (AI). This has primarily been due to the increasingly rapid developments in the field of machine learning (ML) which, in turn, benefited from the confluence of four technological trends: (1) availability of ever-increasing training data sets, (2) comparatively cheap high-performance computing hardware, (3) open source code sharing and access to software for model training or to pre-trained models, and (4) theoretical and practical progress in deep learning and artificial neural networks. As a consequence, there have been significant advancements, say, in natural language processing, image/speech recognition, or autonomous systems. As a result of these developments, AI has now made its way out of academic research into companies and our daily lives.

Already the resulting economic impact is enormous. Practitioners in every sector, from finance or medicine to logistics or administration, have begun using AI or are planning for its introduction. Seemingly not a day goes by without the media reporting on new AI applications and how these will transform economies and societies on a level comparable to the industrial revolution.

However, there still are considerable challenges when it comes to harnessing the full potential of AI in areas or domains outside of fully digitized industries.

A key feature of today's cutting-edge AI technologies is their hunger for resources. This is because modern ML models (deep neural networks) have become incredibly large and complex and involve millions if not billions of adjustable parameters. Their training therefore requires enormous amounts of data and considerable computing infrastructures and therefore energy. Alas, in many industries and application domains, data is still scarce or incomplete and there often is limited access to large-scale high performance computing facilities.

But even if data availability, compute resources, and energy costs are not an issue, model complexity may still pose challenges with respect to explainability, accountability, or trustworthiness of AI solutions which can be dire in settings where regulatory guidelines have to be met or safety guarantees must be ensured.

This is where the paradigm of Informed Machine Learning (Informed ML) comes into play.

In a nutshell, the idea of Informed ML is to systematically leverage additional prior knowledge for the design and training of data-driven AI models. The overall goal is to use reliable background knowledge in order to, on the one hand, reduce model complexity and the need for extensive training data and, on the other hand, increase interpretability and explainability of the decisions made by trained models.

There are of course various possibilities for how to inject what kind of additional knowledge into data-driven learning. It can consist of human expertise, scientific insights, or simple common sense facts, all of which may be represented in different forms, and these representations may enter the ML pipeline at various stages.

The contributions gathered in this volume illustrate the broad range of possibilities when working with different knowledge sources, representations, and integration strategies. They largely assume an application-oriented perspective and discuss working solutions for a wide range of industrial AI applications. We hope readers will find them interesting, get an appreciation for the many practical benefits of Informed ML, and find inspiration for their own work.

Sankt Augustin, Germany                                                    Daniel Schulz
August 2024                                                          Christian Bauckhage

# Contents

**3    AITwin: A Uniform Digital Twin Interface for Artificial
Intelligence Applications** . . . . . . . . . . . . . . . . . . . . . . . . . . . . . . . . . . . . . .    41
Alexander Diedrich, Christian Kühnert, Georg Maier, Joshua
Schraven, and Oliver Niggemann

**Part II    Optimization**

**4    A Regression-Based Predictive Model Hierarchy for
Nonwoven Tensile Strength Inference** . . . . . . . . . . . . . . . . . . . . . . . . .    63
Dario Antweiler, Jan Pablo Burgard, Marc Harmening, Nicole
Marheineke, Andre Schmeißer, Raimund Wegener, and Pascal Welke

# Chapter 1
# Introduction and Overview

Christian Bauckhage, Daniel Schulz, and Laura von Rueden

**Abstract** Informed Machine Learning (Informed ML) refers to the idea of injecting additional prior knowledge into data-driven learning systems. Such knowledge can be given in various forms such as scientific equations or logic rules which provide relevant information about a problem domain or task at hand. Integrating prior knowledge at various stages of the machine learning pipeline can help to improve generalization and trustworthiness. Specifically, Informed ML can help to train models when training data is scarce or to ensure conformity with regulations or safety demands.

In this introductory chapter, we briefly explain the concept of Informed ML, provide an overview of the chapters in this book, and categorize the contributed research and results with respect to a taxonomy of Informed ML.

## 1.1 Introduction to Informed Machine Learning

Over the past couple of years, Artificial Intelligence (AI) has finally found its way into the consciousness of a wider public and into the reporting of the mainstream media. On the one hand, this is not surprising as the capabilities of modern (generative) AI tools are astounding and will likely disrupt societies and economies. On the other hand, we said "finally" because the scientific discipline of AI has a long and venerable history which largely went unnoticed except by its practitioners, science fiction authors, and filmmakers. Yet, a brief look at this history can provide context and motivation for what this book on *Informed Machine Learning* is all about.

C. Bauckhage · D. Schulz (✉) · L. von Rueden
Fraunhofer IAIS, Sankt Augustin, Germany
e-mail: christian.bauckhage@iais.fraunhofer.de; daniel.schulz@iais.fraunhofer.de;
laura.von.rueden@iais.fraunhofer.de

© The Author(s) 2025                                                                                                  1
D. Schulz, C. Bauckhage (eds.), *Informed Machine Learning*,
Cognitive Technologies, https://doi.org/10.1007/978-3-031-83097-6_1

## *1.1.1   Historical Context and Motivation*

Ideas for computational intelligence date back to the 1940s when the first electronic computers became available, and people began researching how to equip these "electronic brains" with thinking capabilities. The roots of Machine Learning (ML) date back to this time, too: McCulloch and Pitts devised a mathematical model of neural computation in 1943, Turing coined the term Machine Learning in his 1948 work on "learning machinery", Hebb thought of associative learning in 1949, and Rosenblatt introduced perceptron learning in 1957. Despite this immediate appearance of the idea of (neural) learning systems, early AI research was dominated by methods based on symbolic logic and logical inference. In the 1950s, pioneers like Shannon, McCarthy, or Minsky thought about computer chess, logical programming languages, and automated theorem proving. The 1960s saw the emergence of knowledge-based systems, the development of a rule-based chatbot (Weizenbaum's ELIZA) and, importantly, the publications of a book by Minsky and Pappert which was largely read as a discouragement of further neural networks research. Indeed, AI research in the 1970s was dominated by work on rule- or knowledge-based expert systems and neurocomputing resurfaced only in the 1980s when the back-propagation algorithm was independently discovered several times and finally allowed for a consistent, data-driven training neural network models.

In the 1990s, there were thus two major paradigms: knowledge-based deduction which largely relied on hand-crafted rules for planning and decision making and example- or data-driven learning which often involved features engineered by experts and was mainly used for pattern recognition. Both approaches worked reasonably well at the time (albeit not well enough for the public to take notice) but seemed to be irreconcilable. Indeed, there were numerous issues pertaining to the problem of the semantic gap between observations (data) and symbolic representations (abstract concepts and their relations) and the question of whether learning-based systems can perform symbolic inference.

These issues remained unresolved until the late 2000s when the availability of massive amounts of data (on the Web), affordable high-performance computing (GPUs), and open-source libraries for neural network training kickstarted what has become known as the deep learning revolution. Ever since, the remarkable capabilities of large-scale end-to-end machine learning models across a wide range of domains, such as computer vision, speech recognition, text understanding, or gaming, have become common lore [4, 14, 19, 31, 34] and deep neural networks have begun to revolutionize engineering and the sciences [6, 7, 20].

These achievements are rooted in systematic big data analysis which allows learning algorithms to draw insights from- or identify pattern in billions of (input/output) examples. However, these achievements also come at a cost.

First, modern (foundation) models require massive amounts of data and compute resources for their training. These are not always available or at least not to everybody. Moreover, insufficient data can hinder the training of well-performing and generalizing models and miss out on constraints or easily explained facts such as those imposed by natural laws or regulatory guidelines, which are essential for ensuring trustworthy Artificial Intelligence (AI) [5].

Second, as machine learning models are becoming more and more complex, demands for explainability and trustworthiness are growing [29]. This poses a challenge for deep learning solutions as massive neural networks are essentially black boxes whose internal decision making processes involving several billions of adjustable parameters are largely intractable even to experts in the field.

This has spurred research into enhancing machine learning models through by means of hybrid approaches which integrating reliable prior knowledge into the data-driven learning process. While one could argue that such an integration of knowledge into learning is common through techniques such as example selection, data labelling, or feature engineering, hybrid learning is supposed to go beyond such measures and to incorporate more profound knowledge and formal representations.

For instance, researchers have explored the inclusion of logic rules [10, 40] and algebraic equations [18, 32] as a means of constraining loss functions. Another example are knowledge graphs which have been utilized to equip neural networks with information about relationships between instances, particularly relevant in image classification [17, 23]. Last but not least, physical simulations are now playing an increasingly important role in enriching training data [8, 21, 27].

To refer to these methods under a single umbrella term, the designation "Informed Machine Learning" (Informed ML) has been proposed [36]. This concept describes the systematic fusion of data-driven and knowledge-driven approaches and is gaining momentum as an avenue for further advancements in Artificial Intelligence.

## 1.1.2 Concept and Taxonomy

From a very abstract point of view, the main idea of Informed ML is to inject additional prior knowledge into data-driven learning as illustrated in Fig. 1.1.

Such prior knowledge is usually specific to the application context and task at hand. For example, fundamental, well established scientific- or medical knowledge can inform the modeling process for applications in the domains of material science or healthcare (e.g., Chaps. 8 and 9 of this volume). Basic knowledge like this often exists independently and in parallel to the practically gathered data samples a machine learning system uses for training and thus constitutes a valuable additional source of information.

Knowledge about an application context or domain is often available as formal representations like (logic) rule bases, equations describing insights in the natural sciences, or knowledge graphs. For example, in Chap. 8, scientific equations from material sciences are used and, in Chap. 9, a knowledge graph is used to improve healthcare analytics.

In Informed ML approaches, such formalized representations of prior knowledge are injected into the ML pipeline. In general, this can happen at various stages

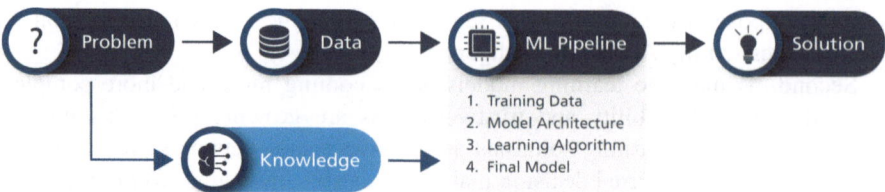

**Fig. 1.1** Schematic illustration of where data independent, prior knowledge can be integrated into the machine learning pipeline. Diagram adapted from [36]

of this pipeline. Knowledge can, for instance, inform the selection of training data, the design of the model architecture, the choice of learning algorithm, or the final model. Further dimensions for categorizing Informed ML approaches are the sources for- and the representations of additional knowledge. The former can be scientific facts and known laws of nature, general world knowledge about history, politics, economy, society and the like, or specific expert knowledge about, say, organizations, products, or markets. The latter typically comprise representations in form of algebraic equations, differential equations, logic rules, invariances, probabilistic relations, knowledge graphs, or simulations of real world phenomena.

Being conceptualized this broadly, Informed ML therefore includes more specific paradigms such as Neuro-Symbolic ML [11, 15, 16] or Neuro-Mechanistic ML [12, 25] which—as their names suggest—focus on hybrid modeling centered around neural networks. Indeed, the above three dimensions of knowledge source, knowledge representation, and knowledge integration are deliberately general and have been used to devise a taxonomy of the field of Informed ML [36]. It resulted from an extensive literature survey of more than 150 scientific reports on hybrid learning and allows for a more fine-grained categorization of how different solutions or frameworks integrate knowledge into various data-driven learning approaches. For instance, Table 1.1 shows how the different chapters of this book can be categorized with respect to this taxonomy.

Looking at this table, it becomes apparent that there exists a wide spectrum of combinations of knowledge sources and representations and stages where knowledge is integrated into the machine learning pipeline. This naturally goes hand in hand with a variety of system designs and processing modules which seems to be in stark contrast to modern end-to-end learning systems which may have different (neural) architectures but are fairly standardized when it comes to information processing and information flow. The obvious question is then what are the particular benefits that make it worthwhile to design and apply Informed ML systems?

### *1.1.3  Benefits*

One of the main benefits of Informed ML is that the use of additional knowledge about what is to be learned can allow for reducing the number adjustable parameters (degrees of freedom) of a machine learning model as well as for restricting the ranges of parameter values; in short, it can help to reduce model sizes and restrict search spaces.

This is of particular interest in practical settings where training data is scarce as the generalization capabilities of very large models typically correlate with the amount of data they have processed during training. At first sight, it seems peculiar to point to situations where data is scarce; after all, we are living in the age of (very) big data and modern foundation models are now being trained on data sets in the petabyte range. However, not all industries and organizations that want to enhance production and business with Artificial Intelligence have such massive data at their disposal. On the contrary, hardly any player outside of the IT sector has access to such vast amounts of data and not everybody can fine-tune available foundation models to their needs. Put differently, lack of data can prevent modern general purpose architectures from generalizing well and performing reliably. Problem specific informed architectures, on the other hand, may achieve these goals from training with substantially less data.

Similar aspects of where Informed ML may lead to improvements are of economical and environmental nature. While modern foundation models whose hundreds of billions of parameters are trained on vast amounts of (multi-modal) data are capable of remarkable feats, there are growing concerns as to the sustainability of this current paradigm. On the one hand, the energy demands for transformer-model training have reached levels which are difficult to justify in times of global warming [39]. On the other hand, there now are signs of diminishing returns of training ever ore complex models with ever growing data sets [33].

Again, an appropriate use of additional knowledge for tailoring learning systems to specific contexts or resources may lead to smaller models with reduced training efforts and thus reduced energy consumption. Moreover, it may even lead to novel training procedures or algorithms which could run on resource efficient hardware such as, say, FPGAs or reemerging analog computers. An example for the latter is found in Chap. 12 which proposes to train simple classifier by means of solving differential equations in a manner that could be implemented using energy efficient analog circuits.

Finally, there are the aspects of explainability, accountability, and trustworthiness of AI models and their alignment with human intention. We all have heard anecdotes of hallucinating large language models or of vision systems which recognize, say, trains because they learned and trains and railroad tracks go together and implicitly infer trains from the presence of tracks. Then there are reinforcement learning systems which were supposed to determine train schedules with minimal risk of accidents and concluded that the best way of avoiding train collisions is not to have trains riding at all. While these examples are silly, they illustrate the potential (or, more daringly, the "importance") of informed learning: fact checking against knowledge bases, carefully curated training data, or expertly formulated learning goals, i.e. the integration of knowledge at different stages of the ML pipeline, can circumvent issues like these.

It is obvious that AI solutions for real world applications in most industrial sectors must be reliable and their decisions must be in line with regulatory guidelines and the kind of reasoning that is explainable to- or interpretable by human experts. Modern end-to-end deep learning poses challenges in these regards. Decisions made by purely data-driven models with (hundreds of) billions of parameters are typically opaque and hardly ever tractable and can lead to unintended results in down-stream processing. This, in turn, may cause accidents or costly mistakes or may even prevent the use of learning-based AI in scenarios where there are legal requirements with respect to the transparency of decision making processes. Informed ML with knowledge-driven models (as in Chap. 4) or knowledge-based data augmentation (as in Chap. 10) can circumvent such shortcomings.

## 1.2   Overview

Above, we emphasized that Informed Machine Learning approaches are typically tailored to specific contexts or problem domains so that there exists a plethora of knowledge-integration methods. The Informed Machine Learning taxonomy in [36] systematically structures the vast landscape of hybrid techniques which integrate data- and knowledge-driven models using the broad categories of knowledge source, knowledge representation, and knowledge integration. These, in turn, are further refined into fifteen subcategories (see Table 1.1) so that the taxonomy covers a wide spectrum of combinations of knowledge sources, representations, and integration strategies. The contributions gathered in this volume emphatically illustrate the variety of possibilities. They report applied- and basic research on Informed Machine Learning and account for various methodologies and the kinds of results they allow for. In the following, we provide a short overview over the chapters of this book and classify their contributions according to the Informed Machine Learning taxonomy as summarized in Table 1.1. Furthermore, we sorted and arranged all chapters after the methods they rely on respectively their area of application. This results into four parts, namely "Digital Twins", "Optimization", "Neural Networks" and "Hybrid Methods".

**Part I: Digital Twins**
In Chap. 2, Wallner et al. [37] are concerned with energy optimal climate control (cooling) for data centers, industrial plants, or office buildings. They describe how to generate data-driven digital twins for cooling systems which can predicting the effects of adjusting control parameters and, when combined with monitoring capabilities, allow operators to make informed decisions for adjustments. Their

**Table 1.1** Overview of book parts and chapters. Each chapter employs a different Informed ML strategy. We categorize them with respect to the taxonomy in [36], which considers knowledge sources, knowledge representation, and stages where knowledge is integrated into the ML pipeline

| | | Informed ML approach | | | | | | | | | | | | | | |
|---|---|---|---|---|---|---|---|---|---|---|---|---|---|---|---|---|
| | | Source | | | Representation | | | | | | | Integration | | | | |
| Part | Chapter | Scientific knowledge | World knowledge | Expert knowledge | Algebraic equations | Differential equations | Simulation results | Spatial invariances | Logic rules | Knowledge graphs | Probabilistic relations | Human feedback | Training data | Hypothesis set | Learning algorithm | Final hypothesis |
| Digital Twins | 2 Optimizing Cooling System Operations with Informed ML and a Digital Twin | | | ✓ | | | | | ✓ | | | ✓ | ✓ | | | |
| | 3 AITwin - A Uniform Digital Twin Interface for Artificial Intelligence Applications | ✓ | | ✓ | | | | | ✓ | ✓ | | | | | ✓ | |
| Optimization | 4 A Regression-Based Predictive Model Hierarchy for Nonwoven Tensile Inference | ✓ | | ✓ | ✓ | | ✓ | | | | | | ✓ | ✓ | | |
| | 5 Machine Learning for Optimizing the Homogeneity of Spunbond Nonwovens | ✓ | | ✓ | | | ✓ | | | | | ✓ | ✓ | | | |
| | 6 Bayesian Inference for Fatigue Strength Estimation | | | ✓ | | | | | | | ✓ | | | ✓ | | |
| | 7 Incorporating Shape Knowledge into Regression Models | ✓ | | ✓ | ✓ | | | | | | | | | | ✓ | |

(continued)

**Table 1.1** (continued)

| Part | Chapter | Source: Scientific knowledge | World knowledge | Expert knowledge | Representation: Algebraic equations | Differential equations | Simulation results | Spatial invariances | Logic rules | Knowledge graphs | Probabilistic relations | Integration: Human feedback | Training data | Hypothesis set | Learning algorithm | Final hypothesis |
|---|---|---|---|---|---|---|---|---|---|---|---|---|---|---|---|---|
| Neural Networks | 8 Predicting Properties of Oxide Glasses Using Informed Neural Networks | ✓ | | ✓ | ✓ | | | ✓ | | ✓ | | | ✓ | ✓ | | ✓ |
| | 9 Graph Neural Networks for Predicting Side Effects and Indications of Drugs Using… | ✓ | | | | | | | | ✓ | | | ✓ | | ✓ | |
| | 10 On the Interplay of Subset Selection and Informed Graph Neural Networks | ✓ | | | ✓ | | ✓ | ✓ | | | | | ✓ | ✓ | | |
| | 11 Informed Machine Learning Aspects for the Multi-Agent Neural Rewriter | | | ✓ | ✓ | | | | | | | | | | ✓ | |
| Hybrid Methods | 12 Training Support Vector Machines by Solving Differential Equations | | | | | ✓ | | | | | | | | | ✓ | |
| | 13 Informed Machine Learning to Maximize Robustness and Computational Performance… | | | ✓ | ✓ | | ✓ | | | | | ✓ | | ✓ | ✓ | |
| | 14 Anomaly Detection in Multi-variate Time Series Using Uncertainty Estimation | ✓ | | ✓ | | | | | | | ✓ | | | ✓ | ✓ | |

solution involves the formalization of expert knowledge, the use of rules and human feedback, and simulated training data.

In Chap. 3, Diederich et al. [9] observe that cyber-physical systems with AI and/or earning components rely on a virtual representation of the underlying real physical system but that different practitioners may opt for different virtual representations. They therefore argue for standardized digital twins and present a corresponding API which takes into account different levels of modeling granularity informed by expert knowledge. Case studies with simulated and real-world examples from industrial process and manufacturing demonstrate the potential of this digital twin framework for AI-based solutions in industry.

**Part II: Optimization**
In Chap. 4, Antweiler et al. [1] focus on a prediction task whose application can be found in practical materials science. In particular, they address the problem of in the context of simulation-based design of non-woven textiles which, to this day, still requires considerable compute resources. They propose a predictive model hierarchy for inferring non-woven tensile strengths behavior which leverages linear or polynomial regression models whose predictions are interpretable to human experts. They find that scientific- and expert knowledge encoded in equations and simulation modules and integrated into the data acquisition and hypothesis class selection stages of the ML pipeline allows for significant speedup over conventional simulations while achieving very reliable results.

Victor et al. [35] report another industrial application of Informed ML from the same sector as in Chap. 4. Their contribution in Chap. 5 describes a learning-based optimization workflow in textile production that improves the homogeneity of spunbond nonwoven products. Their solution involves general scientific- and specific expert knowledge which allows for informed simulations and, consequently, informed training data acquisition.

Weichert et al. [38] present further work on predicting material properties in Chap. 6. They are concerned with the long life fatigue strength of metals which is costly to determine by means of experimental measurements. They therefore propose a ready-to-use experimental and analysis procedure which involves probabilistic learning methods. Their system connects expert knowledge about material behaviors and test setups with historical and newly generated data and achieves the same precision as standard experimental procedures albeit at considerably lower costs.

In a more theory oriented contribution in Chap. 7, Poursanidis et al. [28] show that shape knowledge such as monotonicity or convexity of functions can compensate for insufficient training data. They consider the training of shape-constrained regression models and propose an adaptive feasible-point algorithm which guarantees optimality up to arbitrary precision while being faithful to the constraints. In other words, their work incorporates scientific- and expert knowledge encoded in equations which informs the learning algorithm. Experimental evaluations with respect to manufacturing applications with scare training data show that this leads to better generalizing and better performing models than purely data driven approaches.

**Part III: Neural Networks**
Chapter 8 was contributed by Maier et al. [22] who also consider a practical application in material science. In particular, they observe that Machine Learning of the composition-property relationship of glasses promises to save on expensive trial-and-error approaches in the design stage. They further observe that, despite their considerable sizes, existing datasets on the composition of glasses and their properties only cover only a minuscule fraction of the space of all possible glass compositions. They therefore propose a neural network model which incorporates prior scientific and expert knowledge in various representations into various stages of the learning pipeline. Extensive empirical evaluations show that it achieves results which are consistently better than those of corresponding uninformed models.

In Chap. 9, Sharma et al. [30] are concerned with AI assisted drug development and note that existing learning-based solutions for drug repositioning and side effect prediction large rely on pre-clinical data not really representative for of the real-world situation faced by patients. They therefore work with knowledge graphs based on diagnoses, prescriptions and diagnostic procedures found in repositories of health records and related scientific databases and show that graph neural networks based on such representations allow for an accurate and interpretable prediction of indications and side effects which underlines the potential of Informed ML in healthcare.

In Chap. 10, Breustedt et al. [3] address the problem of predicting properties of molecules and observe that learning with large sets of chemical compound data is still very much limited due to lacking computational resources and missing label information. They therefore discuss the use of domain knowledge in form of equations, simulations, and invariances when training models for predicting atomic energies. Their approach allows them to maximize molecular diversity in training data selection and they introduce a model-agnostic explainer component for graph neural networks based on the rate-distortion explanation framework.

In Chap. 11, Paul et al. [26] use expert knowledge to inform team-optimal policy learning in an agent system for multi-vehicle routing. Their multi-agent reinforcement learning algorithm extends single-vehicle routing solutions through a neural rule rewriter which iteratively rewrites possible solutions and, in the end, enables agents to act and interact in a parallel and conflict-free manner.

**Part IV: Hybrid Methods**
In Chap. 12, Bauckhage and Sifa [2] turn the idea of Physics Informed Machine Learning on its head and consider classifier training by means of differential equation solving. They thus approach one of the most fundamental machine learning problems from a direction which has the potential for implementations on energy efficient hardware (analog computers) and thus hint at how Informed ML may addresses sustainability concerns.

In Chap. 13, Gries [13] focuses on learning algorithms, too. He shows how to apply methods of evolutionary- and surrogate learning for optimizing the numerous fine-grained control parameters of state-of-the-art linear solvers for numerical simulations. This way, he leverages expert knowledge provided in terms of equations, simulations, and feedback and integrates it into hypothesis formulation and learning algorithms. Practical benefits for complex simulations are demonstrated for industrial use cases ranging from fluid dynamics and geological simulations towards structural mechanics and battery aging predictions.

Finally, in Chap. 14, Müller et al. [24] are concerned with anomaly detection in multivariate times series such as occurring in machine maintenance scenarios. Their idea is to incorporates multivariate uncertainties quantified by a Bayesian neural network and expert knowledge in the form of probabilistic relations into a novel anomaly score. Their approach thus integrates scientific- and expert knowledge and considers probabilistic relations during learning. Experimental verification shows that scores separate into normal and anomalous regions when they exploit probabilistic relations between multivariate features and comparisons to recent state-of-the art approaches reveal competitive performance.

## 1.3 Summary

Artificial Intelligence has made considerable progress over the past couple of years. To a large extent, this progress was due to end-to-end machine learning models (deep neural networks) of ever increasing complexity and size whose training requires ever more humongous datasets. This poses considerable challenges for practitioners in industry who want to benefit from these developments.

In many industries and practical settings, data are still scarce, inhomogeneous, or incomplete and compute infrastructures are not always available to the extent needed for training modern foundation models. Moreover, even if data availability, computing resources, and energy costs are of little concern, explainability, accountability, and trustworthiness of large black-box models are often reason for concern.

Informed ML is an attempt to alleviate these kind of practical challenges. The main idea is to inject additional prior knowledge into data-driven learning because this can reduce model complexity and need for extensive training data and, just as well, increase interpretability and explainability of model computations.

Nowadays, there are therefore increased efforts on combining modern models and learning algorithms with suitable representations of problem specific prior knowledge and an attempt has been made to systematically categorize what kinds of prior knowledge and representations thereof to integrate where in the machine learning pipeline to achieve well working solutions across a wide range of practical application scenarios [36].

The contributions gathered in this volume illustrate the broad range of possibilities when working with different knowledge sources, representations, and integration strategies. They are largely application oriented, propose workable solutions to industrial problems such as material property prediction, anomaly detection, routing, or digital twin based process control and, all in all, demonstrate various practical benefits of Informed ML.

# References

1. Antweiler, D., Burgard, J., Harmening, M., Marheineke, N., Schmeißer, A., Wegener, R., Welke, P.: A Regression-based Predictive Model Hierarchy for Nonwoven Tensile Strength Inference. In: D. Schulz, C. Bauckhage (eds.) Informed Machine Learning. Springer (2024)
2. Bauckhage, C., Sifa, R.: Training Support Vector Machines by Solving Differential Equations. In: D. Schulz, C. Bauckhage (eds.) Informed Machine Learning. Springer (2024)
3. Breustedt, N., Climaco, P., Garcke, J., Hamaekers, J., Kutyniok, G., Lorenz, D., Oerder, R., Shukla, C.: On the Interplay of Subset Selection and Informed Graph Neural Networks. In: D. Schulz, C. Bauckhage (eds.) Informed Machine Learning. Springer (2024)
4. Brown, T., et al.: Language Models Are Few-Shot Learners. In: Proc. Neural Information Processing Systems (2020)
5. Brundage, M., et al.: Toward Trustworthy AI Development: Mechanisms for Supporting Verifiable Claims. arXiv preprint arXiv:2004.07213 (2020)
6. Butler, K.T., Davies, D.W., Cartwright, H., Isayev, O., Walsh, A.: Machine Learning for Molecular and Materials Science. Nature **559**(7715), 547–555 (2018)
7. Ching, T., Himmelstein, D.S., Beaulieu-Jones, B.K., Kalinin, A.A., Do, B.T., Way, G.P., Ferrero, E., Agapow, P.M., Zietz, M., Hoffman, M.M.: Opportunities and Obstacles for Deep Learning in Biology and Medicine. Journal of The Royal Society Interface **15**(141), 20170387 (2018)
8. Cully, A., Clune, J., Tarapore, D., Mouret, J.B.: Robots that can adapt like animals. Nature **521**(7553), 503–507 (2015)
9. Diedrich, A., Kühnert, C., Maier, G., Schraven, J., Niggemann, O.: AITwin – A Uniform Digital Twin Interface for Artificial Intelligence Applications. In: D. Schulz, C. Bauckhage (eds.) Informed Machine Learning. Springer (2024)
10. Diligenti, M., Roychowdhury, S., Gori, M.: Integrating Prior Knowledge into Deep Learning. In: Proc. Int. Conf. on Machine Learning and Applications (2017)
11. Dong, T., Bauckhage, C., Jin, H., Li, J., Cremers, O., Speicher, D., Cremers, A., Zimmermann, J.: Imposing Category Trees Onto Word-Embeddings Using A Geometric Construction. In: Proc. Int. Conf. on Learning Representations (2019)
12. Faure, L., Mollet, B., Liebermeister, W., Faulon, J.: A Neural-mechanistic Hybrid Approach Improving the Predictive Power of Genome-scale Metabolic Models. Nature Communications **14**, 4669 (2023)
13. Gries, S.: Informed Machine Learning to Maximize Robustness and Computational Performance of Linear Solvers. In: D. Schulz, C. Bauckhage (eds.) Informed Machine Learning. Springer (2024)
14. Hinton, G., et al.: Deep neural networks for acoustic modeling in speech recognition: The shared views of four research groups. IEEE Signal Processing Magazine **29**(6), 82–97 (2012)
15. Hitzler, P., Eberhart, A., Ebrahimi, M., Kamruzzaman Sarker, M., Zhou, L.: Neuro-symbolic Approaches in Artificial Intelligence. National Science Review **9**(6), nwac035 (2022)
16. Hochreiter, S.: Toward a Broad AI. Communications of the ACM **65**(4), 56–57 (2022)
17. Jiang, C., Xu, H., Liang, X., Lin, L.: Hybrid Knowledge Routed Modules for Large-scale Object Detection. In: Proc. Neural Information Processing Systems (2018)
18. Karpatne, A., Watkins, W., Read, J., Kumar, V.: Physics-guided Neural Networks (PGNN): An Application in Lake Temperature Modeling. arXiv preprint arXiv:1710.11431 (2017)

19. Krizhevsky, A., Sutskever, I., Hinton, G.E.: Imagenet Classification with Deep Convolutional Neural Networks. In: Proc. Neural Information Processing Systems (2012)
20. Kutz, J.N.: Deep Learning in Fluid Dynamics. Journal of Fluid Mechanics **814**, 1–4 (2017)
21. Lee, K.H., Ros, G., Li, J., Gaidon, A.: Spigan: Privileged Adversarial Learning from Simulation. arXiv preprint arXiv:1810.03756 (2018)
22. Maier, G., Hamaekers, J., Martilotti, D.S., Ziebarth, B.: Predicting Properties of Oxide Glasses Using Informed Neural Networks. In: D. Schulz, C. Bauckhage (eds.) Informed Machine Learning. Springer (2024)
23. Marino, K., Salakhutdinov, R., Gupta, A.: The More You Know: Using Knowledge Graphs for Image Classification. arXiv preprint arXiv:1612.04844 (2016)
24. Müller, M., Ernis, G., Mock, M.: Anomaly Detection in Multivariate Time Series Using Uncertainty Estimation. In: D. Schulz, C. Bauckhage (eds.) Informed Machine Learning. Springer (2024)
25. Ororbia, A., Kifer, D.: The Neural Coding Framework for Learning Generative Models. Nature Communications **13**, 2064 (2022)
26. Paul, N., T.Wirtz, Wrobel, S., Kister, A.: Informed Machine Learning Aspects for the Multi-Agent Neural Rewriter. In: D. Schulz, C. Bauckhage (eds.) Informed Machine Learning. Springer (2024)
27. Pfrommer, J., Zimmerling, C., Liu, J., Kärger, L., Henning, F., Beyerer, J.: Optimisation of Manufacturing Process Parameters Using Deep Neural Networks as Surrogate Models. Procedia CiRP **72**, 426–431 (2018)
28. Poursanidis, M., Link, P., Schmid, J., Teicher, U.: Incorporating Shape Knowledge into Regression Models. In: D. Schulz, C. Bauckhage (eds.) Informed Machine Learning. Springer (2024)
29. Roscher, R., Bohn, B., Duarte, M.F., Garcke, J.: Explainable Machine Learning for Scientific Insights and Discoveries. IEEE Access **8**, 42200–42216 (2020)
30. Sharma, J., Lentzen, M., Krix, S., Linden, T., Madan, S., Tran, V., Fröhlich, H.: Knowledge Informed Machine Learning in Healthcare: Graph Neural Networks for Predicting Side Effects and New Indications of Drugs Using Electronic Health Records. In: D. Schulz, C. Bauckhage (eds.) Informed Machine Learning. Springer (2024)
31. Silver, D., et al.: Mastering the Game of Go with Deep Neural Networks and Tree Search. Nature **529**(7587), 484–489 (2016)
32. Stewart, R., Ermon, S.: Label-free Supervision of Neural Networks with Physics and Domain Knowledge. In: Proc. AAAI Conf. on Artificial Intelligence (2017)
33. Thompson, N.C., Greenewald, K., Lee, K., Manso, G.F.: The Computational Limits of Deep Learning. arXiv preprint arXiv:2007.05558 (2022)
34. Vaswani, A., Shazeer, N., Parmar, N., Uszkoreit, J., Jones, L., Gomez, A., Kaiser, L., Polosukhin, I.: Attention Is All You Need. In: Proc. Neural Information Processing Systems (2017)
35. Victor, V., Schmeißer, A., Leitte, H., Gramsch, S.: Machine Learning-based Optimization Workflow of the Homogeneity of Spunbond Nonwovens with Human Validation. In: D. Schulz, C. Bauckhage (eds.) Informed Machine Learning. Springer (2024)
36. von Rueden, L., et al.: Informed Machine Learning–A Taxonomy and Survey of Integrating Prior Knowledge into Learning Systems. IEEE Transactions on Knowledge and Data Engineering **35**(1), 614–633 (2023)
37. Wallner, S., Bernard, T., Kühnert, C.: Optimizing Cooling System Operations with Informed ML and a Digital Twin. In: D. Schulz, C. Bauckhage (eds.) Informed Machine Learning. Springer (2024)
38. Weichert, D., Haedecke, E., Ernis, G., Houben, S., Kister, A., Wrobel, S.: Bayesian Inference for Fatigue Strength Estimation. In: D. Schulz, C. Bauckhage (eds.) Informed Machine Learning. Springer (2024)

39. Wright, D., Igel, C., Samuel, G.: Efficiency is Not Enough: A Critical Perspective of Environmentally Sustainable AI. arXiv preprint arXiv:2309.02065 (2023)
40. Xu, J., Zhang, Z., Friedman, T., Liang, Y., Broeck, G.: A Semantic Loss Function for Deep Learning with Symbolic Knowledge. In: Proc. Int. Conf. on Machine Learning (2018)

# Part I
# Digital Twins

# Chapter 2
# Optimizing Cooling System Operations with Informed ML and a Digital Twin

**Steffen Wallner, Thomas Bernard, and Christian Kühnert**

**Abstract** Today, there are a variety of cooling systems available to serve smaller data centers, industrial plants or office buildings. These are often one-off installations, and in most cases their control parameters are set at the time of installation and are not changed subsequently. In addition, these parameters are set conservatively and are not designed for energy-optimized operation. A digital twin of the plant, including a simulation model, is essential to bring the cooling system closer to energy-optimized operation over its lifetime, but this is not usually the case. One of the main reasons is that digital twins are expensive and time-consuming to create. However, today's cooling systems are extensively equipped with sensors, so this information can be used and the effort to create a digital twin is greatly reduced.

This chapter proposes an approach to generate parts of the digital twin for the cooling system from measured data by using ML methods. In a subsequent step, this digital twin is used to calculate the effects of alternative control parameters, and the results are presented to the operator in an understandable way. Combined with appropriate monitoring this allows the operator to make informed decisions to adjust the control parameters accordingly.

## 2.1 Introduction

Nowadays, there is a vast amount of larger sized mostly individual designed plants for cooling. They consist of components from different manufacturers, are differently dimensioned and are adapted as far as possible to the requirements of their specific location. But it can be assumed that the boundary conditions of a plant change over the course of its lifetime and therefore a large number of these plants have energy-saving potentials. In the use case considered here, there are already several examples. In the last few years, the data center has expanded more and

S. Wallner (✉) · T. Bernard · C. Kühnert
Fraunhofer IOSB, Karlsruhe, Germany
e-mail: steffen.wallner@iosb.fraunhofer.de; thomas.bernard@iosb.fraunhofer.de;
christian.kuehnert@iosb.fraunhofer.de

© The Author(s) 2025
D. Schulz, C. Bauckhage (eds.), *Informed Machine Learning*,
Cognitive Technologies, https://doi.org/10.1007/978-3-031-83097-6_2

more, so the cooling systems needed more energy. Conversely, more people are working from home offices and meetings are held online. This reduces the cooling requirements of the office building.

The unused energy-saving potential results from an operation of the plant that is not optimized to current requirements and environmental conditions. Reasons for this are too rigid control strategies, poorly set control parameters or an intrinsically oversized plant. Hence, to find and raise energy-saving potential, the existing control strategies and parameters must be replaced by better, ideally dynamic ones.

Because these systems are not manufactured in series, the control strategies and parameters cannot be readjusted according to a standard scheme. Hence, to be able to find optimal settings, one needs to use a detailed model of the plant to estimate the impact of new settings in advance. Unfortunately, in practice this is not the case.

Even if a simulation model of the plant exists, the integration into the running operation of the plant is often missing. Therefore, a digital twin is needed which contains not only a simulation model but also the specific configuration of a plant so that the effects of alternative operating parameters can be determined.

Since no plant operator will allow operating parameters to be changed automatically, e.g. since there would be problems with liability if an accident happens, especially when the cooling system is part of a critical infrastructure, an automated application of possibly better operating parameters is not feasible. In summary, an overall concept is needed which makes the optimized operating parameters explainable to the plant operator and in a subsequent step makes their effect verifiable.

## 2.1.1 Related Work

Energy optimization of a cooling system basically means finding the most energy-efficient way to cool down a medium to a certain temperature with as little energy as possible. As outlined in the introduction, the traditional solution is to use model-based optimization. Therefore, a digital twin is required that describes the main functions of the plant. This model incorporates information about the behavior of the cooling system depending on the control parameters, the cooling load, and other influencing factors. The traditional way means that the model itself is based on physical equations. For example, [1] uses energy equations for a cooling tower and a centralized water-cooled chiller to calculate the proper cooling provisioning of a data center. Another example is [6], which uses the model to switch the cooling mode of a data center depending on the water-side economizer.

However, due to the high complexity and heterogeneity of cooling systems the generation of those models is very time consuming [10]. That makes it even more time consuming when one wants to integrate domain knowledge.

Fortunately, in recent years, data-driven methods have become more and more an alternative. The clear advantage of these methods is that models can be trained on historical data. Furthermore, if changes are made to the system a new model can be learned with little effort. In literature, there are already many publications that propose to use Machine Learning and other data-driven methods for optimization. An example is [2], where the authors propose a novel energy optimization method of a multi-cell tower with particle swarm optimization. Within this approach, a neural network is used to predict the overall cooling system power consumption. Hosoz [4] also uses a neural network to predict the power consumption of a cooling tower, but additionally considers several operating conditions. Other publications include the usage of a linear regression model to predict the outlet temperature of a cooling tower [7], the prediction of the energy consumption using a neural network with the overall aim to find optimal operating conditions of the chiller [5] or the combination of clustering and different machine learning models to measure the energy-saving benefits of a chiller system [3]. Finally, in [9] it has been examined to which extent data-driven methods can be used for fault diagnosis in cooling systems.

## 2.1.2  Informed Machine Learning for Cooling System Optimization

Informed Machine Learning refers to the process of training machine learning models while taking into account prior knowledge. Basically, the aim is to generate models with better performance, higher reliability, and more meaningful insights. Within the taxonomy of von Rüden [8], several approaches are characterized on how to integrate prior knowledge into ML, while this chapter focuses mainly on using logic rules to improve ML performance. In detail, this chapter follows three different ways of integrating prior knowledge into the ML model:

- Expert knowledge: Throughout the complete data analysis workflow expert knowledge was needed. Among others, this includes especially the data preparation, model selection, training and validation phase. Furthermore, expert knowledge led to the point that three separate regression models instead of one large model were used.
- Human feedback: The final decision on which cooling system is used is up to the expert. The ML model only makes a proposition.
- Data Engineering knowledge: Throughout the publication AutoML is used to tune the three ML models. Tuning is based on many heuristics which were developed by data engineers in the past.

Therefore, in Sect. 2.3 containing the training of the ML model, not one, but three models distinguishing the three different operation modes are trained. Furthermore, although this is not part of the taxonomy, care was taken to adjust the training data depending on the season, as different modes are used with different frequencies depending on the outside temperature. The adjustment of training data can also be seen as one way to bring prior knowledge into ML models. Finally, it must be noticed that publications proposing to use data-driven methods in cooling systems usually do not consider prior or domain knowledge and are solely based on historical data.

### 2.1.3 Structure

In Sect. 2.2, we will describe the setup of the investigated cooling system itself. In addition to the known characteristics of the plant, we will deal specifically with the three different operating modes and the basic control logic. The operating modes and control logic are the basis for the composition of the simulation model, with parts being trained using AutoML.

Section 2.3 explains the necessary preprocessing of the sensor data for the training phase and how the simulation model is learned. After the simulation model is assembled, we describe the used optimization approach in Sect. 2.4. As part of the overall concept, we show where and how we use the optimization approach in the currently developed assistance system. Finally, Sect. 2.5 summarizes the results and gives an outlook for the further development of the assistance system. As a use case, an existing cooling system in a research facility in the city of Karlsruhe is used.

## 2.2 Cooling System Description and Plant Operation

The research facility in the city of Karlsruhe consists of an office building with a medium-sized data center. To cool the data center as well as the conference rooms, a larger cooling plant is part of the building. The cooling plant itself consumes up to 100 kW of electrical energy to provide the cooling energy using two different generators. In 2019, the annual energy consumption of the cooling plant was 80 MWh which corresponded to a cost of approximately 16,000 € per year.

### 2.2.1 Components of the Cooling System

Figure 2.1 shows the cooling system, which consists of a cooling storage and two cooling generators, namely a cooling tower and a chiller. The demand side of the cooling system consists of a data center and during summer of some conference rooms in the office building.

The cooling reservoir has a volume of 100 m$^3$ and is placed outside, just like the two cooling generators. It has the shape of an upright boiler, including distinguishable temperature levels in the coolant. This means that the cooling reservoir does not have one temperature, but has several layers with different temperatures. For simplicity, within this chapter the average reservoir temperature $\overline{T}_{\text{reservoir}}$ is used.

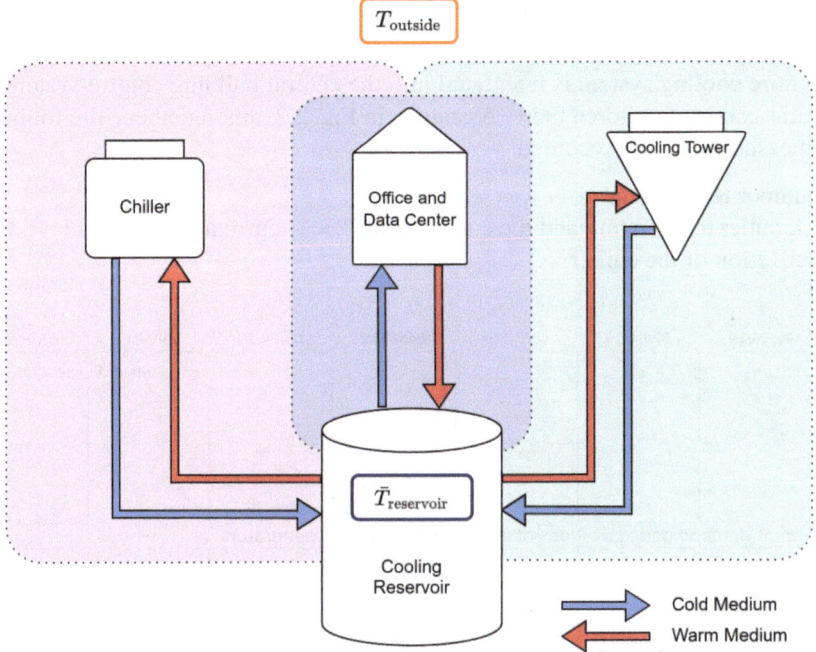

**Fig. 2.1** The principle diagram shows the main components of the cooling system. The chiller, data center and office (load) and the cooling tower have a direct influence on the average reservoir temperature in the cooling reservoir through their in- and outflows. The central global influencing variable on all system components is the outside temperature

The demand side is served directly from the cooling reservoir and the reservoir can be charged individually by the cooling generators. The active chiller has a mean electrical power consumption of 90 kW and, compared to the chiller, a much lower power consumption. Nevertheless, it can only be operated at lower outside temperatures. The electrical power consumption of the cooling tower depends on the speed of the fans and the pump power. Conservatively calculated, the energy consumption of the active cooling tower yields 20 kW.

The operational energy of the chiller is 4.5 times that of the cooling tower. Hence, from an energy-saving point of view, the operation of the cooling tower is generally preferable to that of the chiller. Thus, the essential optimization task is to reduce the operating times of the chiller and compensate them instead by operating the cooling tower.

## 2.2.2 Sensors of the Cooling System

The entire cooling system is integrated into the central building control system and its status can be monitored there. As shown in Fig. 2.2, among others, the following five measurements are recorded:

- Outdoor temperature
- In-, outlet temperature and the current flow $\implies$ current cooling load
- Activation of the chiller

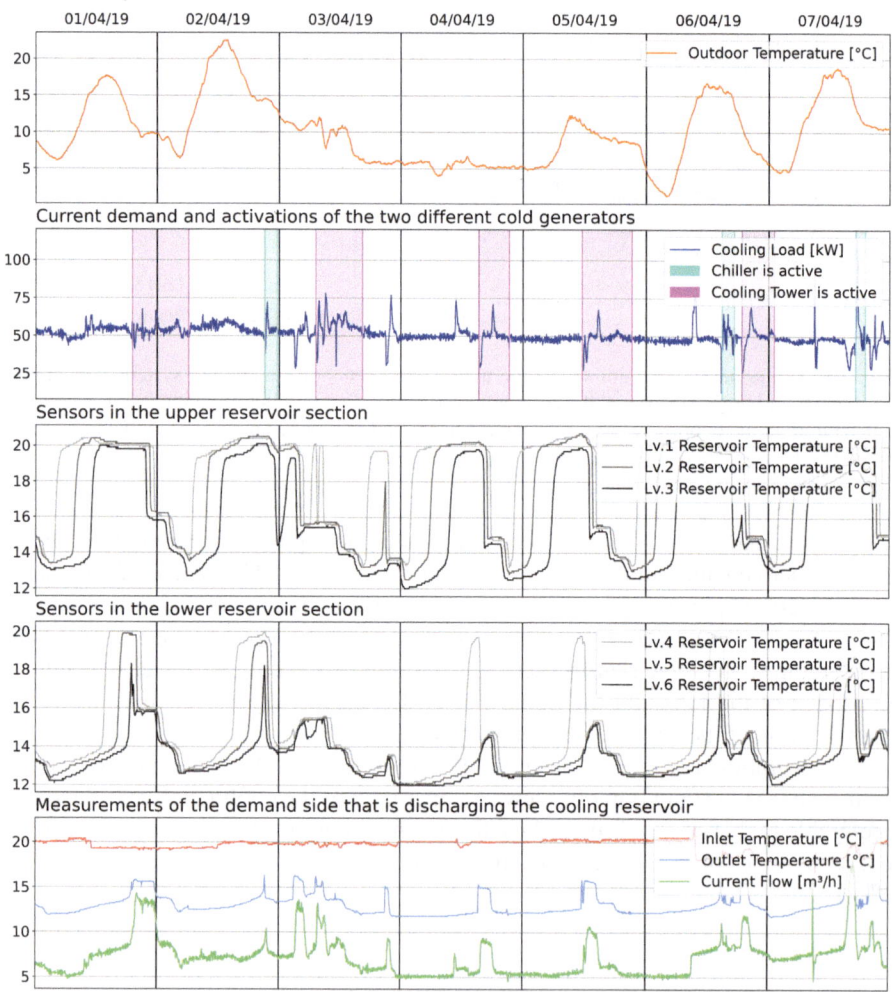

**Fig. 2.2** All relevant measurements of the cooling system. The second subplot, highlights the phases with either chiller (cyan) or cooling tower (magenta) active. The effect of the activation, with a temperature drop can also be seen in the plots below

**Fig. 2.3** The evenly distributed temperature sensors at the cooling reservoir are aggregated to the mean reservoir temperature $\bar{T}_{reservoir} = \frac{1}{6} \sum_{n=1}^{6} T_{Lv.n}$. The mean temperature is sufficient when deriving the model

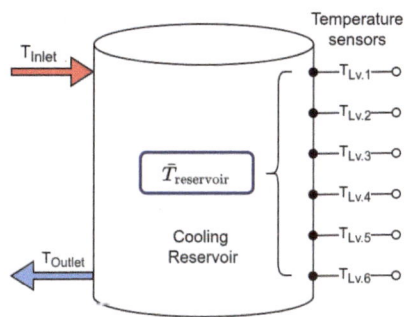

- Current flow of the cooling tower $\implies$ activation of the cooling tower
- Cooling reservoir temperatures at 6 different levels

The different temperature levels in the cooling reservoir are most relevant for the real control logic. If the inlet temperature levels exceed a threshold value, the respective aggregate is activated. For the scope of the publication it is sufficient to use the reservoir mean temperature $\bar{T}_{reservoir}$ as illustrated in Fig. 2.3.

Within the second subplot of Fig. 2.2 it can be seen that the cooling load can be assumed to be nearly constant. This is true for the colder seasons as only the data center has to be supplied, while during the warmer seasons office rooms are additionally cooled.

## 2.2.3 Analysis of the Operation Strategy

To charge the cooling reservoir and thus reduce the mean reservoir temperature, one of the two given cooling units must be activated. In order to decide which of them is used and how long it operates, five adjustable threshold parameters are set in the control of the cooling system. The basic assignment of these parameters with their corresponding measuring points can be seen in Fig. 2.4.

In summary, these are the relevant parameters and their actual purpose regarding the control of the cooling units:

- One parameter to decide which of the units is used (switchpoint temperature)
- Two to decide when to turn a unit on (activation temperature)
- Two to decide when to turn a unit off (deactivation temperature)

Figure 2.5 shows the unit used depending on the so called switchpoint temperature parameter. If the switchpoint temperature is higher than the actual outside temperature, the chiller is used. If the switchpoint temperature is lower than the actual outside temperature, the cooling tower is used.

Furthermore, there is an activation temperature parameter for each unit. If a unit-specific temperature level in the cooling reservoir exceeds this temperature, the corresponding unit is activated. Conversely, the unit runs if a temperature measured

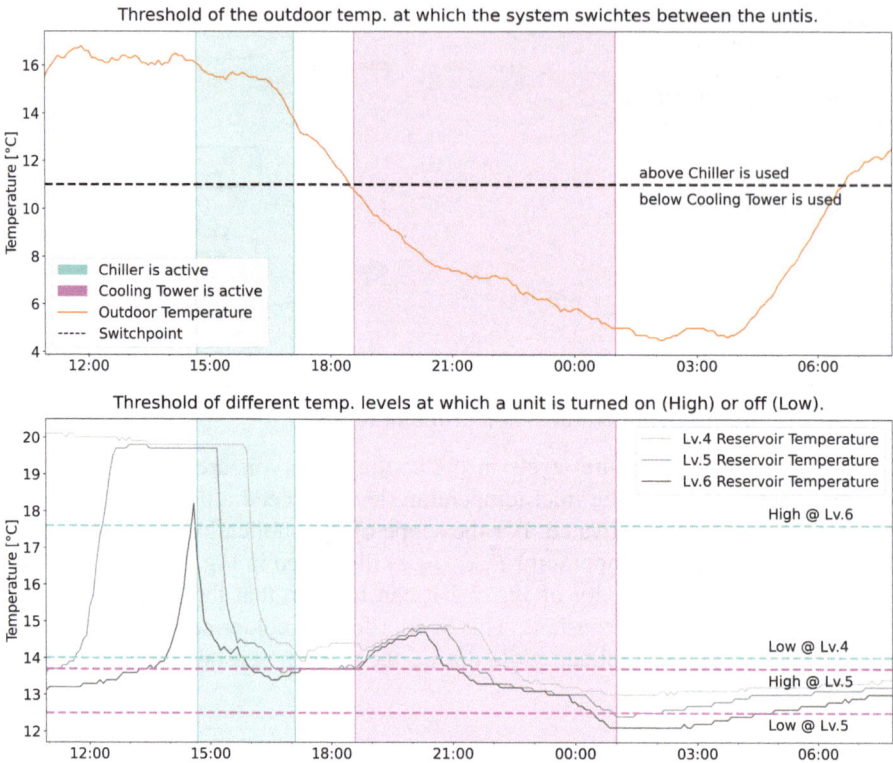

**Fig. 2.4** The upper plot shows the outdoor temperature and depending on its value, if the chiller or the cooling tower is used. The lower plot shows the thresholds and the different temperature levels measured in the cooling tower. The trained simulation model will be used to optimize the switchpoint between the chiller and the cooling tower. Hence, the later learned threshold differs quantitatively from the one shown here

at a certain level in the cooling reservoir is lower than a unit-specific deactivation temperature parameter.

### 2.2.4 Cooling Reserve

Since the data center counts as critical infrastructure, special requirements for the cooling system like a relatively large storage tank have been installed. Furthermore, a large reserve for cooling is set. This reserve ensures that the data center can be cooled and thus operated even in the event of a prolonged failure of the two cooling units.

To calculate the reserve, we define a maximum outlet temperature at the reservoir $T_{\text{outlet,max}}$ at which cooling of the data center is still possible. The lowest subplot in

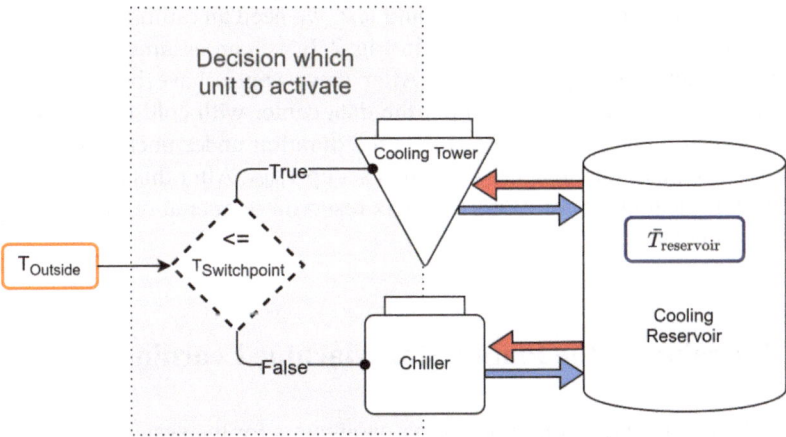

**Fig. 2.5** The switchpoint parameter compared to the outdoor temperature, is solely relevant for deciding which unit is used for charging the cooling reservoir

Fig. 2.2 shows the controller that adjusts the current flow so that there is always a constant inlet temperature $T_{\text{inlet}} = 20\,^\circ\text{C}$.

The inlet temperature can not be exceeded by the outlet temperature so we define now that this constant value is also the maximum temperature $T_{\text{outlet,max}}$ at which the data center can still be cooled. Then the current reserve energy $E_{\text{reserve}}$ results from the difference between the reservoir temperature $\bar{T}_{\text{reservoir}}$ and $T_{\text{outlet,max}}$. That can be calculated as follows:

$$E_{\text{reserve}} = \rho_{\text{water}} \cdot V_{\text{reservoir}} \cdot c_{\text{water}} \cdot (T_{\text{inlet}} - \bar{T}_{\text{reservoir}}) \tag{2.1}$$

$$t_{\text{reserve}} = \frac{E_{\text{reserve}}}{Q_{\text{load}}} \tag{2.2}$$

$$\rho_{\text{water}} = 997 \frac{\text{kg}}{\text{m}^3} \tag{2.3}$$

$$c_{\text{water}} = 4190 \frac{\text{Ws}}{\text{kg} \cdot \text{K}} \tag{2.4}$$

$$\bar{T}_{\text{reservoir}} = 16\,^\circ\text{C} \tag{2.5}$$

Assuming (2.3)–(2.5), since the mean reservoir temperature is still above the highest $T_{\text{outlet}}$ in Fig. 2.2, the current reserve results in

$$E_{\text{reserve}} = 997 \frac{\text{kg}}{\text{m}^3} \cdot 100\,\text{m}^3 \cdot 4190 \frac{\text{Ws}}{\text{kg} \cdot \text{K}} \cdot$$

$$(20\text{–}16\,^\circ\text{C}) \cdot \frac{h}{3600s} \cdot 10^{-3}\text{k} = 464\,\text{kWh}. \tag{2.6}$$

To find out how long this reserve would last, we need an estimate for the cooling load. Referring to the second subplot in Fig. 2.2, we can assume that the cooling load is constant at $Q_{load} = 50$ kW. After these values have been inserted into (2.2), the cooling system could supply the data center with cold for about *9 hours* without activating a cooling unit. This is the duration under unchanged load until the reservoir temperature reaches the actual set up $T_{inlet}$. After this point in time, the reservoir will heat up above its normal max reservoir temperature, which leads to a critical system state.

## 2.3   Modeling of the Plant Using Machine Learning

To be able to calculate optimal operating parameters for the system, the simulation model of the cooling system is generated. The model must be able to consider the critical boundary conditions, meaning that it must be able to calculate the reservoir temperature.

In the following it is described how the cooling system is decomposed into its submodels. Furthermore, the preparation of the training data (Sect. 2.3.2) is described, where the main focus lies on the physical plausibility of the model and in particular the definition of the validity domain. We go into detail about the training procedure (Sect. 2.3.3) and at the end of this section, the recalculation of results with the trained models is shown (Sect. 2.3.4).

### 2.3.1   Submodels of the Cooling System

Figure 2.6 shows the three disjoint operating modes, all influencing the reservoir temperature. To train a separate submodel for each mode, initially the influencing variables need to be set. This results in the following three submodels:

**Load Model**   is used when none of the cooling aggregates is active. The temperature change in the reservoir depends only on the cooling power and the losses through the outer shell of the tank. The cooling power demand is composed of a constant part (data center) and a volatile part (office). It is assumed that the latter correlates with the outside temperature. The losses via the outer shell of the cooling tower depend on the difference between the reservoir temperature and the outside temperature.

**Chiller Model**   is used when the chiller is active. The efficiency of the chiller depends on the difference between the reservoir and the outside temperature. Even if this unit is running, the load that would increase the reservoir temperature acts at the same time.

**Cooling Tower Model**   is used when the cooling tower is active. It can be compared to a passive heat exchanger. As soon as the storage water is warmer

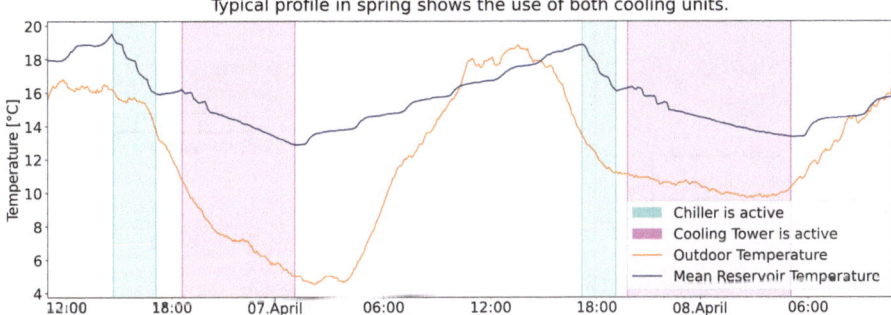

**Fig. 2.6** Operating modes of the cooling system. The reservoir temperature drops in the two modes in which either the chiller (cyan) or the cooling tower (magenta) is active

than the outside temperature, it can release heat energy and cool itself. In this operating mode, the load has an increasing effect on the reservoir temperature.

All three operating modes depend on the same influencing variables. Therefore, all three submodels have the same input variables—current outside and reservoir temperature—and each output the resulting reservoir temperature for the next time step. The overall model of the cooling system for the calculation of the reservoir temperature is composed as described in Fig. 2.7. In Sect. 2.3.3 the machine learning model will be connected to the chiller and cooling tower model. Furthermore, in Sect. 2.3.4 the parameters of the control logic will be varied to find optimal points of operation.

## 2.3.2 Data Processing

To train the submodels from the existing time series data, the data is split based on the respective operating mode, resulting in three time series data sets. Each data set contains fragmented time series, which is filtered again, while only using sequences that have a certain minimum duration and where a certain gap duration is not exceeded. For the load and cooling tower models, the minimum length of the sequences in the time series data is set to 4 hours and the maximum gap to 1 hour. For the chiller model, the sequence needs to be at least 1 hour, since this unit is often briefly activated.

Next, all sequences are filtered again according to the temperature difference between two consecutive time steps. Thus, in the operating mode for the load model, the reservoir temperature should never drop and the difference should always be positive. If one of the generating units is running the reservoir temperature should not rise and the difference should be negative. However, this assumption may be violated. For instance, in the case of the load model, it is possible that the outside temperature and the load are very low. In that case the heat energy loss through the

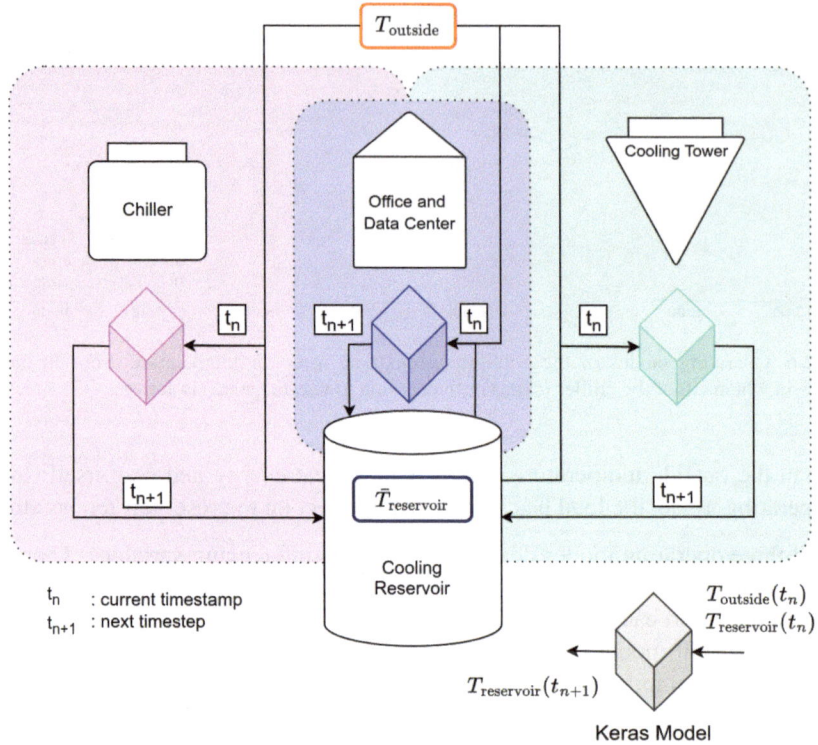

**Fig. 2.7** Linking the measured data to the submodel. Each submodel refers to a different operating mode and therefore they are trained with different data sets. However, the interfaces of the submodels are always the same. The input data is $\bar{T}_{\text{reservoir}}$ and $T_{\text{outside}}$ at time $t_n$ and output data is $\bar{T}_{\text{reservoir}}$ for the next time step $t_{n+1}$

outer shell of the storage is greater than the gains through the waste heat of the data center.

Finally, we resample the data to a 15 minute interval meaning that the models can *only* predict the storage temperature 15 minutes in the future.

Figures 2.8, 2.9 and 2.10 show the resulting set of samples used for training. The slopes of the data sets are quite uniform for each operating mode.

Figure 2.8 shows that the chiller always needs a similar amount of time steps to reach about the same temperature inside the reservoir and its performance seems to be less dependent on the outside temperature—compared to the other two sample sets.

If we look at the samples for the cooling tower in Fig. 2.9, the situation is more difficult. Here the condition when the cooling tower is activated cannot be derived directly from the reservoir temperature.

The reason of the varying start temperature is that the threshold for activating the cooling tower may already have been exceeded, but the outside temperature is

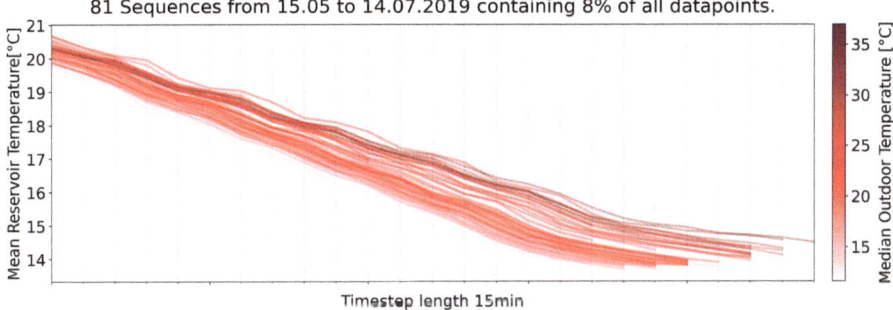

**Fig. 2.8** The time series sequences relevant for training the chiller model. It can be seen that most of them start at 20 °C while the final reservoir temperature does not fall below 14 °C. With higher median outdoor temperature, the slope decreases slightly

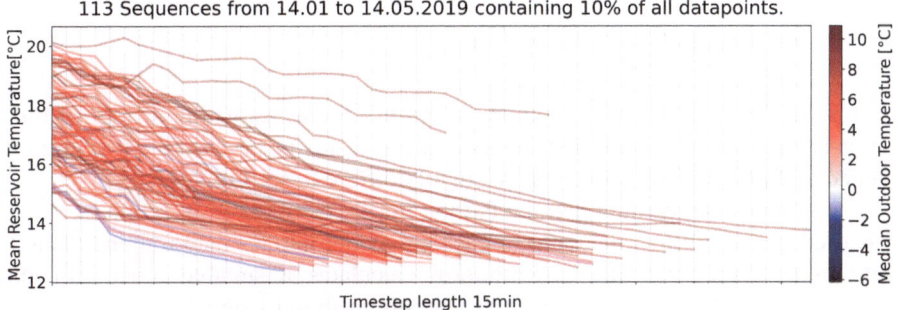

**Fig. 2.9** Time series sequences relevant for training the cooling tower model. The starting temperature varies between 15 and 20 °C while the final mean reservoir temperatures stay above 12 °C. The higher the median outdoor temperature, the longer is the second phase

still above the switchpoint temperature to activate the unit. The reserve temperature continues to rise and at some point, the outdoor temperature could fall again below the switchpoint temperature. The cooling tower is activated but the reservoir temperature is already above the constant activation temperature.

The plot shows that compared to the other models a different training period for the Cooling Tower is used. This is necessary since getting closer to warmer seasons, the cooling tower is activated less often.

Additionally, the time series sequences for the load model have nontrivial initial conditions, see Fig. 2.10. Still, this is not problematic, since the load model will always be enclosed by sequences of the other two submodels.

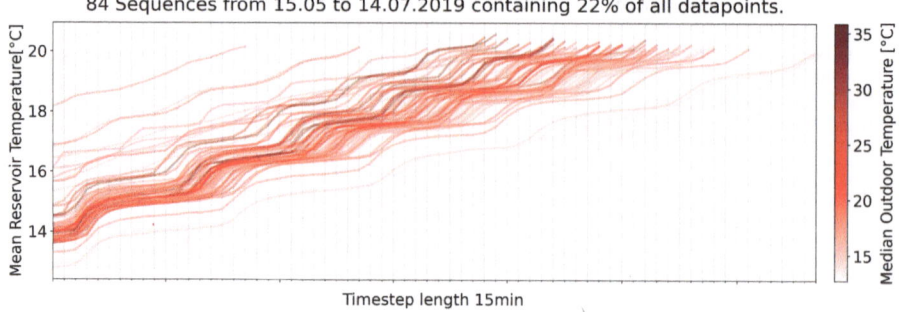

**Fig. 2.10** The time series sequences relevant to the load model training. The starting temperature is between 13 and 18 °C. And sequences do not exceed a max mean reservoir temperature of around 20 °C

### 2.3.3 Training and Plausibility

As for the choice of learning method, an approach that learns the history of time series is possible. Especially for the cooling tower this is, because the first and the second half of a running sequence are slightly different, see Fig. 2.9.

For the machine learning part the Python library `autokeras` in version `1.0.19` was used. After splitting the data for each submodel into test and training dataset, `autokeras.StructuredDataRegressor` calculates a suitable model for the selected approach. You can see the corresponding call in Listing 2.1.

**Listing 2.1** Python code to train a submodel

```
import autokeras as ak
reg = ak.StructuredDataRegressor(max_trials=1)
reg.fit(x_train, y_train, epochs=100)
```

Libraries that support AutoML such as `autokeras` are designed to search the hyperparameter space themselves. The libraries automatically vary the hyperparameters of a model structure to find those that best fit the model to the given training data. Maybe just because the physical phenomena in the three operating modes are not particularly complex, but it was already sufficient for us to directly use the default starting configuration of the hyperparameters that `autokeras` uses by default in its first trial. For all three submodels, even with the standard hyperparameters, the model reaches a mean square error of the mean reservoir temperature below 0.025 K after 100 epochs. Other variations of the hyperparameters did not significantly improve these learning results.

In order to check their physical plausibility, heatmaps of the trained submodels are generated and all combinations of input variables that lie within the training data of a submodel are calculated. Finally, the resulting $(t_{n+1})$ reservoir temperature is calculated and the difference to the input $(t_n)$ reservoir temperature is given in the heatmap.

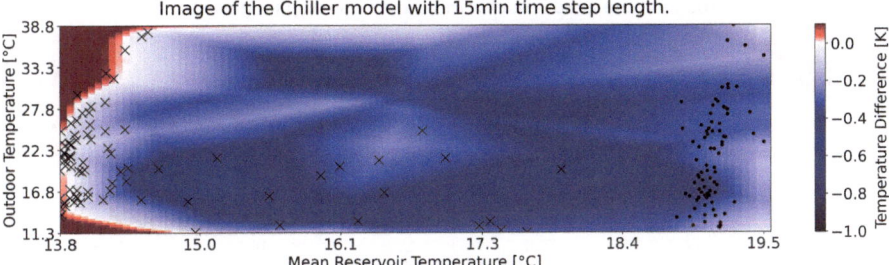

**Fig. 2.11** Blue: Areas in which the chiller model calculates a temperature decrease. Red: Areas that are implausible temperature increases. The start temperatures of the sequences in the training data are marked with · and the end temperatures with ×. The sharp gradients are probably rather fragments since they do not converge during several training runs

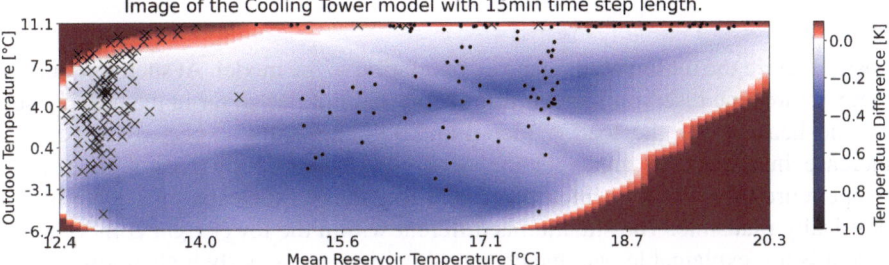

**Fig. 2.12** Blue: areas in which the cooling tower model calculates a temperature decrease. Red: areas are actually implausible temperature increases. The start temperatures of the sequences in the training data are marked with · and the end temperatures with ×. In contrast to the chiller model, larger parts of the plausible temperature increase lie outside the training data. For example the area in the upper right corner

Due to the internal implementation of training in `autokeras`, the heatmaps are slightly different for different runs and the same training data, but converge for each submodel. Thus, it makes sense to discuss the learned submodel from a physical point of view. In Figs. 2.11, 2.12 and 2.13 the set of start and the set of end temperatures of all sequences from the training data are shown too. All other training data are located between these two sets. The markings help to estimate which range of data the model has already seen during the training.

The the heatmap for the chiller model shows (Fig. 2.11) a temperature drop for the most part. However, the dark red areas at the left edge and the upper left corner on the heat map are physically implausible. The reservoir threshold temperature for deactivating the chiller is at the right of these areas. Here the learned model extrapolates since there was no value for those input variables in the training data.

The image for the cooling tower (Fig. 2.12) also shows a temperature reduction for the most part. The absolute temperature reduction is mostly lower than for the chiller. Again, there are areas in the image that make less sense from a physical point of view. In the lower right corner, for example, the temperature changes are

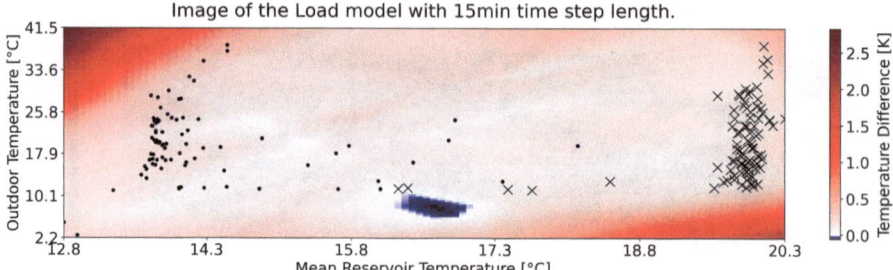

**Fig. 2.13** The load model calculates a plausible temperature increase for almost in the entire area. The start temperatures of the sequences in the training data are marked with · and the end temperatures with ×. The blue areas of very small decreasing temperatures appear sporadically in the trained models. Although they are not fundamentally physically implausible, they are also not learned reliably and stably

positive. Here the reason is again the extrapolation of the model. At such low outside temperatures the reservoir temperature simply did not rise so far in the training set.

The heatmap of the load model (Fig. 2.13) shows the expected temperature increase in almost all areas. In the upper left corner is an area with extreme temperature increase, the which could be a fragment of the exploration, but is also physically plausible. The situation is different within the lower right corner. In that case it is not explainable why the change should be particularly high at low outside temperature and high reservoir temperature.

More of an interest is the area to the left of this lower right corner. Here the temperature difference is negative, which means that the reservoir temperature would fall. In principle, this is physically plausible, because at very low outside temperatures the heat loss through the outer shell of the reservoir can be very high. If this loss is larger or equals to the heat gained by the waste heat from the data center, then the storage temperature decrease remains constant even without activating one of the generator units.

In summary, it can be stated here that the `autokeras` models have in principle probably learned static plant parameters. Not only the load, but also the generating units have a quite constant output. Furthermore, the domain of the model that is physically plausible can be seen. With other learning methods this domain could be perhaps even larger. The advantage of using AutoML here is that we did not have to worry about the limits and prerequisites of the learning procedure in advance.

### 2.3.4 Recalculation of the Entire Cooling System

Each submodel contains plausible output values for a single calculation. Next it is checked how the models behave when they calculate consecutive values. Therefore, as input signal we take the historical time series of the outdoor temperature,

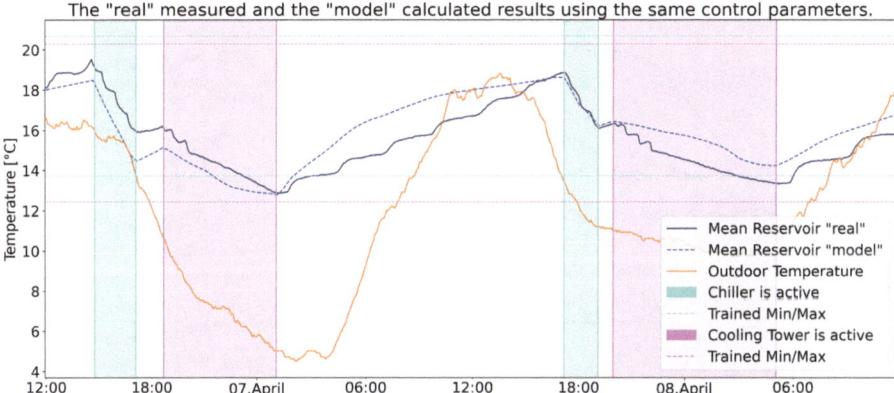

**Fig. 2.14** Result of recalculation of 48 hours with the trained submodels. The boundaries of the chiller model (green) and the boundaries of the cooling tower model (red) show the minimum and maximum temperatures that the respective models have seen in the training data

containing the first value $(t_n)$ of the reservoir temperature as well. Finally, the submodel is used to calculate the reservoir temperature $(t_{n+1})$ as it would be 15 minutes later. The result is then used as the new reservoir temperature in the next iteration. At each iteration, the historical data is used to check which operating mode was present. Accordingly, the submodel matching the mode is used. If, for example, according to the historical data, the chiller was active at the current time, the chiller model is also used for this time step, regardless of the reservoir temperature. As an illustrative example, Fig. 2.14 shows an exemplary behavior of the trained submodels with its boundaries. It is pointed out that other time intervals show a similar behavior of the models.

Furthermore, it should be noted that the submodels are only able to extrapolate to a limited extent. For example, it could occur that during the recalculation a submodel computes a reservoir temperature that is outside of that ever reached in reality. Then the reservoir temperature $(t_{n+1})$ can adopt to physically implausible values. Since there is no temperature control in the recalculation by a proper control, we defined the maximum and minimum reservoir temperatures in the training data as limits. If a model exceeds these limits during the recalculation, the reservoir temperature $(t_{n+1})$ is set to this value, until a new value is calculated that is within the limits.

Since the submodels are sufficiently plausible even for consecutive calculations, they can be combined to form a simulation model of the cooling system. A control logic must then also be implemented and be part of this model. Based on this, the simulation model decides when which unit is switched on and off. The logic for this is shown in Fig. 2.15. In addition to the switchpoint temperature, the four temperature limits are also needed to control when the cooling units are de- and activated. We also extract these temperature limits from the training data. Therefore, we calculate the average from all start and end times of all time series sequence in the data sets.

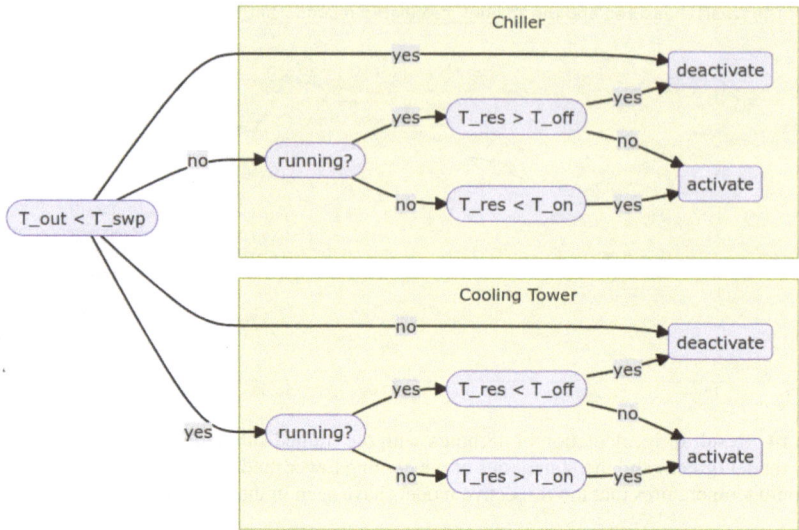

**Fig. 2.15** The basic decision whether to use chiller or cooling tower depends on the outdoor temperature $T_{out}$ compared to the switchpoint temperature $T_{swp}$. Depending on this, it is checked whether the unit is already running and whether the current reservoir temperature $T_{res}$ is below the deactivation temperature $T_{off}$ or above the activation temperature $T_{on}$ of the respective unit. In each control loop, a decision is made for each unit whether to activate or deactivate it exclusively

With this simulation model of the entire cooling system, we can now search for optimal control parameters and make predictions about the reserve of the cooling system.

## 2.4   Optimization Concept

Our optimization concept is embedded into an assistance system that continuously suggests optimized operating strategies. An operating strategy here means setting certain variable control parameters of the cooling system. So our concept is to optimize settings and do not make changes to the underlying control loops. The Optimization is done by searching for the current best value for a particular parameter.

### 2.4.1   Variable Switchpoint Temperature

As mentioned, the plant can generate cooling energy with two different units, the cooling tower unit requiring less energy and the chiller is working at warmer outdoor temperatures. The decision when to switch to which of the two units is defined by the adjustable threshold parameter *switchpoint temperature*, shown in Fig. 2.5.

For this parameter we want to find the current best value and therefore calculate three different variants of future operations. For a defined forecast horizon, the progression of the storage temperature is calculated in every variant with a different switchtpoint temperature in each case. Since the trained models of the components require the outside temperature as input, the current data from a weather forecast service is used here as well. The three variants differ in that they calculate the trend of the reservoir temperature once with:

- the current switchpoint temperature,
- the current switchpoint temperature reduced by 1 K and
- the current switchpoint temperature increased by 1 K.

The results of a calculation for an exemplary day are shown in Fig. 2.16. In order to compare the variants, a balance is made of how long which cooling unit was in operation. These activation times are then multiplied by the respective power input of the unit. At the end, the total required energy for the prediction horizon is obtained for each variant. The value for the switchpoint temperature of the variant with the lowest energy consumption is then, in principle, the one with the optimal operation parameters.

## 2.4.2   Forecast Horizon

A forecast horizon must be defined for the calculation of future operating variants. Since the trained models depend on the outside temperature, initially only the horizon for which reliable weather forecasts are available comes into question.

Another central factor for the length of forecast horizon is the size of the buffer in the storage component of the cooling system. The buffer in the storage is the amount of energy, which can be loaded and unloaded in safe operation minus the reserve. A rather small buffer only allows short-term optimization. If, for example, a buffer can cover the demand for the next 6 hours when fully charged, the reasonable horizon is also limited to 6 hours. Otherwise, since cooling demand could depend at least on the time of day due to the dependence on the outside temperature, the length of the horizon should cover at least one daily cycle. Even if the storage of the plant considered here is not sufficiently dimensioned, we use a 36-hour forecast horizon. For this horizon, the optimum parameters for operation can be found, which must then of course also be set in the cooling plant.

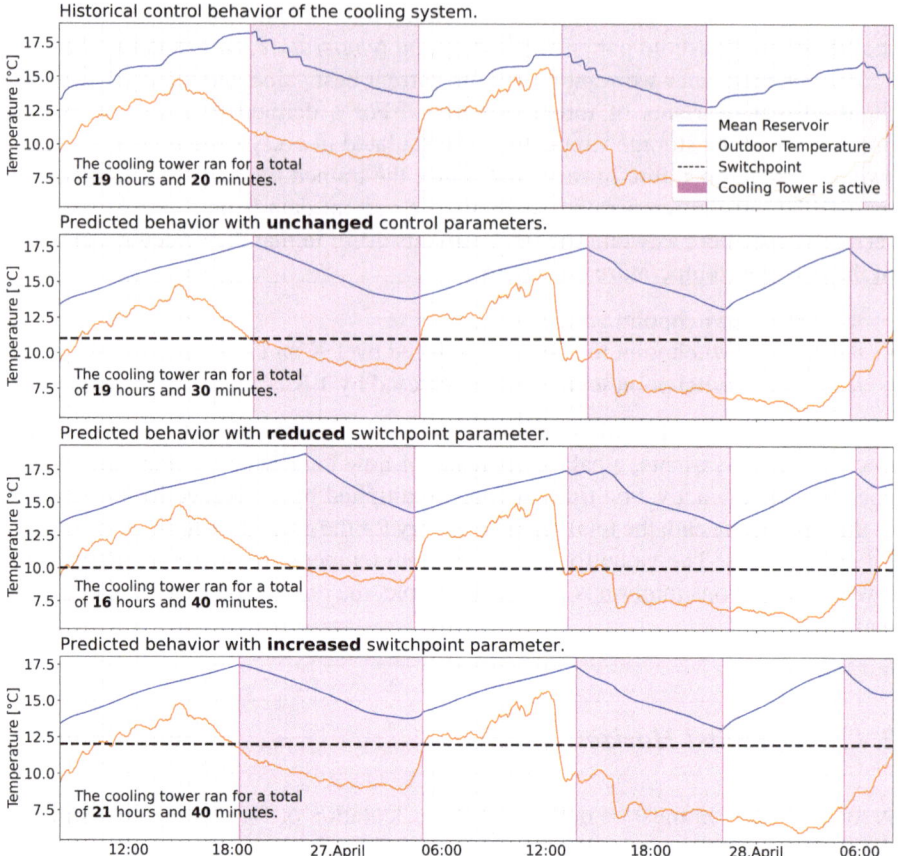

**Fig. 2.16** Result of the calculation of searching the current optimal value of the switchpoint temperature. The top plot shows the historical progression of the storage temperature. It serves to verify the calculated variant of the future operation with unchanged switchpoint temperature in the second subplot. The last two subplots show the future operation with reduced and increased switchpoint temperature parameter. For the evaluation which of the three variants is more favorable, the colored areas are accumulated. In each case they show when a cooling unit was activated. Here, only the cooling tower is used, and it can be said that the most optimal variant is the one with the least activation time of the unit

### 2.4.3 Software Implementation as Assistance System

In critical infrastructures such as the cooling system studied here, setting parameters automatically is very sensitive and demanding to implement. It is hard to define a proper prediction horizon and operators must ensure that the automation does not bring the cooling system into a critical state. This would require extensive test scenarios and comprehensive models, and these were not available here. Within the herein presented application the operator herself must decide on the application of

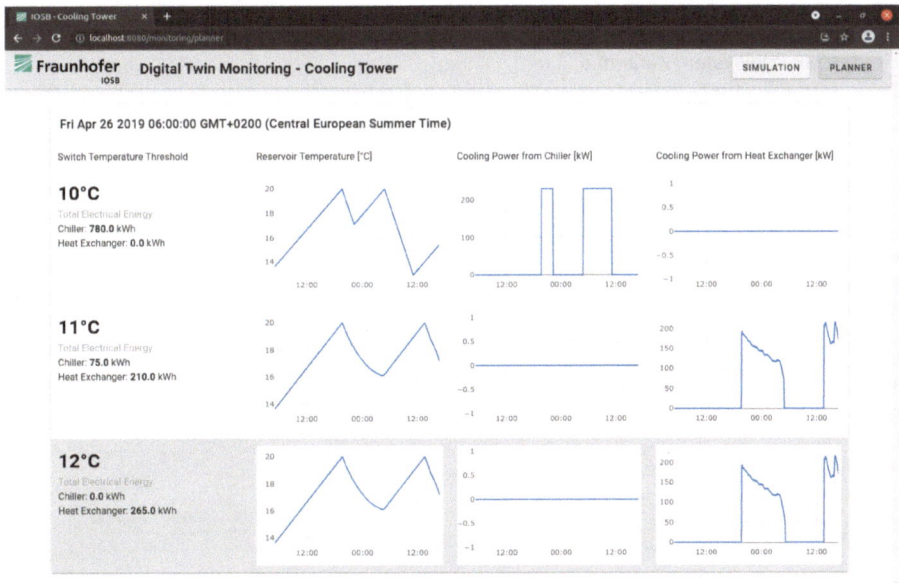

**Fig. 2.17** View of the results from the continuously executed calculation of the future operation variants. It is used by the operator to select the current optimal switchpoint temperature. One can see all calculated variants and see also the total energy consumption per variant. The operator should then set the switchpoint parameter according to the value of the variant with the lowest energy consumption. Here it is highlighted in gray

the optimized control parameters continuously found and proposed to the assistance system. This system must therefore provide the operator with enough information to verify the optimal parameter and let her evaluate the decision in real-time. For implementation, the following steps must be performed in a regularly repeating workflow.

1. Regular automatic calculation of the variants for future operation. This should happen at least once a day or whenever new forecast data is available. In energy system a usual time step length for the prediction is 15 minutes. That means a reasonable lower limit would be to perform the calculation not more than four times per hour.
2. Present the effect of applying these parameters to the operator in a comprehensible way. You can see in Fig. 2.17 this is accomplished by providing an interactive view of the results of a search for optimal parameters.
3. Decision and application of the optimal parameters by the operator.
4. Evaluation by the operator with the aid of monitoring (Fig. 2.18) whether the effect of the new parameters is as expected.

The developed assistance system shows the current state of the real cooling system and additionally the simulation model adapted to the specific plant. Regarding

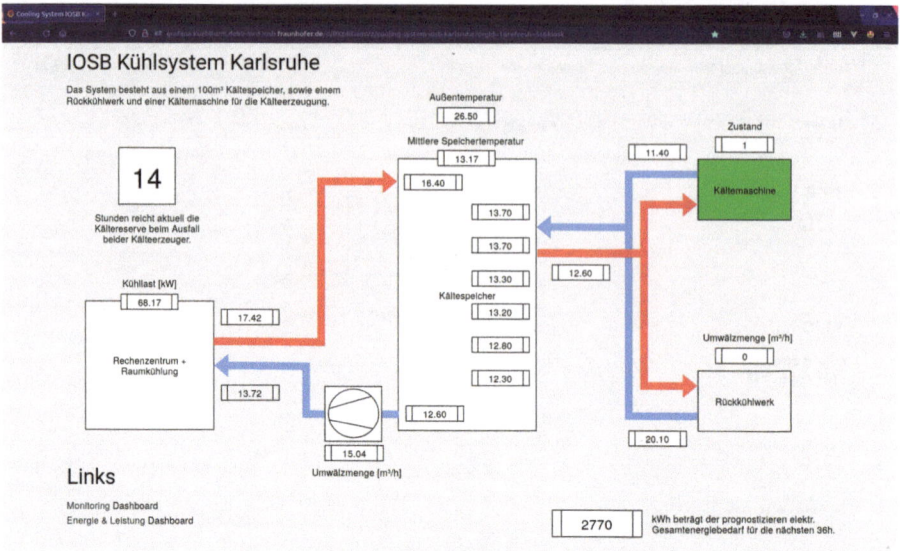

**Fig. 2.18** Screenshot of the interface to monitor the state of the cooling system. It shows the current sensor and meter data and the calculated reserve of the cooling storage in hours. In addition, the predicted energy demand of the cooling system for the current forecast horizon is displayed

potential energy-savings data from 2017 was used. Therefore, the threshold value of the cooling system, see Sect. 2.2.3, was increased from 10 to 18 °C. This resulted in energy-savings of around 18%. With the total costs of the plant of 16,000 € listed above, this results in a savings potential of almost 3000 € per year. In general, as described in this guideline [11], savings of 10–20% can be expected for plants of this size.

## 2.5 Conclusion and Outlook

We have presented an approach for using Informed Machine Learning for optimizing a real-world cooling system. It was shown that there is a way to derive a simulation model from measured data, which can then be used for operation optimization. Informed ML was integrated into the model using several ways. During the complete data analysis workflow, expert knowledge was needed, covering, among others, data selection, data preparation and the decision to use three different regression models instead of one large one. Since the resulting ML model only makes propositions to the operator, human feedback is integrated into the decision-making. Finally, AutoML containing a large number of heuristics was used to train the regression models. All these individual points led to the final generation of a simulation model of the cooling system.

The simulation model was used to reduce the overall energy consumption of the cooling system by performing regular calculations and returning optimal operating parameters. Due to the classification of the cooling system as critical infrastructure, the results were embedded in an assistance system to support the operators in their work. A procedure has thus been described for applying the optimized parameters found to the system. In order to be able to check the effect of the updated parameters later by the operator, corresponding monitoring is also provided.

In terms of operational optimization, there are several opportunities for future research. One first example covers the possibility to improve the cost evaluation by balancing the computed energy consumptions with the volatile electricity prices. In doing so, it is possible that power consumption becomes cheaper or emits less $CO_2$ if energy is only drawn when renewable energy is available. In principle, it would also make sense to vary other control parameters as part of the search and not just the switchpoint temperature.

During modelling, it was still necessary to manually decompose the plant operation into disjoint operating modes in order to identify which submodel needed to be trained. Thinking further, it would be surely conceivable to identify these operating modes automatically from an algorithmic analysis of the available measurement data. This would help to apply the optimization of operating parameters applied in this chapter to a large number of existing plants.

**Acknowledgments** This contribution was supported by the Fraunhofer Cluster of Excellence "Cognitive Internet Technologies".

# References

1. Beitelmal M. H. and Chandrakant D. P.: Model-based approach for optimizing a data center centralized cooling system. In: Hewlett-Packard (HP) Lab Technical Report (2006).
2. Blackburn L. and Tuttle J. and Powell K.: Real-time optimization of multi-cell industrial evaporative cooling towers using machine learning and particle swarm optimization. In: Journal of Cleaner Production (2020) https://doi.org/10.1016/j.jclepro.2020.122175
3. Chun-Wei C. and Chun-Chang L. and Chen-Yu L.: Combine Clustering and Machine Learning for Enhancing the Efficiency of Energy Baseline of Chiller System, Energies (2020) https://doi.org/10.3390/en13174368
4. Hosoz, M. and Ertunc, H.M. and Bulgurcu, Hüseyin: Performance prediction of a cooling tower using artificial neural network. In: Energy Conversion and Management (2007) https://doi.org/10.1016/j.enconman.2006.06.024
5. Jee-Heon K. and Nam-Chul S. and Wonchang C.: X. et al.: Modeling and Optimizing a Chiller System Using a Machine Learning Algorithm, Energies (2019) https://doi.org/10.3390/en12152860
6. Jiajie L. and Zhengwei L.: Model-based optimization of free cooling switchover temperature and cooling tower approach temperature for data center cooling system with water-side economizer. In: Energy and Buildings (2020) https://doi.org/10.1016/j.enbuild.2020.110407
7. Karunamurthy K. et al.: Prediction of Thermal Performance of Cooling Tower of a Chiller Plant Using Machine Learning, Earth and Environmental Science (2020) https://doi.org/10.1088/1755-1315/573/1/012029

8. von Rueden L. et al.: Informed Machine Learning - A Taxonomy and Survey of Integrating Prior Knowledge into Learning Systems. In: IEEE Transactions on Knowledge and Data Engineering, (2021) https://doi.org/10.1109/TKDE.2021.3079836.
9. Tian C. et al.: Chiller Fault Diagnosis Based on Automatic Machine Learning, Front. Energy Res. (2021) https://doi.org/10.3389/fenrg.2021.753732
10. Zhang Q. et al.: A survey on data center cooling systems: Technology, power consumption modeling and control strategy optimization. In: Journal of Systems Architecture (2021) https://doi.org/10.1016/j.sysarc.2021.102253
11. Leitfaden Effizientes Monitoring von Energiedaten im Bereich des Facilitymanagements https://publica-rest.fraunhofer.de/server/api/core/bitstreams/991b3429-be99-4cf8-9a06-ff9e435e09ef/content (last access 6.3.2023)

# Chapter 3
# AITwin: A Uniform Digital Twin Interface for Artificial Intelligence Applications

**Alexander Diedrich, Christian Kühnert, Georg Maier, Joshua Schraven, and Oliver Niggemann**

**Abstract** Cyber-physical systems that integrate machine learning (ML)-based services and methods from the broader field of Artificial Intelligence (AI) rely on a virtual representation of the underlying real physical system. Unfortunately, depending on respective solution approaches, usually similar but rarely the same virtual representation of the physical system is required. Thus, two solutions for the same problem might use different virtual representations. Informed Machine Learning is one technique to integrate expert knowledge into AI applications. It uses techniques to combine an often proprietary and expert-defined virtual representation with data from a real cyber-physical system. But methods for Informed ML have a much higher demand on the virtual representation than, for example, traditional distance-based methods in Machine Learning. Informed ML requires domain specific knowledge, which needs to be represented in some standardized Digital Twin as its virtual representation. Practitioners benefit through some categorization indicating which Digital Twin can be used to acquire a unique virtual representation of a cyber-physical system. Especially, by using a common standardized application programming interface (API). In short: a standardized Digital Twin is needed for AI-based solutions. In this chapter, such an API for Digital Twins for AI solutions is presented and different levels of complexity for Digital Twins are defined. The suggested API is considered as an AI reference model and is verified by using it on several simulated and real examples from the process and manufacturing industries. Additionally, it is compared against currently ongoing research projects.

A. Diedrich
Fraunhofer IOSB-INA, Lemgo, Germany
e-mail: alexander.diedrich@iosb-ina.fraunhofer.de

C. Kühnert (✉) · G. Maier
Fraunhofer IOSB, Karlsruhe, Germany
e-mail: christian.kuehnert@iosb.fraunhofer.de; georg.maier@iosb.fraunhofer.de

J. Schraven · O. Niggemann
Helmut Schmidt University, Hamburg, Germany
e-mail: joshua.schraven@hsu-hh.de; oliver.niggemann@hsu-hh.de

© The Author(s) 2025
D. Schulz, C. Bauckhage (eds.), *Informed Machine Learning*,
Cognitive Technologies, https://doi.org/10.1007/978-3-031-83097-6_3

## 3.1  Introduction

If one wants to implement an AI-solution for some manufacturing or production process, the virtual representation of the cyber-physical system—the Digital Twin— is essential. A Digital Twin is a virtual representation of the (application specific) most important information of some cyber-physical system. In the case of this chapter these might be the dependencies and causalities within a system, the components and their connections within the system, and the available process data and expert knowledge (as far as it is relevant from an artificial intelligence standpoint). Overall, the Digital Twin holds a collection of relevant information about the complete physical system and of each of its entities. This virtual representation can be used for different tasks, reaching from monitoring over fault diagnosis till optimization of the process, depending on which information is included. Therefore, the vast majority of publications considering the Digital Twin have a focus on engineering and process-oriented aspects, covering topics like performing simulation scenarios [24], the continuous enrichment of the Digital Twin during its life cycle [33] or doing different modeling issues such as quality assurance for products [9], engineering chains [24] or detecting optimal meta-levels [32].

Within the area of industry 4.0 (I4.0) the Digital Twin is mainly defined through the so-called Asset Administration Shell (AAS) [15]. The AAS proposes a standardized virtual representation of an I4.0 component which provides an interoperability between different automated industrial systems. Like most other publications covering the Digital Twin, the AAS barely focuses on Artificial Intelligence (AI) and Machine Learning (ML). The growing research area of Informed Machine Learning focuses on combinations between expert knowledge, AI, and ML [19, 31]. Using informed ML, Models will integrate more information of the environment and domain knowledge, meaning that the Digital Twin concept and AI should become ever more compatible. Still, there are two contradictions:

- ML methods are very heterogeneous, meaning that each method comes with a specialized representation. Hence, the question remains how a Digital Twin can provide a general representation to each ML method
- ML and Informed Machine Learning, in particular, requires explicit, by an algorithm processable, knowledge. Knowledge of a process in the Digital Twin such as simulation libraries, unstructured data [29], or executables cannot be used, or are only included implicitly. Still, most publications about Digital Twins refer to this kind of information. Since this chapter is about the development of a Digital Twin for ML/AI applications, no reference is made to this type of knowledge. The conversion of unstructured knowledge into structured knowledge which is applicable for ML is not in the scope of the chapter.

In this chapter we follow the key idea in [21] in which it is claimed that "The intellectual heart of CPS is in studying the joint dynamics of physical processes, software, and networks". In terms of AI this means that the main task of a Digital Twin must be to use models to predict the behavior of the corresponding CPS

**Fig. 3.1** Illustration of the first research question. The Development of a standardized API for Digital Twins

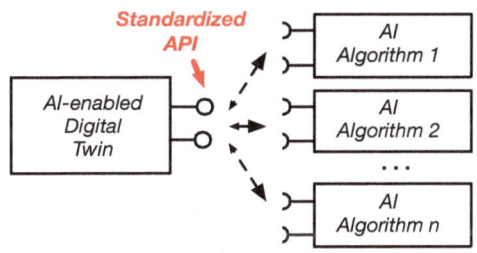

subsystem, comprising cyber- and physical parts. Furthermore, in this chapter the following constraint is added: The prediction knowledge, including domain knowledge, must be explicit and processable by an AI algorithm. In summary, the chapter addresses the following three research questions (RQ):

1. Is it possible to develop a common API of a Digital Twin which is suitable for use with a set of heterogeneous AI algorithms like shown in Fig. 3.1 some AI application shall be able to use this standardized API to get and post training and data, as well as results? Therefore, neither the AI algorithms nor the internal structure of the Digital Twin is addressed in this chapter but a common application programming interface (API) for Digital Twins is presented.
2. Can AI algorithms use a common information base being stored in the Digital Twin? How can AI algorithms in different hierarchies exchange information that way? An algorithm for fault diagnosis, for example, might require information about detected anomalies. How can this information be transmitted and stored using a standardized Digital Twin?
3. Can a Digital Twin be integrated with Machine Learning, Informed Machine Learning, as well as with symbolic AI algorithms? Methods from these different fields have quite different requirements on their input data. Bringing sub-symbolic and symbolic AI together is currently a major topic in the AI community.

### 3.1.1  Related Work

The Digital Twin with its enabling technologies plays a crucial role for success in several areas and is a current field of research. Fuller gives a very comprehensive overview [14] of manufacturing, smart cities and healthcare. He concludes that there is still a lack of Digital Twin reference models, leading to discrepancies between similar projects and thus slowing the progress of this technology. Proposals for Digital Twin reference models have already been made by [1] focusing on risk control and prevention in process plants, by [3] for cloud-based CPS, and by [22], which, among others, developed a web-based Digital Twin of a thermal power plant. An example of a more recent publication that considers Digital Twins in combination with ML is given by Castellani [7], who uses the Digital Twin to generate synthetic datasets for anomaly detection. As one of the first papers in the Industry 4.0 initiative, Löcklin et al. [20] proposes an AAS for data analytics projects.

Section 3.2 analyzes the requirements of AI/ML methods to the corresponding models, Sect. 3.3 presents the main solution approach, deriving the Digital Twin AI reference model, called **AITwin**. Case studies, being the major part in the publication, can be found in Sect. 3.4. A conclusion is given in Sect. 3.5. The results in this publication build on the results presented in [2]. The chapter would improve by explaining more using the taxonomy why one wants to use a digital twin, at least to a large degree because a digital twins provides access to representations of knowledge about the application.

**Informed Machine Learning and the Digital Twin**

Informed Machine Learning refers to the use of prior knowledge and domain expertise to improve the accuracy and efficiency of ML models. Instead of relying solely on data-driven approaches, informed ML integrates prior knowledge into the modeling process e.g. by guiding the selection of relevant features, optimization algorithms, or model architectures. Within [31] a comprehensive overview and taxonomy on how and in which way prior knowledge can be used with ML. Promising results for the interaction of a Digital Twin with informed machine learning algorithms have already been shown in recent publications. Gong et al. [16] generates a physics-informed digital twin and uses it in combination with ML to solve problems in nuclear reactor physics, Jiang et al. [18] uses ML and the Digital Twin to develop a surrogate model for coastal floods, Chakraborty and Adhikari [8] combine a physical and data-driven model to describe multi-timescale dynamical systems. Still, results show that combining ML with its Digital Twin results always in an individual solution for specific application field. The herein presented work tries to close this gap by proposing the Digital Twin reference model AITwin. In particular, the approach for integrating causal dependencies presented in Sect. 3.3.3 is crucial for informed ML, since this information can be directly integrated into the model.

## 3.2 ML/AI and the Digital Twin

Within this section, a classification of different AI methods is performed and requirements for Digital Twin are derived. Like the large number of different methods, there exists also a large number of different formalisms to model and represent domain knowledge as well as the description of the environment. For simplicity, within this chapter the focus is on methods being used mainly for Cyber-Physical-Systems. Therefore, the AI methods are classified into three main categories shown on the left side of Fig. 3.2.

**Fig. 3.2** Complexity of the digital twin in comparison to different levels of ML

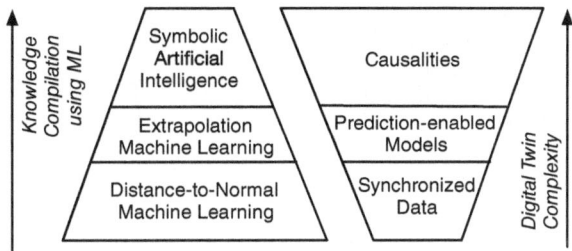

*Distance-to-Normal Machine Learning* Typical examples of this approach are traditional supervised and unsupervised approaches to anomaly detection. In that case the machine learning model stores historical data points and calculates a distance measure between an incoming data sample, e.g. from a production process and the stored points. A classic use case is that historical data points are interpreted as "normal" and samples that are in distance too far away from those samples are classified as "abnormal". To some extent, interpolation can be used by some methods to process data points that are close to previous data points. To implement distance-to-normal machine learning, the digital twin, shown on the right side in Fig. 3.2, must contain a common temporal model to synchronize all data coming for the CPS. Still, no prior knowledge needs to be made available for the ML model, since results a completely data-driven.

*Extrapolation Machine Learning* This subsumes applications such as control and optimization tasks. The use of a Digital Twin for optimization and self-configuration in autonomous systems means that it is used in a closed control loop. A classic example in that case is the use for model predictive control tasks. The Digital Twin has the task to predict output values like resource consumption or component positions which is again used for new input values. It is pointed out, that this extrapolation capability means that new values are predicted from the observed data points far away in time and/or space. In terms of Machine Learning, this can especially cause problems for technical systems if there are only a few individual data points available. In this case, methods that integrate physical and data-based information can bring advantages.

*Symbolic Artificial Intelligence* Compared to the two other categories, symbolic AI is not considered as being part of ML since it is not data-driven. Tasks that cover the fields of process reconfiguration, planning, fault diagnosis or similar need additional causal information of the process. In terms of AI this means that symbolic models are needed: Diagnosis needs at least cause-symptom relationships (e.g. if the pipe leaks, the tank fills slower than expected), planning requires as minimal information the different steps in the process (e.g. a specific actuator moves a component from position 1 to position 2). It needs to be noted that causalities always describe effects resulting from one or several causes. Both, effects and causes, are symbolic predicates in a formal logic. In that case, the Digital Twin must contain such logic-

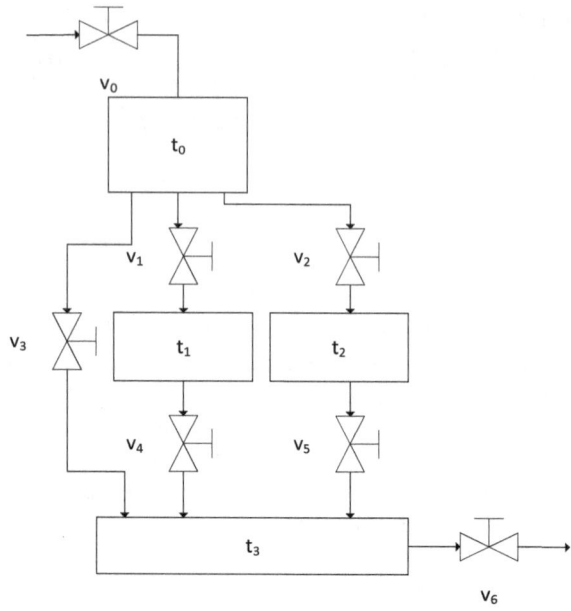

**Fig. 3.3** The introductory example for the different complexity levels of the Digital Twin in form of a four tank model

based models. Furthermore, those models can be for Informed Machine Learning, e.g. by integrating the logic directly into the ML model to solve a specific task.

**Introductory Example**

To illustrate the possibly used ML methods, a running example, as shown in Fig. 3.3 is used. The system is composed of four water tanks $t$, seven valves $p$ with flow sensors, a water source and a water sink (not shown). Possible applications for the prior presented AI methods are:

*Distance-to-Normal Machine Learning* The joined vector of all flows as well as the filling levels and valve positions can be used to detect anomalies in the system. Only the synchronized data stream is needed.

*Extrapolation Machine Learning* The joined vector of all flows can be used to predict some sort of consumption profile. In combination with domain knowledge (e.g. how to valves can the controlled) this information can be used to optimize the system configuration.

*Symbolic Artificial Intelligence* The switching of a valve leading to the overflow of a container can be modeled. In a subsequent step, this model can be used for an AI to perform an diagnosis and to identify the root cause of an error.

## 3.3   AI Reference Model

Within this section the Digital Twin  AI reference model, the *AITwin*, is derived. In general, the AI reference model provides a set of interfaces in different levels of abstractions that some AI method may use to obtain its input data and post its outputs. The actual implementations of such an AI Twin may differ and may include logical knowledge bases, simulations, time-series data, or graphs. Most importantly, though, the API shall be general enough such that it is irrelevant what exact process (logical model, simulation etc.) has generated such data. In case of the introductory example of the four-tank system the AI Twin would provide synchronized time-series data such as the valve-set points, water level in the tanks, and flow-rates through the pipes.

### *3.3.1   Synchronized Data*

Time is one main feature of physical processes. However, software does not inherently handle the concept of time. Within physical processes parallel events can occur at arbitrary points in time. Often, the time span between commands is defined through some measure, but through variations in the executing hardware, the physical time span can vary. It is therefore necessary to establish a common time model of the Digital Twin . Following the idea from [21], this leaves two options:

*Cyberizing the Physical*  A signal values $x$ as $x(t, k)$ is defined where $t \in \mathbf{R}$ is the physical time and $k \in \mathbf{N}$ the number of an ordered sequence of events/instructions in the software.

*Physicalizing the Cyber*  For time synchronization in distributed systems the Precision Time Protocol [17] allows for a consistent concept of time, especially in controllers. An overview of industrial data acquisition for real-time system are given in [13]. This means that software can refer to precise points in time.

"Cyberizing the Physical" has the drawback that most ML algorithms need to be adapted, as they need a clear time model and concept of causality. Out of this reason, "Physicalizing the Cyber" is preferred. All information $\mathbf{x}(t)$ refers to a unique point in time $t$, which means that the underlying technical systems contain a time synchronization between all devices. Timing refers first of all to the measurement signals while prior knowledge plays a subordinate role.

To obtain synchronized data the Digital Twin  needs two interfaces:

1. The function *getData(i,t)* returns the i'th signal $x \in \mathbf{R}$ at time point $t$. Here i is the i'th signal of total number $\mathcal{X}$ of available signals in the digital twin.
2. The function *getData(t)* returns all signals $\mathbf{x} \in \mathcal{X}$ at a time point $t$. While $\mathcal{X}$ is the set of all possible values $\mathbf{x}(t)$ and vector $\mathbf{x}(t)$ records the complete observable state of the physical system (Fig. 3.4).

**Fig. 3.4** Interface needed for
the AI twin reference model
for synchronized data

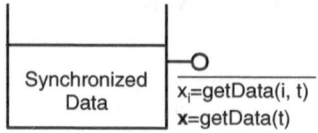

## 3.3.2 Prediction-Enabled Models

Prediction models can be structured in different ways. On the one hand, there are the purely data-driven models which cover the majority of use cases, but also expert systems or simulation models are possible. Regarding the Digital Twin it is important that all models cover the same API. Especially with data-driven ML models there are a number of difficulties when applied to a CPS, like for manufacturing or similar. One main challenge is that usual data intensive ML algorithms like deep learning, often fail or deliver wrong results even if they generate very good results in other domains. The reasons are:

- Data in production systems is comprised of repetitive patterns—mirroring the repetitive structure of typical production processes, this leads to sufficient data but insufficient information.
- In production data only small numbers of samples include failure information or other problems.
- Physical system data seldom contains multiple instances of data from the same failure.

Thus, in many cases learned models from CPS have only limited extrapolation capabilities. Even if learned models predict outputs for new input values, prediction accuracy quickly decreases when values are far away from observations in time or distance. Hence, the proposed AI reference model, not only needs to extrapolate data, but also needs to provide information about the quality of extrapolation. Specifically, probabilistic models should be used such that predictions are not only implemented as a point estimator, but contain the probability that the prediction is correct. In the case of the introductory example, the extrapolation could be implemented by control theoretic methods, that vary the set-point of a valve depending on some tank level. If the tank's water level changes the valve would be adjusted accordingly in order to keep a stable state, for example. Within the chapter, extrapolation is distinguished between so-called static and dynamic predictors. Static predictors make predictions only based on the current value, while in contrary, dynamic predictors makes predictions based not only on the current but on previous values, e.g. by looking at history table. Both are explained in the following.

**Static Predictions**
Condition-monitoring or anomaly detection are examples of static predicts, since they only use signal values $\mathbf{x(t)} \in \mathbf{R}^n$ at some point in time $t$. Thereby, the assumption is that the data sequence with respect to time does not contain any

information. It needs to be noted that this assumption is true for many CPSs, even for systems which have a dynamic nature them-self. Three typical fields of application can be derived from this:

- Anomaly Detection: For a new data point $\mathbf{x} \in \mathbf{R}^n$ its probability $p(\mathbf{x}|X)$ given some historical data $X$ is computed.
- Prediction: Given a feature vector $\mathbf{x} \in \mathbf{R}^n$, those entries are used to predict a label $\mathbf{y}$ with its probability.
- Optimization: The partial gradient over the signals can be created by extrapolating in the vicinity of an operation point. Hence, the Digital Twin needs no special functionality.

**Dynamic Predictions**
Once information is inside a sequence of signal values over time a different situation arises. Here, three typical fields of application can be derived:

- Anomaly Detection: Given historical data $X = \{\mathbf{x}(t = t_0), \mathbf{x}(t = t_1), \ldots, \mathbf{x}(t = t_k)\}$, for a time step $\Delta t$ a new data vector $\mathbf{x}(t = t_k + \Delta t)$ is computed with its probability, such that $\mathbf{x}(t = t_k + \Delta t)$ is the logical continuation of the series.
- Prediction: Provided historical data $X = \{\mathbf{x}(t = t_0), \mathbf{x}(t = t_1), \ldots, \mathbf{x}(t = t_k)\}$ and time step $\Delta t$ the next value $\mathbf{x}(t = t_k + \Delta t)$ is computed with its probability distribution.
- Optimization: The temporal gradient over the signals is be created by extrapolating in the vicinity of the time point $\Delta t$. Therefore, no special functionality of the Digital Twin is needed here.

For reliable predictions, the model should not only capture normal behavior but also different failure modes. Hence, the Digital Twin must comprise this knowledge and be enabled to set some components to failure states. Figure 3.6 presents the resulting API methods with extrapolation. In the static case the extrapolation is solved by a function *extrapolateStatic(x')*, where **x'** is a partially filled vector with signals. The function outputs the missing values **x** with the probability vector, with *null* entries if not computable. This solves the three above discussed use cases. The function *extrapolateDynamic(X, $\Delta t$)* with **X** being a sequence of historical signals up to time $t$ is defined for the static use case. The function outputs the next signal vector $\mathbf{x}(t + \Delta t)$ with associated probabilities, both with *null* entries if not computable. Again, this solves the three discussed use cases. Different failure modes are set using the function *setFailedComps(C')* with **C'** being a subset of all components within the physical system. The functions *extrapolateStatic* and *extrapolateDynamic* predict the component's behavior assuming that the components are not working correctly. Function *getComps()* returns the names of all components.

### 3.3.3  Causalities

For ML algorithms and especially for the emerging topic of informed ML the integration of physical information, such as the knowledge of causal dependencies, is becoming more and more important. When causal dependencies are included in the Digital Twin , logical rules or knowledge graphs can be derived and according to [31], used to be integrated into the ML model. Unfortunately, concerning causality, until now there is no definition of causality that is accepted by the majority of researchers. Therefore, within this section, initially, we analyze the specific usage of causality models in two main AI applications and from there we derive a suitable definition of causality for the Digital Twin . To illustrate, in the introductory example, causalities could be represented through a graph-based structures or logical frameworks [11]. Such graph-based structures describe which component's output influences which other components. This is necessary information for tasks such as fault diagnosis and planning.

**Application 1—Consistency-Based Diagnosis**
Diagnosis is the task of computing a fault based on incomplete symptoms, e.g. observations. Therefore, knowledge about causalities between faults and symptoms must be available since failures can often lead to complex symptom patterns. More specific, operators often face complex alarms but fail to identify the true fault. Models often predict only normal behavior, while consistency-based diagnosis (CBD) [10] often uses so-called weak fault models. In many cases, CBD uses partial system models such as

$$OK(C_{i_1}) \wedge \ldots \wedge OK(C_{i_k}) \rightarrow (s_1 \rightarrow s_2 \wedge \ldots \wedge s_l), k, l \in \mathbf{N}, \tag{3.1}$$

where $OK(C)$ denotes a correct functioning of component $C$ and $s_i \in \{0, 1\}$ is a binary/symbolic representation of the current system status. This means, if components $C_{i_1}, \ldots, C_{i_k}$ work correctly, the implication describes the normal system causality. CBD then redraws OK-assumptions until no contradictions between predictions and observation are left.

**Application 2—Planning**
Planning describes the procedure to compute a sequence of process steps to transform raw objects into some given final object. Let $P$ be the set of all possible objects. Then a process step $q_i$ is an implication $q_i : x \rightarrow y$, which takes $u$ input objects $x \in P^u$ and transforms them into $v$ output objects $y \in P^v$. We denote the set of all process steps as $Q$.

**Fig. 3.5** Causality concept of AITwin, $s_i$s are system states, $C_j$ are components, $p_k$ are products

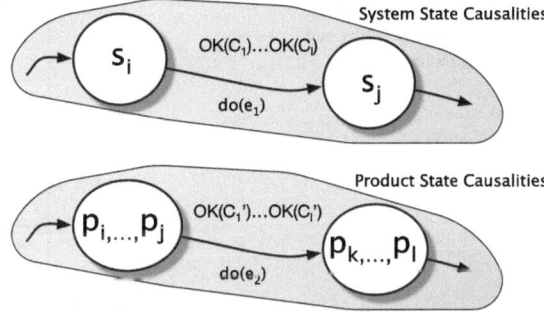

### 3.3.3.1  System and Product State Causalities

From those two applications, a common causality model for the Digital Twin can be defined. Figure 3.5 shows two distinct types of causalities, namely system state causalities and product state causalities. Both are explained in the following.

**System State Causalities**
System state causality models describe the transition from one system state $s_i \in S$ to another system state $s_j \in S$. The occurrence of some event $e_1 \in \{0, 1\}$ triggers this transition, while the event $e_1$ denotes Pearl's **do** operator [30]. This means, that if $e_1$ is forced to be one, the event is triggered.

A system state $s \in S$ corresponds to a subset of $\mathcal{X}$, with $\mathcal{X}$ being the set of all possible values $\mathbf{x}(t)$. An event $e_1$ is defined as the crossing of some threshold in $\mathcal{X}$. The thresholds correspond to a number of equations of the form $\mathbf{f} \cdot \mathbf{x}(t) < \mathbf{c}, \mathbf{f}, \mathbf{x}(t), \mathbf{c} \in \mathbf{R}^n$ defining a halfspace of $X$. Such a system state causality is only valid under the condition that specific components $C_1, \ldots, C_l$ are functioning correctly, f.e. $OK(C_1) \wedge \ldots \wedge OK(C_l)$ is true (see Sect. 3.3.2). With this information the diagnosis (CBD) task from above can be implemented, such that consistencies between predictions $\mathbf{x}(t)$ from the predictions API level and causalities predicting system states $s_i$ can be checked as $s_i$ is a subspace of $\mathcal{X}$.

**Product State Causalities**
Product state causalities model the transition from one set of objects $p_i, \ldots, p_j$ to another set of objects $p_k, \ldots, p_l$. A product $p_i$ might be any kind of complete or partial product or some raw material.

The occurrence of an event $e_2$ triggers this transition, while empty events are possible. Events are interpreted as defined above. A system state causality is only valid under the condition that specific components $C'_1, \ldots, C'_l$, see Sect. 3.3.2, are functioning correctly, e.g. $OK(C'_1) \wedge \ldots \wedge OK(C'_l)$ is true. The resulting API extensions are shown in Fig. 3.6. Using this information the planning task from above can be implemented.

Using an implication API allows for the handling of causalities in form of transitions or implications: The function getSystemCausalities($s_i$) returns all causalities

**Fig. 3.6** The complete Digital Twin AI reference model AITwin

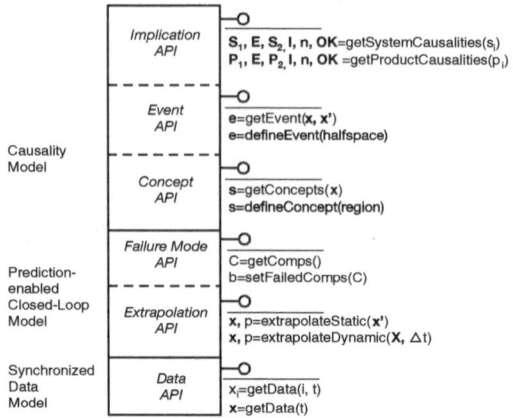

$s_1, e, s_2, I, n, OK$. Where $s_1$ denotes the starting state, $s_2$ the end state, $e$ is the event, $n$ the name of the causality and $I$ stores additional information such as probabilities, timing etc. $OK$ is the set of $OK$ assumptions $OK = OK(C_1) \wedge \ldots \wedge OK(C_l)$ which are a precondition for the validity of the causality. The function getProductCausalities($p_i$) returns all causalities $p_1 = (p_1^1, \ldots, p_1^l), e, p_2 = (p_2^1, \ldots, p_2^k), I, n, OK$ where $p_1$ denotes the original objects, $e$ is the event, $p_2$ denotes the resulting objects, $n, I, OK$ are defined as above.

$e = $ defineEvent(halfspace) defines an event $e$ which corresponds to a halfspace. The parameter *halfspace* is defined using a mathematical inequality in the language MathML [27]. The function *getEvent*($\mathbf{x}, \mathbf{x}'$) returns all events which are triggered when moving from the operation point $\mathbf{x} \in \mathcal{X}$ to $\mathbf{x}' \in \mathcal{X}$.

$s = $ defineConcept(region) defines a system state $s$, namely a concept, as a causal convex region *region*, meaning that there is no violation of causality. The parameter *region* is part of in $\mathcal{X}$ using a mathematical inequality in the language MathML [27]. The function *getConcepts*($\mathbf{x}$) returns all system states $s_i$ (aka concepts) with $\mathbf{x} \in s_i$.

### 3.3.4 The AITwin Reference Model

The complete AITwin reference model is shown in Fig. 3.6. The focus on those parts that are required by AI algorithms and further information about specific domains must be added. It has to be noted that the AITwin can be used in parallel with different system granularities. Note that the three levels on the right of Fig. 3.2 correspond to the three levels on the left of Fig. 3.6 and form a holistic concept: Data $\mathbf{x}$ from the "Synchronized Data Model"-API is used also in the "Prediction-enabled Closed-Loop Model"-API. These data points may further be correlated to system states from the "Causality Model"-API. Note that all APIs use the same components for modeling failure modes.

Besides the APIs, the Digital Twin is based on three state spaces necessary for the Digital Twin definition: $\mathcal{X}$ denotes all possible signal values, $P$ denotes all possible object/product configurations and $COMPS$ is the set of components which may fail. Further, two derived state spaces exist, which normally are defined via API methods during the operation phase: $S$ is a set of system states where $s \in S$ is a subset of $\mathcal{X}$ and $E$ is the set of all events where $e \in E$ is a halfspace of $\mathcal{X}$.

## 3.4 Evaluation

The aim of this section is to focus on known use cases from literature as well as one real-world process. This includes the four tank model from control engineering, which has already been presented in Sect. 3.2, the Tennessee Eastman Process [12] and the SECOM [28] data set. Finally, as a real-word use case, the Digital Twin is evaluated on a sensor-based sorting system.

### 3.4.1 Applying AITwin to a Four Tank Model

To evaluate AITwin the API is used on the four tank process introduced in Sect. 3.2. Therefore, especially the *getData(t)* method to have synchronized data and the *extrapolateStatic(x')* method have been used to detect anomalies in the measurements. Abnormal data has been inserted via the *setFailedComps(C)* API.

In a first step, data describing the normal behavior of the process is extracted by using the *getData(t)* method from the Digital Twin. As a subsequent step these samples are used to perform the training of the model by using an unsupervised learning algorithm. To evaluate the trained model, non-normal data is extracted, again using the *getData(t)* function.

### 3.4.2 Applying AITwin to Tennessee Eastman Process

As second use case, the AITwin API is applied to symbolic AI tasks. Therefore, concepts are extracted from the Digital Twin via the "concept" API and causalities via the "implication" API. This information can be used to create a model for Consistency-Based Diagnoses (see Sect. 3.3.3). Failures can be added via the "failure" API. As Table 3.1 shows the results, most failures can be identified correctly.

Therefore, two quantitative simulations of the Tennessee Eastman process where used. (1) The implementation of Downs et al. [12] and (2) an implementation of the alarm management benchmark developed by Manca [26].

**Table 3.1** Results for the
simulated Tennessee Eastman
Process [12]

| IDV | Fault isolated | Injected fault |
|-----|----------------|----------------|
| 1 | *not ok* | Feed ratio changed |
| 6 | *ok* | Pipe A feed loss |
| 8 | *ok* | Feed ratio changed |
| 13 | *not ok* | Reactor kinetics fault |
| 14 | *ok* | Reactor cooling fault |
| 15 | *ok* | Condenser cooling fault |

**Table 3.2** Experimental results with the Tennessee Eastman process by Manca. $|\omega'|$ denotes the size of the diagnosis set

| Runs | Avg. # injected faulty $OBS$ | $|\omega'|$ |
|------|------------------------------|-------------|
| 100 | 13.16 | 2.05 |

In summary, the proposed simulation in [12] has the possibility to generate 20 different injected faults. Still, it is assumed that the instrumentation does not exactly identify all faults and some fault modes are not due to the fault of one component. Hence, Table 3.1 shows the results for the remaining six experiments.

In Table 3.1, the *ok* means that the that the minimal cardinality diagnosis found faulty component, while the *not ok* means that the faulty component has not been detected by using the diagnosis algorithms. In summary, the AI-solution is able to detect all faults which are identifiable as the cause of the process disturbance. Since no observations are available at the inputs, the change of input ratios can only be detected indirectly.

Regarding the AITwin reference model, the *getData(t), setFailedComps(),* and *getSystemCausalities()* functions were used. *getData()* is used to obtain simulated process information from the simulation, *setFailedComps()* returns binary residuals for each component, and *getSystemCausalities()* returns expert-defined causality rules formulated in predicate logic.

Using the another simulation developed by Manca [26], we were able to evaluate more fault modes than by the implementation described by Downs et al. Additionally, we used a different implementation of the AITwin API: *getComps()* is implemented through expert knowledge denoting the time and place when faults occur. *setFailedComps()* is implemented through a simulation. The system causalities are learned data-driven using *getSystemCausalities()* to obtain a Granger Causality measurement. The results in Table 3.2 show that through Granger Causality we can obtain predicate logic rules suitable for consistency-based diagnosis and symbolic AI.

It is evident that, although on average 13 faults were injected (with a high standard deviation), the average diagnosis size is small compared to the number of components. It must be noted, however, that the number of components contained in the rules is less than those in the simulation, due to the constraints explained above.

**Table 3.3** Results with
systems S1, S2, S3, S4 of the
developed benchmark

| System | Runs | Avg. # injected faulty $OBS$ | $|\omega'|$ |
|--------|------|------------------------------|-------------|
| S1     | 100  | 1.09                         | 3.66        |
| S2     | 100  | 1.9                          | 7.7         |
| S3     | 100  | 2.41                         | 4.5         |
| S4     | 100  | 0.52                         | 1.38        |

We have used the same API functions as above to automatically generate rules
and diagnose systems S1, S2, S3, and S4 of the benchmark developed by Balzereit
et al. [5]. The results are presented in Table 3.3.

It can be seen that although the average number of injected faults was low (albeit
with a high standard deviation), the number of possible faults components is large
compared to the system size. This can be explained by the connected nature of the
tanks systems, where injected faults propagate almost instantaneous.

### 3.4.3 Applying AITwin to a Quality Assurance Example

A real manufacturing process of semiconductors is used as an application example
for the use of the AITwin model for product quality assurance. This process is
continuously monitored using various sensors and measuring points, so that 591
characteristics are recorded and documented for the evaluation of the product
quality. These data points are included in the SECOM dataset [28] for 1567
instances along with the quality assessment (pass or fail). As [28] describes, not
all included features have the same relevance. Therefore, the features should be
evaluated by an intelligent system depending on their impact on the product quality.
Furthermore, causal relationships between the identified main features could be
detected. Arif et al. [4] uses a ML approach to train a prediction model that classifies
this dataset with an accuracy of >90%. With the API introduced in Sect. 3.3.3, the
prediction model can be connected to the digital twin in a standardized way. Similar
to the sensor-based sorting system described in Sect. 3.4.4, the collection of sensor
data is done by the function *getData(t)*. Once the prediction model is trained, it
becomes part of the Digital Twin and can be called through the Extrapolation API.
By using the function *extrapolateStatic($x'$)* the predicted product quality can be be
retrieved, while **x'** describes in the input parameters needed for the prediction of the
quality.

### 3.4.4 Applying AITwin to a Sensor-Based Sorting System

As a real world example on how to apply the AITwin model, a lab-scale sensor-
based sorting system is used. Sensor-based sorting is a single particle separation

**Fig. 3.7** Sensor-based
sorting system used to
evaluate the AITwin API at
Fraunhofer IOSB

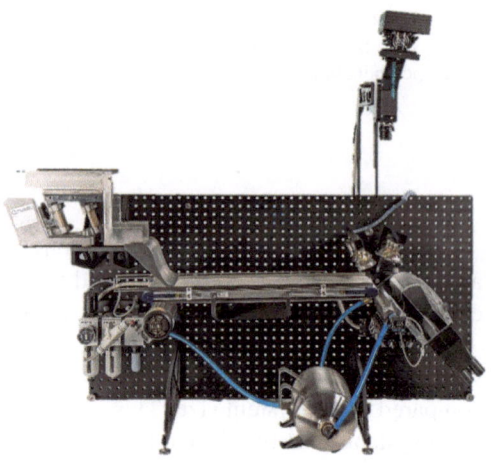

technology that finds wide application in the fields of mining, recycling, and
the processing of agricultural products and foodstuffs. The goal is to remove
residues, for instance low-quality entities, from a product stream to be recovered.
The functional principle, as for instance described in [25] in more detail, can
be summarized as follows. A material stream is observed by one or multiple
imaging, line-scanning sensors. The sensor data is then processed with the goal
to calculate a sorting decision per particle. The sorting decision is then executed by
an actuator that handles the actual physical separation. Most commonly, an array of
fast switching pneumatic valves is used for this purpose. An image of the considered
laboratory scale sensor-based sorting system is given in Fig. 3.7. The system is used
to sort small quantities of bulk materials, in our case roasted coffee beans according
to their quality. The system is equipped with a vibratory feeder, a conveyor belt,
a hyperspectral short-wavelength infrared (SWIR) line-scan camera and an array
of pneumatic valves. The coffee bean quality is determined using the camera and
an array of pneumatic valves is used to blow out the low quality roasted beans.
Since the conveyor belt runs on high speed, in our case approximately $1\frac{m}{s}$, the
attached valves need to execute the sorting decision in high speed as well. To ensure
optimum sorting quality, the beans need to be evenly distributed on the conveyor belt
perpendicular to the transport direction. Whenever this is not the case, the following
two problems may arise:

- The valves on the valve bar operate unevenly. This leads to the point that
  individual valves have to be replaced more frequently since some valves have
  reached their maximum switching life cycle.
- Due to the fact that the beans are discharged from the conveyor belt at high speed,
  if only a small number of valves is operating, the sorting quality is reduced since
  the valve can not open and close as fast as low quality beans pass by.

To remedy these two problems, a ML model, namely an Autoencoder, is used to
monitor the lateral distribution of particles using the sensor data as well as the

behavior of the valve bar, respectively checking if the bean distribution on the conveyor belt and the activation of valves follows an equal distribution. Therefore, the Autoencoder is trained on a data set containing the observed positions and valve behavior with equally distributed beans. In a subsequent step this model is used to monitor the distribution and raise an alarm if the valve bar is unevenly stressed.

In terms of Fig. 3.2, the ML-task corresponds to distance-to-normal learning. Since the digital twin of the sensor-based sorting system has already implemented some features of the AITwin reference model, for training the model, the function *getData(t)* is used to collect the training data. Therefore, an operator monitors whether the beans are equally distributed. After training, the Autoencoder itself becomes a part of the Digital Twin . Therefore, the function *extrapolateStatic($\mathbf{x'}$)* is implemented and used to ask the current ML model results, e.g., if the valves are evenly stressed. In that case, $\mathbf{x'}$ describes the opening and closing of the individual valves over a certain time interval. As mentioned after training, the Autoencoder becomes a part of the AITwin and can be queried by using the function *getComps()*.

## 3.5   Discussion and Future Work

Within the introduction, three research questions focusing on the Digital Twin for AI based solutions were identified as key challenges. In short, the challenges were the development a common API, a common information storage and whether an interaction between symbolic and sub-symbolic AI is possible. As an answer to those challenges, a common Digital Twin reference model, called the AITwin, has been defined. Therefore, in Sect. 3.3, initially the conceptual requirements of CPS with respect to timing models were analyzed. This was expanded in Sect. 3.3.2 by analyzing the conceptual requirements of CPS with respect to machine learning models. Section 3.3.3 continued with the analysis of the conceptual requirements of CPS with respect to symbolic AI models and Informed Machine Learning. Finally, all sections led to the point that a common application programming interface (API) could be derived in Sect. 3.3.4. It is concluded that the proposed and developed AITwin reference model could answer all three key challenges defined in the introduction.

In summary, the authors do not claim that following the ideas of the developed reference models is the ultimate definition of a Digital Twin for AI solutions. But we hope it is still a helpful standardized interface which practitioners may use to build compatible AI applications over several hierarchies of abstractions. Still, it can be assumed, that in the future Digital Twins, as well as AI, will be one of the center-points of future CPS architectures. Nonetheless, lots of work still needs to be done to bring both worlds together and to harmonize them. When looking at some public research projects it can be identified that their results may easily benefit from adherence to the AITwin model. For examples, the results from Li [23], could have used many of the methods to predict the surface roughness in extrusion-based additive manufacturing. Similarly, in [6] an on-line defect recognition may have benefited from an AITwin. Planning and diagnosis tasks may also derive most of their needed information from the AITwin. Thus far, descriptions of machines and about most expert knowledge must be implemented individually and often using proprietary methods. Using the AITwin, instead of having several incompatible data models, domain knowledge, interfaces, and formats, it is possible to use a single Digital Twin with its common interfaces and thus integrate a multitude of disparate algorithms. However there are still some applications whose requirements may not be met specifically. For example, applications making explicit use of simulations. The API is designed in such a way that simulations are only included implicitly. This is sufficient for many industrial AI use-cases, for example, reinforcement learning applications would prefer a more simulation oriented API.

**Acknowledgments** This contribution was supported by the Fraunhofer Cluster of Excellence "Cognitive Internet Technologies".

# References

1. Bevilacqua et al., M.: Digital twin reference model development to prevent operators' risk in process plants. Sustainability **12**(3), 1088 (2020)
2. Niggemann et al., O.: A generic digitaltwin model for artificial intelligence applications. In: IEEE International Conference on Industrial Cyber-Physical Systems (ICPS) (2021)
3. Alam, K.M., El Saddik, A.: C2PS: A digital twin architecture reference model for the cloud-based cyber-physical systems. IEEE Access (2017)
4. Arif, F., Suryana, N., Hussin, B.: A data mining approach for developing quality prediction model in multi-stage manufacturing. International Journal of Computer Applications **69**(22), 35–40 (2013)
5. Balzereit, K., Diedrich, A., Ginster, J., Windmann, S., Niggemann, O.: An ensemble of benchmarks for the evaluation of AI methods for fault handling in CPPS. In: 19th IEEE International Conference on Industrial Informatics (2021)
6. Caggiano, A., Zhang, J., Alfieri, V., Caiazzo, F., Gao, R., Teti, R.: Machine learning-based image processing for on-line defect recognition in additive manufacturing. CIRP Annals **68**(1), 451–454 (2019)
7. Castellani, A., Schmitt, S., Squartini, S.: Real-world anomaly detection by using digital twin systems and weakly supervised learning. IEEE Transactions on Industrial Informatics (2021)
8. Chakraborty, S., Adhikari, S.: Machine learning based digital twin for dynamical systems with multiple time-scales. Computers and Structures **243**, 106410 (2021)
9. Cheng, D.J., Zhang, J., Hu, Z.T., Xu, S.H., Fang, X.F.: A digital twin-driven approach for on-line controlling quality of marine diesel engine critical parts. International Journal of Precision Engineering and Manufacturing **21**(10), 1821–1841 (2020)
10. Diedrich, A., Niggemann, O.: Model-based diagnosis of hybrid systems using satisfiability modulo theory. In: Thirty-Third AAAI Conference on Artificial Intelligence (AAAI-19) (2019)
11. Diedrich, A., Niggemann, O.: On residual-based diagnosis of physical systems. Engineering Applications of Artificial Intelligence **109**, 104636 (2022)
12. Downs, J.J., Vogel, E.F.: A plant-wide industrial process control problem. Computers & chemical engineering **17**(3), 245–255 (1993)
13. Flammini, A., Ferrari, P.: Clock synchronization of distributed, real-time, industrial data acquisition systems. In: M. Vadursi (ed.) Data Acquisition, chap. 3. IntechOpen, Rijeka (2010)

14. Fuller, A., Fan, Z., Day, C.: Digital twin: Enabling technology, challenges and open research. arXiv preprint arXiv:1911.01276 (2019)
15. German Institute for Standardization: Reference architecture model industrie 4.0 (rami4.0); din spec 91345 (2016)
16. Gong, H., Cheng, S., Chen, Z., Li, Q.: Data-enabled physics-informed machine learning for reduced-order modeling digital twin: Application to nuclear reactor physics. Nuclear Science and Engineering **196**(6), 668–693 (2022)
17. https://www.nist.gov/el/intelligent-systems-division-73500/ieee-1588. Accessed 10/10/22
18. Jiang, P., Meinert, N., Jordão, H., Weisser, C., Holgate, S.J., Lavin, A., Lutjens, B., Newman, D., Wainwright, H.M., Walker, C., Barnard, P.L.: Digital twin earth – coasts: Developing a fast and physics-informed surrogate model for coastal floods via neural operators (2021)
19. Karniadakis, G.E., Kevrekidis, I.G., Lu, L., Perdikaris, P., Wang, S., Yang, L.: Physics-informed machine learning. Nature Reviews Physics **3**(6), 422–440 (2021)
20. Löcklin, A., Vietz, H., White, D., Ruppert, T., Jazdi, N., Weyrich, M.: Data administration shell for data-science-driven development. Procedia CIRP pp. 115–120 (2021)
21. Lee, E.A.: CPS foundations. In: Proceedings of the 47th Design Automation Conference, DAC '10. ACM, New York, NY, USA (2010)
22. Lei, Z., Zhou, H., Hu, W., Liu, G.P., Guan, S., Feng, X.: Towards a web-based digital twin thermal power plant. IEEE Transactions on Industrial Informatics (2021)
23. Li, Z., Zhang, Z., Shi, J., Wu, D.: Prediction of surface roughness in extrusion-based additive manufacturing with machine learning. Robotics and Computer-Integrated Manufacturing **57**, 488–495 (2019)
24. Madni, A.M., Madni, C.C., Lucero, S.D.: Leveraging digital twin technology in model-based systems engineering. Systems **7**(1) (2019)
25. Maier, G., Pfaff, F., Pieper, C., Gruna, R., Noack, B., Kruggel-Emden, H., Längle, T., Hanebeck, U.D., Wirtz, S., Scherer, V., Beyerer, J.: Experimental evaluation of a novel sensor-based sorting approach featuring predictive real-time multiobject tracking. IEEE Transactions on Industrial Electronics **68**(2), 1548–1559 (2021)
26. Manca, G.: "tennessee-eastman-process" alarm management dataset. IEEE Dataport (2020). https://doi.org/10.21227/326k-qr90
27. https://www.w3.org/TR/REC-MathML. Accessed 10/10/22
28. Michael McCann, A.J.: SECOM data set, UCI machine learning repository. Accessed 10/10/22 (2008). URL https://archive.ics.uci.edu/ml/datasets/SECOM
29. Panzner, M., von Enzberg, S., Meyer, M., Dumitrescu, R.: Characterization of usage data with the help of data classifications. Journal of the Knowledge Economy pp. 1–22 (2022)
30. Pearl, J.: Causality: Models, Reasoning and Inference, 2nd edn. Cambridge University Press, New York, NY, USA (2009)
31. von Rueden, L., Mayer, S., Beckh, K., Georgiev, B., Giesselbach, S., Heese, R., Kirsch, B., Walczak, M., Pfrommer, J., Pick, A., Ramamurthy, R., Garcke, J., Bauckhage, C., Schuecker, J.: Informed Machine Learning – A Taxonomy and Survey of Integrating Prior Knowledge into Learning Systems. IEEE Trans. on Knowledge and Data Engineering **35**(1), 614–633 (2023)
32. Stark, R., Damerau, T.: Digital twin. In: L. Laperrire, G. Reinhart (eds.) CIRP encyclopedia of production engineering. Springer Publishing Company (2019)
33. Tao, F., Qi, Q., Wang, L., Nee, A.: Digital twins and cyber-physical systems toward smart manufacturing and industry 4.0: Correlation and comparison. Engineering **5**(4) (2019)

# Part II
# Optimization

# Chapter 4
# A Regression-Based Predictive Model Hierarchy for Nonwoven Tensile Strength Inference

Dario Antweiler, Jan Pablo Burgard, Marc Harmening, Nicole Marheineke, Andre Schmeißer, Raimund Wegener, and Pascal Welke

**Abstract** Nonwoven materials, characterized by a random fiber structure, are essential for various applications including insulation and filtering. An industrial long-term goal is to establish a framework for the simulation-based design of nonwovens. Due to the random structures, simulations of material properties on fiber network level are computational expensive. We propose a predictive model hierarchy for inferring an important material property—the nonwoven tensile strength behavior. The model hierarchy is built using regression-based approaches, including linear and polynomial models, which provide interpretable results. This allows for significant speedup (six orders of magnitude) over the conventional simulations, while achieving good prediction results ($R^2 = 0.95$). The proposed models open the application to nonwoven material design, as they provide accurate and cost-effective surrogates for predicting material properties. In this way, our work serves as a proof of concept.

D. Antweiler (✉)
Fraunhofer IAIS, Sankt Augustin, Germany
e-mail: dario.antweiler@iais.fraunhofer.de

J. P. Burgard · M. Harmening · N. Marheineke
Trier University, Trier, Germany
e-mail: burgardj@uni-trier.de; harmening@uni-trier.de; marheineke@uni-trier.de

A. Schmeißer · R. Wegener
Fraunhofer ITWM, Kaiserslautern, Germany
e-mail: andre.schmeisser@itwm.fraunhofer.de; raimund.wegener@itwm.fraunhofer.de

P. Welke
University of Bonn, Bonn, Germany
e-mail: welke@cs.uni-bonn.de

© The Author(s) 2025
D. Schulz, C. Bauckhage (eds.), *Informed Machine Learning*,
Cognitive Technologies, https://doi.org/10.1007/978-3-031-83097-6_4

63

## 4.1  Introduction

The efficient prediction of material properties based on production parameters is a common goal for many industrial applications. This includes the nonwoven airlay manufacturing, which serves as practical basis for this chapter. Nonwovens are characterized by a random fiber structure that is usually bonded by thermal, chemical, or mechanical procedures. Their low-cost production makes them a suitable choice for many fabrics, such as filters, insulation materials or hygiene products [7]. Predicting nonwoven properties from production parameters enables nonwoven material design by providing insight into the effects of individual parameters. In order to avoid costly experimental testing, this mainly involves simulation-based approaches, which, however, often suffer from high computational effort [33]. More recently, machine learning approaches have gained ground in this field, as they allow comparatively efficient predictions, see [2]. In particular, the integration of prior knowledge into the training process, termed "Informed Machine Learning", proved to be beneficial in terms of training speed and quality of final predictive models [29]. This chapter demonstrates the use of machine learning approaches for predicting nonwoven material properties. We focus, as an example, on the tensile strength behavior of airlay fabricated nonwovens (see Fig. 4.1a–c), for which we develop and propose a predictive model hierarchy driven by simulation data. With this goal in mind, we begin with a brief discussion of related literature in the field of nonwoven tensile strength simulations and machine learning approaches, and then explain the novelty and the structure of this chapter.

(a)                     (b)                     (c)                     (d)

**Fig. 4.1** Nonwoven airlay manufacturing and property testing: (**a**) simulated fiber dynamics and laydown in turbulent airflow (process of Airlay-K12 by machine manufacturer AUTEFA Solutions), (**b**) fiber laydown zone, (**c**) final product, (**d**) tensile strength test for a material sample (photo by IDEAL Automotive). Image adapted from [14]

### 4.1.1 Literature Overview

Mapping production parameters to the tensile strength behavior of nonwovens requires the simulation of the underlying production process and the mechanical behavior of the resulting fiber structure. There are many approaches to virtual generation of fiber structures, coming from statistical analysis and stochastic geometry [23, 31] or three-dimensional volume imaging covering microscopy and X-ray tomography [10, 24]. However, the challenge is to model the underlying production process, which was done by [14] for nonwoven airlay production. The authors introduced a chain of mathematical models coupled by parameter identification to deal with the computational complexity that arises from several thousand airlaid fibers in a complex machine geometry. The models cover a highly turbulent fiber suspension flow, a stochastic surrogate for the fiber laydown on a moving conveyor belt, and a bonding process mimicking the thermobonding. The suitability of such model hierarchies for virtual nonwoven generation is topic in [33].

Various approaches in the literature deal with the simulation of the mechanic behavior of nonwovens. A common procedure is to treat the nonwoven material as a continuum, which allows the use of finite element methods [8, 12]. In these approaches, the behavior of individual fibers is not considered, but instead knowledge of the statistical fiber orientation is incorporated to account for the randomness in the material web. In contrast, there are approaches that consider the mechanical behavior at the fiber network level, cf. [15, 19]. Kufner et al. [19] described the material's structure as an elastic Cosserat network. As resolving the behavior of the individual beam-type fibers in an industrial-size virtual material sample is too complex, additional homogenization techniques are necessary [20, 27]. Harmening et al. [15] modeled the fiber structure as truss with nonlinear elastic behavior and reduced the applied stress to the forces at the individual fiber joints, which mainly determine the nonwovens' tensile strength behavior. A problem-tailored data reduction strategy and a singularly perturbed regularization approach enable simulations with industrial-size samples. The approach underlying this chapter handles the problem-inherent multiscales (interplay of deterministic structural effects at macro-scale and random fiber alignment at micro-scale) and realizes the randomness in the fiber structure generation by Monte-Carlo simulations.

In the field of Machine Learning, there is much literature on woven material prediction, while modeling approaches for nonwovens are rare. Early work used simple neural network architectures, which prevent any interpretability of the results, to predict the strength of yarns [6] or worsted fabrics [11]. Abou-Nassif [1] investigated neural networks and linear regression for predicting the tensile strength of woven fabrics, limiting the work to seven training samples. Eltayib et al. [9] used linear regression to predict tear strength of fabrics based on yarn count, yarn tenacity and fabric liner density. The approach in [28] deploys multiple regression models to predict different material properties of woven fabrics, but heavily relies on huge datasets and extensive manual feature selection by domain experts. Due to the high computational cost of generating training data and due to their specialization on weaving features, these approaches cannot be applied to property prediction of nonwovens. For nonwovens, Rahnama et al. [26] proposed a feed-forward neural network based on a numerical propagation model to compare heat and moisture propagation through different nonwoven fabrics. Chen et al. [4] integrated simple logical rules developed by domain experts into a neural network to predict elongation at break, but the paper is limited to a single test example. Investigating nonwoven features and developing a stretch algorithm, we employed linear regression for tensile strength prediction in [2], which yields promising, accurate and interpretable results.

Aside from (non-)woven manufacturing, there are other works that address the prediction of material properties from production parameters. Related to our work are those that integrate prior knowledge about the underlying physical mechanics into the data, the model architecture, or the loss term being optimized. For example, Karpatne et al. [18] integrated physical knowledge about feature dependencies into a neural network as additional loss terms. Lu et al. [21] presented an approach in which knowledge about underlying material mechanics was incorporated into a machine learning approach as algebraic formulas. However, with the handcrafted neural network architectures, they are not able to provide interpretable results. Recent research on combining machine learning and simulation approaches in a more general context can be found in [29, 30]. Within the proposed taxonomy, our approach can be contextualized as the integration of (i) algebraic equations and (ii) simulation results from scientific and domain knowledge sources.

## 4.1.2 New Regression-Based Predictive Model Hierarchy

The model-based simulation framework underlying this chapter goes back to [14] for virtual fiber structure generation and to [15] for tensile strength computation. Its evaluation yields a tuple consisting of (utilized) production parameters, a random fiber graph and an associated stress-strain curve indicating the relationship between strain and stress during nonwoven's tensile strength testing. To account for the randomness in the fiber structure generation, Monte-Carlo simulations are required, which multiply the already high time requirements. This makes nonwoven material design impossible in practice and motivates a predictive surrogate. Following our ideas and strategies developed in [2], we propose a new regression-based model hierarchy for the prediction of the nonwovens' stress-strain behavior from production parameters (see Fig. 4.2). Once trained, the regression models are characterized by efficient evaluations allowing for significant speedup, while providing good, interpretable results, as we will show.

The *tensile strength model-based simulation framework* (TSS-model) at the top of the model hierarchy is built on a first principle-oriented model chain. It serves as ground truth for predictions and provides the required datasets for Machine Learning. By considering linear regression, two approaches have been proposed

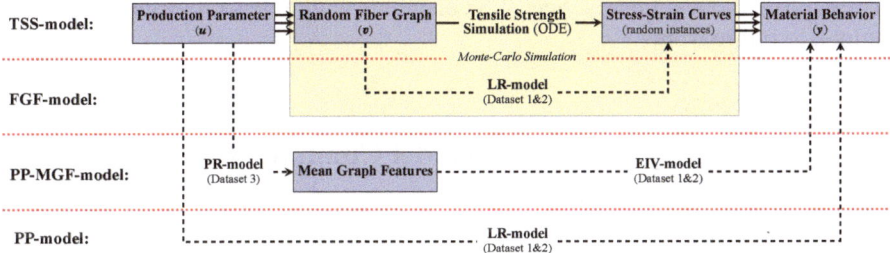

**Fig. 4.2** Predictive model hierarchy: Mappings from the production parameters to the associated tensile strength behavior. Predictive relations are indicated with dashed lines and simulations procedures with solid lines. For predictions, we employ linear regression (LR), polynomial regression (PR) and an errors-in-variables model (EIV)

in [2] that allow to circumvent the high computational effort associated to the TSS-model: The *fiber graph feature-based predictive model* (FGF-model) samples multiple fiber graphs, extracts associated graph and stretch features, and uses them to predict the stress-strain curve for each fiber graph. The *production parameter-based predictive model* (PP-model) predicts directly the mean stress-strain curve based solely on production parameters. The FGF-model provides better predictions, but Monte-Carlo simulations are necessary to derive expectations and variances from individual fiber graph features. This is accompanied by a computational overhead required to generate random fiber graph samples. The fact that the purely linear PP-model performs worse suggests some nonlinear relationships between production parameters and fiber graph features. In this chapter we introduce the novel *production parameter and mean graph feature-based predictive model* (PP-MGF-model) as a compromise between the established ones. The PP-MGF-model intercepts the nonlinearities by predicting the mean graph features using polynomial regression. Then, these (artificial) features are used as additional explanatory variables for predicting the stress-strain curves with a linear (errors-in-variables) model, in order to recover the good quality of the FGF-model. Its main advantage is that the model provides a good predictive quality without requiring Monte-Carlo simulations.

### 4.1.3  Structure

The structure of this chapter is based on the regression-based predictive model hierarchy depicted in Fig. 4.2. Section 4.2 outlines the TSS-model, by introducing the first principle-oriented model chain, and lays the foundations for predictions. Section 4.3 discusses the FGF-model and the PP-model originating from [2] and presents a performance study with focus on predictive quality. Section 4.4 introduces the new PP-MGF-model, which is numerically investigated in comparison to the established ones. Finally, Sect. 4.5 concludes with a general discussion and an outlook to future work.

## 4.2 First Principle Oriented Model Chain for Dataset Generation

The TSS-model is a first principle-oriented model chain that covers fiber graph generation and tensile strength simulation, see [14, 15]. It maps from an input set of 28 (production) parameters to a random stress-strain curve instance as output. In this chapter, we restrict to practically relevant production processes that are characterized by 4 parameters and refer to them as 4-parametric (production) process class. The resulting stress-strain curves obey a similar behavior that motivates a 2-parametric labeling. We refer to this restriction as stress-strain curve class. To improve the predictions in machine learning we consider additional fiber graph features. In this section we explain the TSS-model (Sect. 4.2.1) and introduce input (production parameters $u$, Sect. 4.2.2), output (stress-strain characteristics $y$, Sect. 4.2.3), and auxiliary variables (random fiber graph features $v$, Sect. 4.2.4), before we describe the generation of the datasets used for training and testing our regression models in Sect. 4.2.5. Readers with focus on the predictive models may skip this rather technical and mathematically extensive section and think of it as a black box for data generation.

### 4.2.1 Fiber Graph Generation and Tensile Strength Simulation

The TSS-model involves a stochastic fiber lay-down model (A) with graph generation (B) and an ordinary differential system for tensile strength testing (C). The model parameters are specified in Sect. 4.2.2.

(A) A nonwoven material is the image of fibers deposited onto a moving conveyor belt. Consider a cubic reference material volume $\mathcal{V}_R$ over the nonwoven height $H$ with base area $w_R^2$ and let $T_R$ be the time needed to produce it. A deposited fiber of length $L$ is identified with the lay-down time $T$ and the planar coordinates $(X, Y)$ of one of its end points. It contributes to $\mathcal{V}_R$, if $X - x_B(T) \in [-w_R/2, w_R/2]$ is satisfied, where $x_B$ accounts for the motion of the conveyor belt. In the three-dimensional web a fiber is modeled in terms of the curve $\eta^{(X,Y,T)} : [0, L] \to \mathcal{V}_R$,

$$d\eta_s = R(\eta_s \cdot e_x + x_B(T)) \cdot \tau_s \, ds, \qquad \eta_0 = (X - x_B(T))e_x + Y e_y + r(X)e_z,$$

$$R(x) = \frac{1}{\sqrt{1 + r'(x)^2}} [I + (\sqrt{1 + r'(x)^2} - 1)e_y \otimes e_y + r'(x)(e_z \otimes e_x - e_x \otimes e_z)],$$

$$r(x) = H \int_{-\infty}^{x} g(\bar{x}) \, d\bar{x}$$

with $X$ distributed according to the lay-down probability density function $g$ as well as $Y \sim \mathcal{U}([-w_R/2, w_R/2])$ and $T \sim \mathcal{U}([0, T_R])$ uniformly distributed. The system above is based on the stochastic Stratonovich differential system

$$d\boldsymbol{\xi}_s = \boldsymbol{\tau}_s \, ds, \qquad d\boldsymbol{\tau}_s = -\frac{1}{B+1}[\boldsymbol{\Pi}_s(B) \cdot \nabla\Sigma(\boldsymbol{\xi}_s) \, ds + A \, \boldsymbol{\Pi}_s(\sqrt{B}) \circ d\boldsymbol{w}_s]$$

with unit tensor $\boldsymbol{I}$, projection $\boldsymbol{\Pi}_s(x) = \boldsymbol{n}_{1,s} \otimes \boldsymbol{n}_{1,s} + x \, \boldsymbol{n}_{2,s} \otimes \boldsymbol{n}_{2,s}$ as well as $\boldsymbol{\xi}_0 = 0$ and $\boldsymbol{\tau}_0$ uniformly distributed in the unit circle spanned by $\boldsymbol{e}_x$ and $\boldsymbol{e}_y$. The stochastic lay-down model for position and orientation $((\boldsymbol{\xi}, \boldsymbol{\tau}) : [0, L] \to \mathbb{R}^3 \times \mathbb{S}^2)$ with unit sphere $\mathbb{S}^2 \subset \mathbb{R}^3$ describes the path of a deposited fiber onto the $\boldsymbol{e}_x$-$\boldsymbol{e}_y$ plane. In the modeling for the fiber tangent $\boldsymbol{\tau}$, the drift term prescribes the characteristic coiling behavior with the potential $\Sigma$, while the white noise term with the Wiener process $(\boldsymbol{w} : [0, L] \to \mathbb{R}^3)$ and the amplitude $A$ accounts for fluctuations in the lay-down process. Anisotropic behavior is indicated by the parameter $B \in [0, 1]$ with the local orthonormal triad $\{\boldsymbol{\tau}, \boldsymbol{n}_1, \boldsymbol{n}_2\}$, $\boldsymbol{n}_1 \in \text{span}\{\boldsymbol{e}_x, \boldsymbol{e}_y\}$. The typical nestling behavior of the fiber on the ramp-like contour surface of the nonwoven is modeled by the curve $\boldsymbol{\eta}$. The contour line $r$ of the fiber material in machine direction is described by means of the joint probability density function $g$ of the deposited material. A fiber end point lies on the associated contour surface and the fiber orientation is aligned to it due to the local rotation $\boldsymbol{R}(x) \in \text{SO}(3)$.

(B) Our considerations are restricted to the embedded test material volume $\mathcal{V} \subset \mathcal{V}_R$ with smaller base $w^2$, $w = w_R - 2L$, to exclude lateral boundary effects. The random fiber web is consolidated by adhesive joints resulting from thermobonding. Let $\boldsymbol{\eta}_h$ denote the discretized fiber, i.e., a set of discrete fiber points. An adhesive joint $\boldsymbol{a}$ to be formed between two fibers $\boldsymbol{\eta}_h$ and $\tilde{\boldsymbol{\eta}}_h$ is modeled as

$$\boldsymbol{a} = \frac{1}{2}(\boldsymbol{q}^\star + \tilde{\boldsymbol{q}}^\star)$$

$$\text{if} \quad \|\boldsymbol{q}^\star - \tilde{\boldsymbol{q}}^\star\|_2 < \kappa, \quad (\boldsymbol{q}^\star, \tilde{\boldsymbol{q}}^\star) = \underset{(\boldsymbol{q},\tilde{\boldsymbol{q}})\in\eta_h\times\tilde{\eta}_h}{\text{argmin}} \|\boldsymbol{q} - \tilde{\boldsymbol{q}}\|_2$$

with contact threshold $\kappa > 0$. The adhesive joint takes the place of the fiber points in contact within the respective fibers. As the minimizer might be not unique, we use the first minimizer found for practical reasons. Since the fibers lie rather straight, cf. [14], we assume at most one contact between each fiber pair. If more fibers are involved in a contact, the resulting adhesive joint is centered between the respective fiber points in contact. The resulting adhered fiber structure is considered as a connected graph $G = (V, E)$ with the nodes $V$ representing adhesive joints as well as fiber ends and the edges $E$ indicating fiber connections between them. The graph is supplemented by the node positions $\boldsymbol{p}_0 : V \to \mathbb{R}^3$ and the edge-associated fiber lengths $l : E \to \mathbb{R}_{\geq 0}$.

(C) The tensile strength test is modeled as differential system on the node positions $p : V \times [0, 1] \to \mathbb{R}^3$, initialized with $p(\cdot, 0) = p_0$,

$$p(v, 0) = p_0(v), \quad \forall v \in V_l, \qquad p(v, t) = p_0(v) + t\, h e_3, \qquad \forall v \in V_u$$

$$\varepsilon\, \partial_t p(v, t) = \sum_{e \in \delta(v)} f_e^v(t), \qquad\qquad\qquad \forall v \in V \setminus (V_l \cup V_u)$$

$$f_{e=\{v,v'\}}^v(t) = \frac{p(v', t) - p(v, t)}{d(e, t)} N\left(\frac{d(e, t) - l(e)}{l(e)}\right)$$

with $\delta(v) \subset E$ incident edges of node $v$. For fixed lower face $V_l$, the upper face $V_u$ of the fiber structure is linearly shifted away in (vertical) $e_3$-direction (with maximal displacement $h > 0$). In the interior nodes of the graph the acting traction forces are balanced by a friction term with $\varepsilon > 0$. The force amplitude $N$ depends on the relative strain of the fiber connection $e$ with respect to its length $l(e)$, where $d(e, t)$ denotes the Euclidean distance between its endpoints, $d(e, t) = ||p(v, t) - p(v', t)||_2$. It reflects Hooke's law in the stretched state and is taken as zero in the unstretched state. The characterizing stress-strain relation for the fiber structure (with initial height $H$) is then given by $(\epsilon(t), T(p(\cdot, t)))$, $t \in [0, 1]$,

$$\epsilon(t) = \frac{h}{H} t, \qquad T(p(\cdot, t)) = -\sum_{v \in V_u} \sum_{e \in \delta(v)} f_e^v(t) \cdot e_3.$$

### 4.2.2  Production Process Class

An airlaid nonwoven typically consists of two fiber types for which the TSS-model has 28 input parameters in total: Each fiber type is characterized by length $L_f$, line density $(\rho A)_f$, cross-sectional weighted elasticity modulus $(EA)_f$ and lay-down probability density $g_f$ considered as normally distributed $g_f \sim \mathcal{N}(\mu_f, \sigma_f^2)$, $f = 1, 2$. The joint probability density is then $g = \beta_n g_1 + (1 - \beta_n)g_2$ with fiber number fraction $\beta_n$ determined by mass fraction $\beta$. For technical reasons, we use a compact support $\text{supp}(g) = [x_l, x_r]$. The production plant is characterized by conveyor belt width $b$ and speed $v_B$ as well as mass rate $\dot{m}$. The nonwoven sample is specified by height $H$ and width $w$. Production time $T_R$, trace curve $x_B$ and number of deposited fibers per type $n_f$, for $f = 1, 2$, are resulting quantities. The laydown is parameterized regarding diffusion $A$, anisotropy $B$ and bending potential $\Sigma$ expressed by the three standard deviations $\sigma_x$, $\sigma_y$, $\sigma_z$ in $e_x$, $e_y$, $e_z$-directions. The bonding considers fiber discretization length $\Delta s$ and contact threshold $\kappa$. The strength test is parametrized by adhesive thickness $z$ for upper and lower structure faces, friction-associated regularization $\varepsilon$ as well as traction function $N$ with a regularization parameter $\delta$. Note that the displacement $h$ in the strength test belongs to the input quantities.

**Table 4.1** Characteristic dimensionless input parameters for TSS-model. Values for an industrial airlay process (mixture of solid (PES) and bi-component (PES/PET) fibers in plant K12, cf. scenario in [14]). Referential values in SI units: $w = 1.0 \cdot 10^{-2}$ m, $v_B = 3.3 \cdot 10^{-2}$ m/s, $(EA)_1 = 1.0$ N

| Description | Symbol | | Value |
|---|---|---|---|
| Fiber length | $L_1/w$, | $L_2/L_1$ | 5.5, 1.0 |
| Fiber number | $\alpha_1 w^2/v_B$, | $\alpha_2/\alpha_1$ | 1150, 0.65 |
| Elasticity modulus | $(EA)_2/(EA)_1$ | | 1.0 |
| Lay-down pdf mean | $\mu_1/w$, | $\mu_2/w$ | 0, 0 |
| Lay-down pdf std | $\sigma_1/w$, | $\sigma_2/\sigma_1$ | 2.0, 1.0 |
| Support joint lay-down pdf | $x_l/\sigma_1$, | $x_r/\sigma_1$ | − 5.0, 5.0 |
| Nonwoven sample height | $H/w$ | | 6.0 |
| Bending potential (std) | $\sigma_y/w, \sigma_x/\sigma_y, \sigma_z/\sigma_y$ | | 2.0, 0.75, 0.075 |
| Diffusion | $A\sqrt{\sigma_y}$ | | $2.8 \cdot 10^{-2}$ |
| Anisotropy | $B$ | | $3.0 \cdot 10^{-1}$ |
| Fiber discretization | $\Delta s/w$ | | $2.75 \cdot 10^{-2}$ |
| Contact threshold | $\kappa/w$ | | $2.6 \cdot 10^{-2}$ (calibrated) |
| Adhesive thickness at faces | $z/w$ | | $6.0 \cdot 10^{-2}$ |
| Friction regularization | $\varepsilon$ | | $1 \cdot 10^{-7}$ |
| Traction regularization | $\delta$ | | $1 \cdot 10^{-4}$ |

**Table 4.2** Input $u$ (4-parametric production process class) for Machine Learning. Parameter ranges for dataset used in ML approach and respective values in industrial scenario, Table 4.1. The values of all other parameters (ratios) are taken from Table 4.1

| Symbol | Range | Industrial value | Effect |
|---|---|---|---|
| $\hat{\alpha} = \alpha_1 w^2/v_B$ | [1000, 1515] | 1150 | Amount of fibers |
| $\hat{\sigma} = \sigma_1/w$ | [1.0, 5.0] | 2.0 | Laydown behavior |
| $\hat{\sigma}_y = \sigma_y/w$ | [1.0, 5.0] | 2.0 | Laydown behavior |
| $\hat{\kappa} = \kappa/w$ | $[2.8, 3.0] \cdot 10^{-2}$ | $2.6 \cdot 10^{-2}$ | Bonding |

Since the parameters $(\rho A)_1$, $(\rho A)_2$, $\beta = \beta_1/\beta_2$, $\dot{m}$ and $b$ only occur in the quantities $\alpha_f = \beta_f \dot{m}/((\rho A)_f L_f b)$, for $f = 1, 2$, indicating the number of fibers for each type deposited per second and meter in cross direction on the conveyor belt, three parameters can be eliminated. Making the model dimensionless with nonwoven sample width $w$, conveyor belt speed $v_B$ and elasticity modulus $(EA)_1$ reduces the set of input parameters by further three. The resulting dimensionless numbers are mainly formulated as ratios, cf. Table 4.1. Note that the strength test is stated in dimensionless form to incorporate the friction-associated (dimensionless) regularization parameter $\varepsilon \ll 1$ that ensures a unique solution.

In this chapter, we consider a 4-parametric production process class. The process class is motivated from the industrial test setting in [14]: We adopt all industrial values—except for $u = (\hat{\alpha}, \hat{\sigma}, \hat{\sigma}_y, \hat{\kappa}) \in \mathbb{R}^4_+$. These four inputs affect the fiber amount in the nonwoven (sample), the fiber laydown behavior and the bonding (i.e., fiber graph topology). By varying them in a certain regime, a broad variety of practically relevant airlay scenarios are covered, see Table 4.2 for the parameter ranges underlying our dataset for Machine Learning. Note that the larger chosen $\hat{k}$ ensures a stronger bonding and hence a denser fiber structure than in the industrial test case.

### 4.2.3 Stress-Strain Curve Class

The stress-strain curves of the nonwovens obtained by the 4-parametric production process class show a similar pattern and allow for a 2-parametric labeling, $y = (\alpha, \beta)$. The observed output curves are constant at a stress close to zero up to a threshold value $\alpha$ of applied strain, above which they increase quadratically with coefficient $\beta$, see Fig. 4.3. The behavior results from more and more fibers coming under strain and thus contributing to the tensile strength, neglecting plastic effects and fiber tearing. Hence, we model the relation between strain and stress for a nonwoven sample by

$$T_y(\epsilon) = \begin{cases} 0, & \epsilon < \alpha \\ \beta(\epsilon - \alpha)^2, & \epsilon \geq \alpha \end{cases}, \qquad y = (\alpha, \beta) \in \mathbb{R}_+^2 , \qquad (4.1)$$

where $\epsilon$ refers to the relative strain applied to the sample and $T_y : \mathbb{R}_+ \to \mathbb{R}_+$ to the resulting reacting force.

The approximation of the stress-strain curve by the constant-quadratic ansatz enables a straightforward machine learning modeling approach with only two output parameters $y = (\alpha, \beta)$ as labels for prediction—instead of a complex output curve. The general tensile strength behavior can be characterized using the joint distribution of $\alpha$ and $\beta$. To draw conclusions about the randomness of the material, the constant-quadratic ansatz can be used to compute, for example, the mean stress and the associated variance at individual strain levels.

**Fig. 4.3** Stress-strain curves obtained for fixed parameter setting by TSS-model

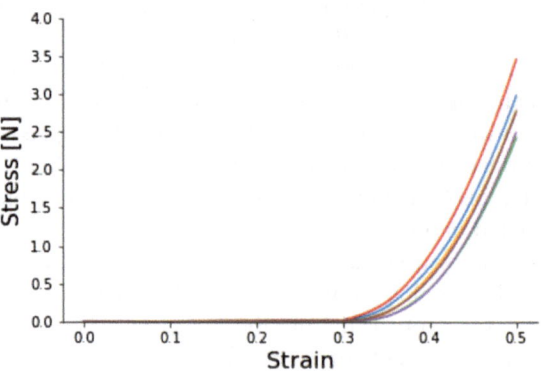

### 4.2.4   Fiber Graph Features

The use of fiber graph features for predicting tensile strength has turned out to be advantageous in Machine Learning. According to [2] we use two groups of features: *topological graph features* representing the fiber structure connectivity which likely affects the nonwoven's tensile strength and *stretch features*, which are obtained by a heuristic stretching algorithm based on elongation of the nonwoven samples, allowing only vertical displacements of nodes and no strain on the individual fibers. The identification of the features and the stretching algorithm originate from [2].

**Topological Graph Features** The graph feature set contains the numbers of nodes $|V|$ and edges $|E|$, maximum node degree $d_{max} = \max_{v \in V} \delta(v)$, total fiber lengths $L_{\text{fiber}} = \sum_{e \in E} l(e)$ as well as the numbers $|V_u|$, $|V_l|$ of upper and lower face nodes. Moreover, to encode the graph connectivity several path and length-associated features are considered, see Fig. 4.4. Let $L_1(P) = \text{len}(P)$ denote the edge count, $L_2(P) = \sum_{e \in P} l(e)$ and $L_3(P) = \sum_{e \in P} d(e)$ the fiber and Euclidean lengths for a path $P$. Of interest are the shortest paths connecting the upper and lower faces—in terms of edge count $P_1$ and fiber length $P_2$, i.e., $P_1 = \text{argmin}\,\{L_1(P_{uv}) \mid u \in V_l, v \in V_u\}$ and $P_2 = \text{argmin}\,\{L_2(P_{uv}) \mid u \in V_l, v \in V_u\}$, so we include $L_1(P_1)$, $L_2(P_2)$ and $L_3(P_2)$ to the feature set. In addition, we consider mean, median and sum of differences between fiber and Euclidean length over all edges $\{l(e) - d(e) \mid e \in E\}$ and the size of a minimum cut $C_{min}$, i.e., edge set with minimum cardinality disconnecting $V_u$ from $V_l$ when removed.

**Stretch Features** The stretch features obtained from the stretching algorithm [2] provide information about the nonwoven behavior under vertical tensile loading. A

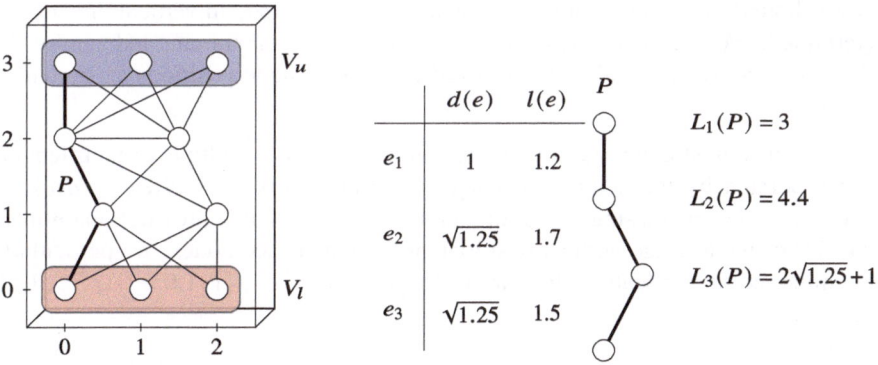

**Fig. 4.4** Illustration of some graph features. Fiber graph with $|V| = 10$ nodes, $|E| = 17$ edges. Face sets $V_u$ and $V_l$ are colored in blue, and red, respectively. For path $P$, length variants $L_1$, $L_2$ and $L_3$ are based on edge-wise fiber lengths $l(e)$ and Euclidean lengths $d(e)$. A minimum cut $C_{min}$ separates all nodes above the value 1.5 from all nodes below that value, $|C_{min}| = 4$ [2]

graph $G$ with node positions $p$ and edge-associated fiber lengths $l$ is called a *valid instance* if the following length constraint is satisfied,

$$l(\{v, w\}) \geq \|p(v) - p(w)\|_2 = d(\{v, w\}) \quad \forall \{v, w\} \in E.$$

The stretching algorithm (Algorithm 1) addresses the question of how far the nonwoven sample can be stretched vertically without stretching any fibers, i.e., maximizing the sum of height coordinates of the upper face nodes $V_u$, while fixing the positions of the lower face nodes $V_l$ and keeping the instance valid. For computational reasons the algorithm assumes that the fiber nodes (outside $V_l$) can only move freely in the vertical (third) dimension while their horizontal position is fixed. Given a valid instance $(G, p, l)$ and $V_l \subseteq V$ the *ZStretch* problem reads:

$$\max \sum_{v \in V_u} \tilde{p}_3(v)$$

$$\text{subject to:} \quad \tilde{p}(v) = p(v) \quad \forall v \in V_l,$$
$$\tilde{p}_1(v) = p_1(v) \quad \forall v \in V,$$
$$\tilde{p}_2(v) = p_2(v) \quad \forall v \in V,$$
$$d(\{v, w\}) \leq l(\{v, w\}) \quad \forall \{v, w\} \in E,$$

where $p_i = p \cdot e_i, i = 1, 2, 3$, denote the spatial coordinates. The optimization problem certainly ignores many real-world structure properties, e.g., fiber intertwining, or the fact that fiber nodes can in reality move in all three dimensions to allow further stretching of the nonwoven sample in the third dimension. But due to its simplicity it can be solved in $O(|E| \log(|V|))$ run time by Algorithm 1. As a result, a lower bound to the maximum movement of any fiber node in vertical direction is determined. We use mean, standard deviation, median, maximum, and sum of the differences between initial and optimized upper face node positions as stretching features.

As extension, stretching of the individual fibers up to a multiple of their lengths is incorporated by weakening the length constraint to $\tilde{l}(e) = cl(e) \geq d(e)$ for some $c > 1$. For increasing values of $c$ and a fixed graph, this provides a nonlinear behavior of the average vertical positions of the upper face nodes. We particularly determine the stretch features for various length factors, $c \in \{1, 1.05, 1.1, \ldots, 1.5\}$, see Table 4.3.

### 4.2.5 Dataset

The machine learning dataset is generated using the TSS-model. The combinations of input production parameters are randomly selected from a range that yields

**Table 4.3** Input features for regression models

| Set | Symbols | Description |
|---|---|---|
| Param | $\boldsymbol{u}$ | Four parameters for production process |
| | $|V|$ | Number of nodes |
| | $|E|$ | Number of edges |
| | $d_{max}$ | Maximum node degree |
| | $L_{fiber}$ | Total fiber lengths |
| | $|V_u|$ | Number of upper face nodes |
| Graph | $|V_l|$ | Number of lower face nodes |
| | $L_1(P_1)$ | Minimal edge count of all paths from $V_u$ to $V_l$ |
| | $L_2(P_2)$ | Minimal fiber lengths of all paths from $V_u$ to $V_l$ |
| | $L_3(P_2)$ | Euclidean length of weighted shortest path $P_2$ |
| | $D_1, D_2, D_3$ | {mean, median, sum} of differences between edge-wise fiber and Euclidean lengths |
| | $|C_{min}|$ | Size of minimum edge cut separating $V_u$ and $V_l$ |
| Stretch | $S_1^c, S_2^c, S_3^c, S_4^c, S_5^c$ | {mean, std, median, max, sum} of stretching distance for $c \in \{1, 1.05, 1.1, \ldots, 1.5\}$ |

---

**Algorithm 1** Graph stretching algorithm

---

**Input**: a valid instance $(G, \boldsymbol{p}, l)$ and $V_l \neq \emptyset$
**Output**: a valid instance $(G, \tilde{\boldsymbol{p}}, l)$ that maximizes the objective of ZStretch

set $\tilde{\boldsymbol{p}}(v) = \boldsymbol{p}(v) \quad \forall v \in V$
set $V_\perp = V_l$ and $B = \mathcal{N}(V_\perp)$ where $\mathcal{N}()$ refers to the neighbor nodes
**for** $v = \text{argmin}_{w \in B} \, maxMove(w, V_\perp)$ **do**
  pop $v$ from $B$
  $\tilde{p}_3(v) = maxMove(w, V_\perp) + p_3(v)$
  add $v$ to $V_\perp$
  $B = B \cup \mathcal{N}(v) \setminus V_\perp$
**end**

---

**Algorithm 2** maxMove subroutine

---

**Input**: a node $v \in V$ and $V_\perp \subseteq V$
**Output**: the largest $h$ such that $\tilde{\boldsymbol{p}}(v) = \boldsymbol{p}(v) + h\boldsymbol{e}_3$ satisfies $\|\tilde{\boldsymbol{p}}(v) - \boldsymbol{p}(w)\|_2 \leq l(\{v, w\}) \quad \forall w \in \mathcal{N}(v) \cap V_\perp$

**forall the** $\{v, w\} \in E$ **do**
  find the largest $h$ s.t.
  $\tilde{\boldsymbol{p}}(v) = \boldsymbol{p}(v) + h\boldsymbol{e}_3$
  satisfies
  $\|\tilde{\boldsymbol{p}}(v) - \boldsymbol{p}(w)\|_2 \leq l(v, w) \quad \forall w \in \mathcal{N}(v) \cap V_\perp$
**end**

---

reasonable fiber structures, cf. Table 4.2. While the generation of fiber graphs and accompanying features is fast, the computation of the stress-strain curves is very time-consuming as it requires solving large-scale dynamical systems on the individual fiber structure samples. To account for the systems' stiffness, we employ

**Table 4.4** Composition of dataset

|                       | Set 1: fully labeled | Set 2: single labeled | Set 3: unlabeled | Total |
|-----------------------|----------------------|-----------------------|------------------|-------|
| Graphs                | 6×25                 | 37×25                 | 2.000 × 1        | 3.075 |
| Stress-strain curves  | 6×25                 | 37×1                  | –                | 187   |

an implicit Euler scheme with variable step size control. For the resulting nonlinear equation systems, we use an exact Newton method with analytical Jacobian and Armijo's line search. An explicit Euler step provides a suitable initial guess for warm start. The ODE-solver typically requires between 24 and 48 hours for a single instance, making it the bottleneck for building datasets. Both the fiber graph generation as well as the tensile strength simulations are performed in parallel on a machine with 88 CPU cores (Intel(R) Xeon(R) CPU E5-2699 v4 @ 2.20 GHz) and 792 GB RAM running Ubuntu 18.04.6 using Matlab (R2019a).

For 43 parameter combinations, we generate 25 sample graphs each, totaling 1.075 graphs. On average, each graph contains 51.507 nodes (standard deviation ±2.182) and 198.744 edges (±29.996). We randomly select six of our parameter combinations and compute the 25 stress-strain curves associated to the graphs (set 1, fully labeled), while for all other combinations we compute only a single stress-strain curve for one of the corresponding samples (set 2, single labeled) because of the high cost of the ODE-solver. The dataset thus includes 187 supervised/monitored samples ($6 \times 25$ samples + $37 \times 1$ sample) across 43 different parameter combinations. The graphs and corresponding stress-strain curves serve as ground truth examples for supervised learning. Additionally, set 3 (unlabeled) contains 2.000 graphs for 2.000 parameter combinations, cf. Table 4.4. Given an unseen parameter combination, our goal is to predict the average behavior as well as a range of deviation of the resulting stress-strain curves as close as possible to the ground truth. The data that we generated and used for the results shown in this chapter is available for download at https://github.com/pwelke/random-nonwoven-fibers.

## 4.3  Linear Regression-Based Predictive Models

This section deals with the two multivariate linear regression models recently proposed by [2]: the PP-model (production parameter-based) and the FGF-model (fiber graph feature-based). We explain the underlying modeling ideas (Sect. 4.3.1) and discuss the advantages of the model variants by means of a performance study (Sect. 4.3.2). For this purpose, we investigate the goodness of fit for the prediction of the mean stress-strain curves as an example. Both models aim at avoiding the high computational effort associated to the TSS-model. In the following the TSS-model is represented by the random field $S : \mathcal{U}_{ad} \times \Omega \rightarrow \mathcal{Y}$ with set of admissible production parameters $\mathcal{U}_{ad} \subset \mathbb{R}^n$, $n = 4$ (cf. Table 4.2) and set of stress-strain curve parametrizations $\mathcal{Y} \subset \mathbb{R}^r$, $r = 2$, cf. (4.1).

### 4.3.1 Linear Regression and Monte Carlo Simulations

**PP-model** The PP-model directly relates the production parameters to the stress-strain curve parametrizations using multiple multivariate linear regression, cf. [17]. Given $k \in \mathbb{N}$ observation pairs $\{(u_i, y_i)\}_{i=1}^{k}$, consisting of input (production parameters) $u_i \in \mathbb{R}^n$ and output (random stress-strain curve parametrizations) $y_i = \mathcal{S}(u_i, \omega_i) \in \mathbb{R}^r$, the model assumes the relation

$$y_i = b_P + B_{P,1}^T u_i + \varepsilon_{P,i} \quad \text{for } i = 1, \ldots, k,$$

where the errors $\varepsilon_{P,i} : \Omega \to \mathbb{R}^r$ account for the stochastic nature of the tensile strength simulation framework. They are assumed to be independent and identically distributed (i.i.d.) with $\mathrm{E}[\varepsilon_{P,i}] = 0$ and $\mathrm{Cov}[\varepsilon_{P,i}] = \Sigma_P \in \mathbb{R}^{r \times r}$. The task is to identify the unknown intercept $b_P \in \mathbb{R}^r$ and regression coefficients $B_{P,1} \in \mathbb{R}^{n \times r}$. The regression model for $B_P = [b_P, B_{P,1}^T]^T \in \mathbb{R}^{1+n \times r}$ can be summarized as

$$Y = U B_P + E_P,$$

$$Y = \begin{bmatrix} y_1^T \\ y_2^T \\ \vdots \\ y_k^T \end{bmatrix}, \qquad U = \begin{bmatrix} 1 \ u_1^T \\ 1 \ u_2^T \\ \vdots \ \vdots \\ 1 \ u_k^T \end{bmatrix}, \qquad E_P = \begin{bmatrix} \varepsilon_{P,1}^T \\ \varepsilon_{P,2}^T \\ \vdots \\ \varepsilon_{P,k}^T \end{bmatrix}$$

with response $Y \in \mathbb{R}^{k \times r}$, design matrix $U \in \mathbb{R}^{k \times 1+n}$, and error matrix $E_P \in \mathbb{R}^{k \times r}$. By the assumptions on the individual errors $\mathrm{Cov}[(E_P)_{.,i}, (E_P)_{.,j}] = (\Sigma_P)_{i,j} I_k$ holds true, for $i, j = 1, \ldots, r$ and the identity $I_k \in \mathbb{R}^{k \times k}$. Thus, the individual observations are independent, but correlations between the responses are allowed.

A linear, unbiased estimator of $B_P$ is the well-known least-squares estimator $\widehat{B}_P = (U^T U)^{-1} U^T Y$. Considering the decomposition $\widehat{B}_P = [\widehat{b}_P, \widehat{B}_{P,1}^T]^T$ yields the predictor

$$\widehat{y}_P(u) = \widehat{b}_P + \widehat{B}_{P,1}^T u, \quad \text{with} \quad \widehat{y}_P : \mathbb{R}^n \to \mathbb{R}^r \tag{4.2}$$

that maps the production parameters to the associated mean stress-strain curve parametrization. If the errors are assumed to be multivariate normally distributed, i.e., $\varepsilon_{P,i} \sim \mathcal{N}(0, \Sigma_P)$, the maximum likelihood estimator of the covariance matrix $\Sigma_P$ is given by

$$\widehat{\Sigma}_P = \frac{1}{k} \widehat{E}_P^T \widehat{E}_P, \qquad \widehat{E}_P = Y - U\widehat{B}_P, \tag{4.3}$$

where $\widehat{E}_P$ are the residuals between actual observation and prediction.

The PP-model approximates the TSS-model as $S \approx \widehat{S}_P$, which is specified by $\widehat{S}_P(\boldsymbol{u}, \cdot) \sim \mathcal{N}(\widehat{\boldsymbol{y}}_P(\boldsymbol{u}), \widehat{\boldsymbol{\Sigma}}_P)$ for all $\boldsymbol{u} \in \mathcal{U}_{ad}$. Note that $\widehat{\boldsymbol{y}}_P$ only predicts the mean stress-strain curve parametrization. However, if we use the additional distributional assumptions, we can resample multiple stress-strain curve parametrizations and insert them in the constant-quadratic ansatz (4.1). Averaging over the resulting curves yields then a prediction of the mean stress-strain curve.

**FGF-model** The FGF-model, unlike the PP-model, builds on predicting the stress-strain curve parametrizations for individual fiber graphs. This requires the generation of random fiber structure samples, from each of which $m \in \mathbb{N}$ features (i.e., combinations of production parameters, topological graph and stretch features, as listed in Table 4.3) are extracted. We view fiber graph generation and feature extraction as a random field $\mathcal{M} : \mathcal{U}_{ad} \times \Omega \to \mathbb{R}^m$. Then, the FGF-model relates production parameters and (fiber graph) features to the associated stress-strain curve parametrizations for which we again consider a multiple multivariate linear regression model. Given $k \in \mathbb{N}$ observation tuples $\{(\boldsymbol{u}_i, \boldsymbol{v}_i, \boldsymbol{y}_i)\}_{i=1}^{k}$, consisting of production parameters $\boldsymbol{u}_i \in \mathbb{R}^n$, random features $\boldsymbol{v}_i = \mathcal{M}(\boldsymbol{u}_i, \omega_i) \in \mathbb{R}^m$ and associated[1] stress-strain curve parametrizations $\boldsymbol{y}_i = S(\boldsymbol{u}_i, \omega_i) \in \mathbb{R}^r$, it reads

$$\boldsymbol{y}_i = \boldsymbol{b}_F + \boldsymbol{B}_{F,1}^T \boldsymbol{u}_i + \boldsymbol{B}_{F,2}^T \boldsymbol{v}_i + \boldsymbol{\varepsilon}_{F,i} \quad \text{for } i = 1, \ldots, k$$

with intercept $\boldsymbol{b}_F \in \mathbb{R}^r$ and regression coefficients $\boldsymbol{B}_{F,1} \in \mathbb{R}^{n \times r}$, $\boldsymbol{B}_{F,2} \in \mathbb{R}^{m \times r}$. The assumptions on the errors are the same as those of the PP-model, with covariance matrix $\text{Cov}[\boldsymbol{\varepsilon}_{F,i}] = \boldsymbol{\Sigma}_F$. However, the errors are here motivated as simple regression errors and not as sampling errors, as the FGF-model describes the input-output behavior of the deterministic tensile strength simulations. The model can be summarized as

$$\boldsymbol{Y} = \boldsymbol{W} \boldsymbol{B}_F + \boldsymbol{E}_F$$

with $\boldsymbol{B}_F = [\boldsymbol{b}_F, \boldsymbol{B}_{F,1}^T, \boldsymbol{B}_{F,2}^T]^T$ and $\boldsymbol{W} = [\boldsymbol{U}, \boldsymbol{V}]$ using $\boldsymbol{V} = [\boldsymbol{v}_1, \ldots, \boldsymbol{v}_k]^T$. Again, the task is to identify intercept and regression coefficients, for which the unbiased linear least-squares estimator is given by $\widehat{\boldsymbol{B}}_F = (\boldsymbol{W}^T \boldsymbol{W})^{-1} \boldsymbol{W}^T \boldsymbol{Y}$. Thus, we obtain the linear predictor

$$\widehat{\boldsymbol{y}}_F(\boldsymbol{u}, \boldsymbol{v}) = \widehat{\boldsymbol{b}}_F + \widehat{\boldsymbol{B}}_{F,1}^T \boldsymbol{u} + \widehat{\boldsymbol{B}}_{F,2}^T \boldsymbol{v} \quad \text{with} \quad \widehat{\boldsymbol{y}}_F : \mathbb{R}^n \times \mathbb{R}^m \to \mathbb{R}^r \qquad (4.4)$$

that maps a given set of production parameters and fiber graph features to the stress-strain curve parametrization associated to the respective fiber graph.

---

[1] The explicit usage of a fixed production parameter combination $\boldsymbol{u}_i$ together with a fixed probabilistic state $\omega_i$ emphasizes that graph features and stress-strain curve parametrization are obtained from the same fiber graph sample (by simulation and feature extraction).

The FGF-model approximates the TSS-model by the coupling of the predictor $\widehat{y}_F$ with the random field $\mathcal{M}$, i.e., $S \approx \widehat{S}_F$ where $\widehat{S}_F(u, \cdot) \sim \widehat{y}_F(u, \mathcal{M}(u, \cdot))$ for all $u \in \mathcal{U}_{ad}$. As we have no analytical insights in the behavior of $\mathcal{M}$, this coupling has to be treated as a stochastic black box. To obtain a predictor of the mean stress-strain curve we have to conduct Monte-Carlo simulations where we repeatedly sample fiber graphs, predict their stress-strain curves and average over the results.

**Remark 4.1** The case where only fiber graph features are used for predictions in the FGF-model can be covered by choosing $\widehat{B}_{F,1} = 0$. The corresponding least-squares estimator is $[\widehat{b}_F, \widehat{B}_{F,2}] = (\tilde{V}^T \tilde{V})^{-1} \tilde{V}^T Y$ with design matrix $\tilde{V} = [1, V] \in \mathbb{R}^{k \times 1+m}$ containing an extra column of ones to account for the intercept term.

**Remark 4.2** We state the closed form solution of the least-square estimator for convenience. In practice, we avoid solving the ill-conditioned normal equations and instead solve the associated least-squares optimization problem via pseudo inverse by means of a singular value decomposition.

### 4.3.2  Numerical Results

**Experimental Setting** To assess the predictive quality for inferring the mean stress-strain curves, we perform a leave-one-out cross-validation (LOOCV) across all 43 production parameter combinations contained in the labeled datasets 1&2 (Table 4.4). In each run, we separate the data into a training set containing the samples of 42 parameter combinations and a test set containing the samples of the remaining parameter combination. Hence, the test set always contains 25 fiber graph samples, where either one (set 2) or all of them (set 1) are labeled with a stress-strain curve parametrization. The training set is used to fit the PP-model and the FGF-model. For the latter one, we compare different combinations of the feature groups (listed in Table 4.3). During inference, the fitted models use the production parameters without/with fiber graph features as input to predict the mean stress-strain curve parametrizations (PP-model) or the stress-strain curve parametrizations associated to individual fiber graphs (FGF-model). For the FGF-model, the stress-strain curves of the 25 fiber graphs in the test set are reconstructed using the constant-quadratic ansatz (4.1). Averaging these curves yields a mean stress-strain curve prediction. For the PP-model, the procedure differs slightly. As it only predicts mean stress-strain curve parametrizations we use the covariance estimate (4.3) on top of the predicted mean parametrization (4.2) to resample 1.000 stress-strain curve parametrizations. Then, averaging over the associated stress-strain curves that are reconstructed by means of the constant-quadratic ansatz provides a prediction of the mean stress-strain curve.

For each production parameter combination, we compare the predicted mean curve to the ground truth curve. Using parameter combinations with multiple training samples (set 1), we compare the means of the predicted and the ground truth curves. For the single-sample parameter combinations (set 2) we take all 25 (mainly unlabeled) graph samples and check how much the ground truth curve of the single labeled sample deviates from the mean of the predicted curves that our model produces. For assessment, we use the test set to calculate the coefficient of determination, $R^2$, and the adjusted coefficient of determination, $\bar{R}^2$, between the means of predicted and ground truth curve evaluations. This provides a measure of the model fit that is independent of the strain. Thus, to compute the $R^2$ values, we evaluate each curve (predicted and ground truth) at $K$ (=1.000) equally distanced strain points in the interval $[0, 0.5]$ and take the means over the values at each strain point. In every run, this yields the predicted means $\hat{\bar{y}}_1, \ldots, \hat{\bar{y}}_K$ as well as the observed means $\bar{y}_1, \ldots, \bar{y}_K$, from which we compute the (adjusted) coefficient of determination through

$$R^2 = 1 - \frac{\sum_{i=1}^K \bar{y}_i - \hat{\bar{y}}_i}{\sum_{i=1}^K \bar{y}_i - \bar{\bar{y}}} \quad \text{and} \quad \bar{R}^2 = 1 - (1 - R^2)\frac{K - 1}{K - m - 1},$$

where $\bar{\bar{y}} = \sum_{i=1}^K \bar{y}_i$ and $m$ is the number of features used by the predictive model. Given the variability of the samples within the same parameter combination, this validation provides a robust estimation of the model quality. While the $R^2$ value is a default evaluation score for regression tasks, we supplement it with the $\bar{R}^2$ value which penalizes for larger numbers of selected attributes within a model. Furthermore, we perform an Optimal Transport (OT) optimization between the sets of curves embedded in $\mathbb{R}^K$. It computes a mapping between two sets of points, that is minimal in terms of total work, i.e. transportation of mass. For optimization, the Wasserstein distance for discrete distributions is used. In comparison to the median $R^2$ score, the OT score penalizes substantial differences between individual predicted and actual curves to a larger degree. With this additional score, we can adequately assess the difference in distribution between prediction and ground truth curves.

As baseline, in each run of the LOOCV we also compare the (mean) ground truth curve to the curve obtained by feeding the constant-quadratic ansatz with the means of the parametrizations in the training set. Computing the corresponding $R^2$ values yields a simple comparative value to beat. Moreover, we include a comparison of the ground truth curves to that obtained by means of the best found constant-quadratic curve fits to get an idea of the suitability of the utilized stress-strain curve model class. Corresponding code and experimental data are available at https://github.com/pwelke/random-nonwoven-fibers and as a reproducible run on CodeOcean https://codeocean.com/capsule/7514050/tree/v1 [3].

**Results and Discussion** The main results for the prediction of the mean stress-strain curves are illustrated in Table 4.5. It reports the median (adjusted) coefficients of determination, $R^2$ and $\bar{R}^2$, observed during the LOOCV. Most importantly, the results show that the constant-quadratic ansatz is a well-chosen approximation for the ground truth stress-strain curves, which is expressed by a median coefficient of determination that is very close to 1. Further, we note that both, the PP-model

**Table 4.5** Regression results for the baseline, the constant-quadratic ansatz, the PP-model and the FGF-model. Listed are the medians of the $R^2$ and $\bar{R}^2$ values observed during the LOOCV as well as the OTLoss

| Model | Feature set | Median $R^2$ ↑ | Median $\bar{R}^2$ ↑ | OTLoss ↓ |
|---|---|---|---|---|
| Baseline | – | 0.3928 | – | – |
| Constant-quadratic | – | **0.999927** | – | – |
| PP-model | Param | 0.7967 | 0.7958 | 292.56 |
| FGF-model | Stretch | 0.9730 | 0.9714 | 111.77 |
| FGF-model | Graph | 0.9737 | 0.9733 | 99.65 |
| FGF-model | Param + stretch | 0.9723 | 0.9705 | 85.24 |
| FGF-model | Param + graph | 0.9717 | 0.9712 | 82.62 |
| FGF-model | Graph + stretch | **0.9760** | **0.9742** | **71.44** |
| FGF-model | Param + graph + stretch | 0.9778 | 0.9761 | 85.71 |

and the FGF-model, outperform the identified baseline by a clear margin. The fiber graph feature-based approach in particular works surprisingly well and delivers significant improvements over the simple production parameter-based approach. With regard to different feature set combinations, it should be emphasized that a union of topological graph and stretch features already achieves a remarkable performance with a median coefficient of determination of $R^2 = 0.9760$, calculated between the mean predicted and the mean ground truth curves (highlighted in bold in Table 4.5). This indicates that the topological and geometric structure of the fiber graphs already encodes much of the tensile strength behavior under vertical load. It should be noted, that we also compared a lasso and ridge regularization for the parameter estimation, leading to no significant change in results.

A major advantage of using simple regression models for prediction is the interpretability of the individual regression coefficients. In the following, we investigate the feature importance exemplarily for the FGF-model (graph + stretch) that uses the union of graph and stretch features for prediction. The regression weights observed during the 43-fold LOOCV are displayed in Fig. 4.5. Apparently, high impact features differ between $\alpha$ and $\beta$ prediction. Generally, stretch features display a large impact, especially $S_1^c$ (mean) and $S_5^c$ (sum) for larger values of the overstretching factor $c$. Examining the graph feature importance discloses the following relationships: For the prediction of $\alpha$ the negative regression coefficient values with respect to $|E|$ indicate that as the number of edges increases, the quadratic behavior of the stress-strain curves sets in earlier ($\alpha$ is smaller). In line with that, the positive coefficient values for the prediction of $\beta$ indicate that an increase in $|E|$ also causes the quadratic incline to grow quicker ($\beta$ is bigger). This underlines the intuition that more fiber connections result in firmer materials (higher tensile strength). Similar relationships can be observed for the maximum degree $d_{\max}$ and the size of the minimum edge cut $|C|_{\min}$, as higher feature values are likely to represent a higher fiber structure connectivity. Opposed to that $L_3(P_2)$, the Euclidean length along the weighted shortest path in terms of fiber length, exhibits

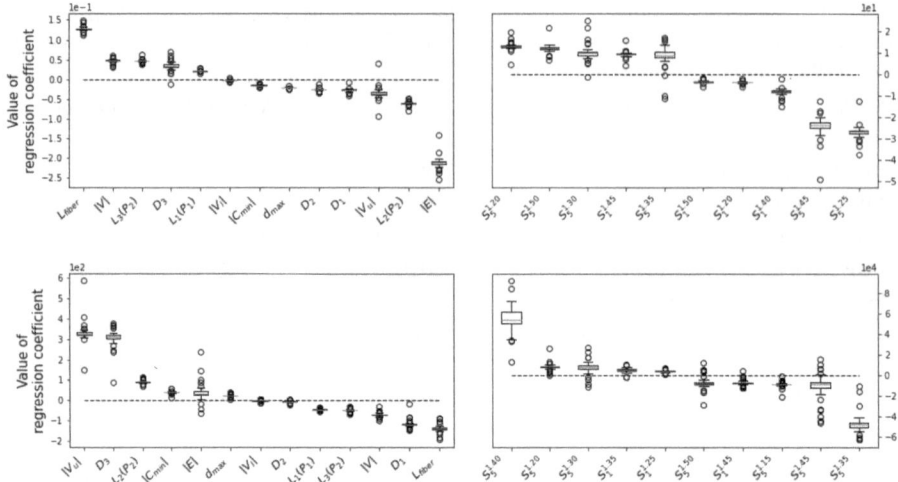

**Fig. 4.5** Feature importance values for the FGF-model (graph + stretch) to predict $\alpha$ (top) and $\beta$ (bottom) with $y = (\alpha, \beta)$. Topological graph (left) and stretch features (right). To reduce visual clutter, we display the five stretch features with biggest and smallest mean values, respectively. For reasons of comparison, the explanatory variables are scaled before training (min-max scaling)

positive coefficient values for the prediction of $\alpha$ and negative coefficient values for the prediction of $\beta$ (reversed effect). An explanation is that at high $L_3(P_2)$ values, the shortest path has more leeway to be pulled apart during the tensile strength experiment without contributing to the tensile strength (lower tensile strength).

The same applies to $L_1(P_1)$, the length of the shortest path in terms of edge count. Comparable interpretations cannot be made for all features, since some of the them exhibit reciprocal relationships. Overall, the coefficients are stable over different parameter combinations, indicating a robust model fitting.

In comparison to the TSS-model, the regression models achieve a significant speedup. The time needed to compute a stress-strain curve for a sample generated by an unseen parameter combination is reduced by the FGF-model by more than three orders of magnitude, from 24 to 48 hours to two minutes per sample. As both workflows can be executed in a parallelized fashion, the speedup is of a factor greater than 1.000.

## 4.4 Sequential Predictive Regression Model

The prediction quality of the FGF-model is significantly better than that of the PP-model, but is brought by a costly underlying Monte-Carlo simulation procedure. Thus, the performance of the FGF-model crucially depends on the fiber graph generation. The remarkable difference in the prediction quality suggests some nonlinear relations between production parameters and fiber graph features. Such relationships are overlooked when using a purely linear model, as is the case with the PP-model. To capture the nonlinearities, we propose an intermediate multivariate polynomial regression model to infer mean topological graph features. This is a fairly straightforward approach, cf. [22, 25]. Alternatives may include multivariate adaptive regression splines (MARS) [13] or radial basis functions [32]. However, in our application, we observe already very good predictive results with polynomials of a total degree up to 5. To enable the prediction of mean stress-strain curve parametrizations we couple the intermediate model with an errors-in-variables model (Sect. 4.4.1). The quality of the resulting predictive pipeline, referred to as production parameter and mean fiber graph feature-based predictive model (PP-MGF-model), is investigated in comparison to the previously discussed models in Sect. 4.4.2. The PP-MGF-model is new, in view of the existing literature, and represents a good compromise between the efficiency of the PP-model and the predictive accuracy of the FGF-model.

### 4.4.1 Coupled Polynomial Regression and Errors-In-Variabels Model

The underlying assumption of the intermediate multivariate polynomial regression model is that the production parameters and the individual graph features obey a perturbed polynomial relation. Let $\Gamma_{g,n}$ denote the set of $n$-dimensional multi-indices up to total degree $g \in \mathbb{N}$ with cardinality $l$, i.e.,

$$\Gamma_{g,n} = \left\{ \boldsymbol{\gamma} \in \mathbb{N}_0^n : |\boldsymbol{\gamma}| = \sum_{j=1}^n \gamma_j \leq g \right\}, \qquad l = |\Gamma_{g,n}| = \binom{n+g}{g}$$

and assume an arbitrary enumeration $\boldsymbol{\gamma}_1, \ldots, \boldsymbol{\gamma}_l$ of the multi-indices such that $\boldsymbol{\gamma}_1 = (0, \ldots, 0)$. Then the multivariate polynomials $q_j : \mathbb{R}^n \to \mathbb{R}$, $j = 1, \ldots, m$, are defined through

$$q_j(\boldsymbol{u}) = \sum_{i=1}^l c_{ij} \boldsymbol{u}^{\boldsymbol{\gamma}_i}, \quad \text{where } \boldsymbol{u}^{\boldsymbol{\gamma}_i} = \prod_{r=1}^n u_r^{\gamma_{ir}}$$

with polynomial coefficients $c_{ij} \in \mathbb{R}$ and factor interactions between the individual production parameters $\boldsymbol{u}^{\boldsymbol{\gamma}_i} \in \mathbb{R}$. Given $s \in \mathbb{N}$ observation pairs $\{(\boldsymbol{u}_i, \boldsymbol{v}_i)\}_{i=1}^s$, consisting of production parameters $\boldsymbol{u}_i \in \mathbb{R}^n$ and random fiber graph features $\boldsymbol{v}_i = M(\boldsymbol{u}_i, \omega_i) \in \mathbb{R}^m$, the intermediate polynomial regression model assumes the relation

$$\boldsymbol{v}_i = \boldsymbol{q}(\boldsymbol{u}_i) + \boldsymbol{\varepsilon}_{\mathrm{R},i}, \qquad \boldsymbol{q}(\boldsymbol{u}) = (q_1(\boldsymbol{u}), \ldots, q_m(\boldsymbol{u}))^T = \boldsymbol{C}^T (\boldsymbol{u}^{\boldsymbol{\gamma}_1}, \ldots, \boldsymbol{u}^{\boldsymbol{\gamma}_l})^T$$

with (unknown) coefficient matrix $\boldsymbol{C} \in \mathbb{R}^{l \times m}$, $(\boldsymbol{C})_{i,j} = c_{ij}$. Analogously to the basic linear regression model, the errors $\boldsymbol{\varepsilon}_{\mathrm{R},i}$ are assumed to be i.i.d. with

$E[\boldsymbol{\varepsilon}_{R,i}] = \mathbf{0}$ and $Cov[\boldsymbol{\varepsilon}_{R,i}] = \boldsymbol{\Sigma}_R$, for $i = 1, \ldots, s$. The task is to identify $C$ in order to simultaneously fit a multivariate polynomial of (total) degree $g$ for each of the $m$ (fiber graph) features. It is convenient to think of the $l$ possible factor interactions as independent variables. This allows to reformulate the model as a multiple multivariate linear regression model, since linearity is only required with respect to the regression coefficients $c_{ij}$. Thus, let $x_{ij} = \boldsymbol{u}_i^{\gamma_j}$ be the set of explanatory variables, then we get

$$V = XC + E_R$$

with design matrix $X \in \mathbb{R}^{s \times l}$ ($(X)_{i,j} = x_{ij}$, containing the factor interactions), response matrix $V \in \mathbb{R}^{s \times m}$ and error matrix $E_R \in \mathbb{R}^{s \times m}$. Note that no intercept must be included for setting up the design matrix $X$, since $x_{i1} = \boldsymbol{u}_i^{\gamma_1} = 1$ for $i = 1, \ldots, s$. Especially, for the case $g = 1$ the polynomial regression model includes the classic multiple multivariate linear regression model with intercept. In view of the reformulation, an adequate estimator for $C$ is given by the least-squares estimator $\widehat{C} = (X^T X)^{-1} X^T V$ which provides a (non-linear) predictor of the mean fiber graph features for given combinations of production parameters

$$\widehat{v}(\boldsymbol{u}) = \widehat{C}^T (\boldsymbol{u}^{\gamma_1}, \ldots, \boldsymbol{u}^{\gamma_l})^T. \tag{4.5}$$

The objective is now to predict the expected stress-strain curve parametrizations based on the production parameters and mean fiber graph features. Assuming a linear relation, which has been shown to be accurate in the case of the FGF-model, the functional relation is as follows

$$y_i = b + B_1^T \boldsymbol{u}_i + B_2^T \widehat{v}(\boldsymbol{u}_i) + \boldsymbol{\varepsilon}_i, \quad \text{for} \quad i = 1, \ldots, k. \tag{4.6}$$

Here $\boldsymbol{\varepsilon}_i : \Omega \rightarrow \mathbb{R}^r$ models the deviation from the mean parametrization caused by the stochastic nature of the simulation framework. To fit the relationship (4.6) to data, we replace the predictor $\widehat{v}$ with the variable $\bar{v}$ representing the mean features. In this context, we note that sampling data to fit the model using the TSS-model is not feasible, because the mean graph features $\bar{v}$ are not directly observable. Instead, we only have access to observations tuples $\{(\boldsymbol{u}_i, \boldsymbol{v}_i, y_i)\}_{i=1}^k$ composed of production parameters $\boldsymbol{u}_i \in \mathbb{R}^n$, random features $\boldsymbol{v}_i = M(\boldsymbol{u}_i, \omega_i) \in \mathbb{R}^m$ and associated stress-strain curve parametrizations $y_i = S(\boldsymbol{u}_i, \omega_i) \in \mathbb{R}^r$. However, the fiber graph features can be thought of as perturbed realizations of $\bar{v}$, i.e., $\boldsymbol{v}_i = \bar{v}_i + \delta_i$. Thereby, $\delta_i \in \mathbb{R}^m$ represents the error of measuring $\bar{v}_i$. Thus, in addition to the conventional errors in the regression equation, we assume errors in the explanatory variables as well. This results in the usage of the generalized errors-in-variables model [16] which assumes the relation

$$y_i = b + B_1^T \boldsymbol{u}_i + B_2^T \bar{v}_i + \boldsymbol{\varepsilon}_i, \tag{4.7a}$$

$$\boldsymbol{v}_i = \bar{v}_i + \delta_i. \tag{4.7b}$$

In Eq. (4.7) the observable variables are $y_i$, $u_i$ and $v_i$, whereas $\bar{v}_i$ is referred to as latent variable. Analogously to the multivariate linear regression model, the joint errors $\psi_i = (\delta_i^T, \varepsilon_i^T)^T$ are assumed to be i.i.d. with $\mathrm{E}[\psi_i] = \mathbf{0}$ and $\mathrm{Cov}[\psi_i] = \Sigma$, for $i = 1, \ldots, k$. Then the task is to estimate $B = [b, B_1^T, B_2^T]^T$. By applying the conventional least-squares estimator $\widehat{B} = (W^T W)^{-1} W^T Y$ (cf. Sect. 4.3.1), we neglect the measurement error described by (4.7b) during estimation. Even though it is well known that the least-squares estimator is not a consistent estimator for $B$ in the errors-in-variables model, it gives good results for prediction [5, 16].

Eventually, the coupling of the feature predictor (4.5) from the polynomial regression model with the fitted errors-in-variables model yields a mapping from the production parameters to the mean stress-strain curve parametrizations. It is determined by the intercept $\widehat{b}$ and the coefficient matrices $\widehat{B}_1, \widehat{B}_2, \widehat{C}$ according to the previous explanations and results in the (nonlinear) predictor

$$\widehat{y}(u) = \widehat{b} + \widehat{B}_1^T u + \widehat{B}_2^T \widehat{C}^T (u^{\gamma_1}, \ldots, u^{\gamma_l})^T, \quad \text{with} \quad \widehat{y} : \mathbb{R}^n \to \mathbb{R}^r. \quad (4.8)$$

A very convenient property of this coupling is that we can use different datasets for fitting the polynomial regression model and for fitting the errors-in-variables model, cf. Fig. 4.2. Particularly, since tensile strength simulations (computational bottleneck) are not necessary, the dataset for fitting the polynomial regression model can be chosen much larger ($s \gg k$). This is appropriate in order to account for the larger number of explanatory variables.

Conclusively, to approximate the input-output behavior of the TSS-model, we need distributional assumptions for the joint error behavior. Again, we rely on a multivariate normal distribution (similar to the PP-model) and determine a covariance estimate $\widehat{\Sigma}$ analogously to (4.3). Then, the PP-MGF-model behaves as $S \approx \widehat{S}$, where $\widehat{S}(u, \cdot) \sim \mathcal{N}(\widehat{y}(u), \widehat{\Sigma})$ for all $u \in \mathcal{U}_{ad}$. Predicting the mean stress-strain curve requires resampling, as it is the case for the PP-model.

### 4.4.2 Numerical Results

**Experimental Setting** Using dataset 3 (Table 4.4), we investigate the relation between production parameters and topological graph features by means of a 5-fold cross validation. Therefore, the set is randomly divided in 5 subsets, containing 400 samples each. In each run, one of the subsets is used as test set, while the remaining ones are used for training. We train the multivariate polynomial regression model for a total degree of $g \in \{1, \ldots, 5\}$, using a least-squares estimator for fitting the regression coefficients. To assess the model quality, we compare the median adjusted coefficients of determination, $\bar{R}^2$, observed throughout the cross-validation.

Subsequently, we investigate the quality of the PP-MGF-model for predicting the mean stress-strain curves. To achieve a fair comparison with regard to the PP-model and the FGF-model, we again perform a leave-one-out cross-validation

**Table 4.6** Results of the 5-fold cross-validation: Median of $\bar{R}^2$ values for the prediction of the topological graph features and for different polynomial degrees. Highest score per feature is highlighted in bold

| Feature | Degree 1 | Degree 2 | Degree 3 | Degree 4 | Degree 5 |
|---------|----------|----------|----------|----------|----------|
| $|V|$ | 0.8526 | 0.9698 | **0.9812** | 0.9809 | 0.9741 |
| $|E|$ | 0.9552 | 0.9852 | **0.9893** | 0.9892 | 0.9868 |
| $|V_u|$ | 0.8465 | 0.9344 | **0.9460** | 0.9460 | 0.9361 |
| $|V_l|$ | 0.8596 | 0.9416 | **0.9499** | 0.9466 | 0.9350 |
| $d_{max}$ | 0.4303 | 0.5861 | 0.6241 | **0.6352** | 0.5567 |
| $L_{fiber}$ | 0.9551 | 0.9823 | **0.9883** | 0.9880 | 0.9852 |
| $L_1(P_1)$ | 0.8819 | 0.9448 | **0.9469** | 0.9423 | 0.9329 |
| $L_2(P_2)$ | 0.8799 | 0.9403 | 0.9393 | **0.9406** | 0.9261 |
| $L_3(P_2)$ | 0.8761 | 0.9356 | **0.9369** | 0.9358 | 0.9190 |
| $D_1$ | 0.8776 | 0.9536 | 0.9680 | **0.9687** | 0.9437 |
| $D_2$ | 0.7835 | 0.9292 | 0.9638 | **0.9652** | 0.8153 |
| $D_3$ | 0.9460 | 0.9842 | **0.9897** | 0.9891 | 0.9851 |
| $|C_{min}|$ | 0.8494 | 0.9150 | 0.9311 | **0.9395** | 0.9223 |

(LOOCV) across the 43 production parameter combinations (dataset 1&2), as described in Sect. 4.3.2. To train the polynomial regression model, we use all fiber graphs (labeled and unlabeled) associated to the training set. Thereby, we test polynomial relations of the degree $g \in \{2, \ldots, 6\}$. To train the errors-in-variables model, we use the labeled training data only. For both models, we employ a least-squares fit. During inference on the test set, the fitted models use the production parameter combinations as input in order to predict the mean stress-strain curve parametrizations. To obtain a prediction of the mean stress-strain curve we resample 1.000 stress-strain curve parametrizations, reconstruct the associated curves using the constant-quadratic ansatz and then average over them (similar to the PP-model). In comparing the predicted mean stress-strain curves to the ground truth curves, we follow the descriptions from Sect. 4.3.2.

**Results and Discussion** The main results of the described 5-fold cross validation are summarized in Table 4.6. We observe that the adjusted $\bar{R}^2$ values, acting as a measure of model fit, peak for a degree of 3 and 4. Further increasing the polynomial degree for regression leads to a deterioration in terms of the adjusted $\bar{R}^2$ value. Since the case of polynomial degree 1 resembles the linear model, an improvement by moving to a higher polynomial degree is apparent. The high $\bar{R}^2$ values, which are even above 0.9 in most cases, are particularly astonishing and justify the use of a polynomial model for the mean fiber graph feature prediction.

The results regarding the 43-fold LOOCV are summarized in Table 4.7. The predictive results of the PP-MGF-model outperform that of the PP-model by a clear margin and almost reach the predictive quality of the FGF-model. We note that this is achieved without the need of a Monte-Carlo simulation procedure. Particularly, a polynomial fit of total degree 5 works best for the relation between production

**Table 4.7** Regression results of the LOOCV comparing the PP-model, the FGF-model and the PP-MGF-model: Median of observed $R^2$ and $\bar{R}^2$ values as well as OT loss. Best performing model is highlighted in bold

| Model | Approach | Median $R^2$ ↑ | Median $\bar{R}^2$ ↑ | OTLoss ↓ |
|---|---|---|---|---|
| PP-model | Param | 0.7967 | 0.7958 | 292.56 |
| PP-MGF-model | Degree 2 | 0.9282 | 0.9258 | 125.37 |
| PP-MGF-model | Degree 3 | 0.9428 | 0.9396 | 130.01 |
| PP-MGF-model | Degree 4 | 0.9427 | 0.9372 | 121.63 |
| PP-MGF-model | Degree 5 | **0.9584** | **0.9515** | **93.43** |
| PP-MGF-model | Degree 6 | 0.9572 | 0.9447 | 110.16 |
| FGF-model | Param + graph | 0.9717 | 0.9712 | 82.62 |

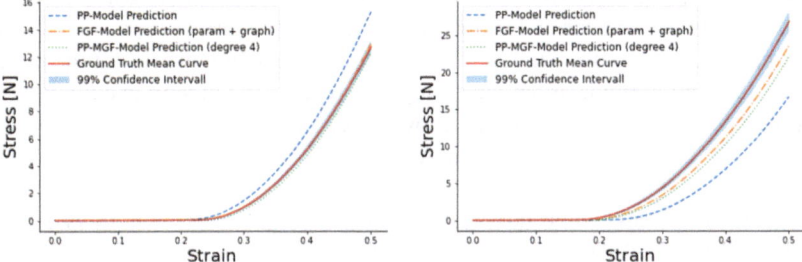

**Fig. 4.6** Mean stress-strain curve predictions resulting from our predictive model hierarchy. The exemplarily illustrated instances are observed during the LOOCV for two fully labeled test sets (belonging to set 1)

parameters and topological graph features. For higher degrees, we observe a deterioration of the $\bar{R}^2$ value, which is probably related to overfitting. We note that we also tested fitting the errors-in-variables model by means of a generalized total-least-squares estimation. However, this did not improve the prediction quality, for which the results presented are limited to the use of a conventional least-squares estimator.

In comparison to the TSS-model, the time needed to compute a stress-strain curve for a sample generated by an unseen parameter combination is reduced by more than six orders of magnitude, from 24–48 hours to 10 milliseconds per sample. In that, the PP-MGF-model is similar to the PP-model and more than three orders of magnitude better than the FGF-model in terms of computation time. However, note that the training of the PP-MGF-model is slightly more expensive than that of the other regression models, since it depends on a large amount of additional graph samples to fit the nonlinear relations between production parameters and graph features.

Summing up, the PP-MGF-model is cheap to evaluate and has excellent predictive quality, making it suitable for nonwoven material design. To conclude our discussion we refer to Fig. 4.6 which shows the predicted mean stress-strain curves of all models included in the predictive model hierarchy. Although isolated instances also led to other predictive gradations, the plots are representative of the observable results of the models and reflect the results of the numerical experiments performed.

## 4.5   Conclusion and Future Work

This chapter demonstrates the power of Informed Machine Learning in predicting material properties. We developed a regression-based model hierarchy for predicting the tensile strength behavior of nonwovens. While direct linear regression on the production parameters lacks accuracy (PP-model) and linear regression using individual fiber graphs requires time-consuming Monte-Carlo simulations (FGF-model), a coupling of a polynomial model with a linear (errors-in-variables) model (PP-MGF-model) has proven to be a good compromise combining the best of both model variants. By reducing the computation time by several orders of magnitude, a high accuracy of the prediction results (compared to the ground truth) is achieved. Thus, the PP-MGF-model promises to be of great benefit as a surrogate model for nonwoven material design, which is a field for further work. Our approach incorporates extensive domain knowledge into the modeling process at the points of training data and hypothesis set via simulation results and algebraic equations from scientific and expert sources. To our knowledge, our approach is completely new in the context of nonwoven material design.

**Acknowledgments** This contribution was supported by the Fraunhofer Cluster of Excellence "Cognitive Internet Technologies".

## References

1. Abou-Nassif, G.A.: Predicting the tensile and air permeability properties of woven fabrics using artificial neural network and linear regression models. Journal of Textile Science & Engineering **5**(5) (2015). https://doi.org/10.4172/2165-8064.1000209
2. Antweiler, D., Harmening, M., Marheineke, N., Schmeißer, A., Wegener, R., Welke, P.: Graph-based tensile strength approximation of random nonwoven materials by interpretable regression. Machine Learning with Applications **8**, 100288 (2022). https://doi.org/10.1016/j.mlwa.2022.100288
3. Antweiler, D., Harmening, M., Marheineke, N., Schmeißer, A., Wegener, R., Welke, P.: Machine learning framework to predict nonwoven material properties from fiber graph representations. Software Impacts **14**, 100423 (2022). https://doi.org/10.1016/j.simpa.2022.100423
4. Chen, T., Li, L., Koehl, L., Vroman, P., Zeng, X.: A soft computing approach to model the structure–property relations of nonwoven fabrics. Journal of Applied Polymer Science **103**(1), 442–450 (2007). https://doi.org/10.1002/app.24909
5. Cheng, C.L., Van Ness, J.W.: Statistical Regression with Measurement Error. John Wiley & Sons (1999)
6. Cheng, L., Adams, D.L.: Yarn strength prediction using neural networks: Part I: Fiber properties and yarn strength relationship. Textile Research Journal **65**(9), 495–500 (1995). https://doi.org/10.1177/004051759506500901
7. Das, D., Pourdeyhimi, B.: Composite Nonwoven Materials: Structure, Properties and Applications. Elsevier Science (2014)

8. Demirci, E., Acar, M., Pourdeyhimi, B., Silberschmidt, V.: Finite element modelling of thermally bonded bicomponent fibre nonwovens: Tensile behaviour. Computational Materials Science **50**(4), 1286–1291 (2011). https://doi.org/10.1016/j.commatsci.2010.02.039
9. Eltayib, H.E., Ali, A.H.M., Ishag, I.A.: The prediction of tear strength of plain weave fabric using linear regression models. International Journal of Advanced Engineering Research and Science **3**(11), 151–154 (2016). https://doi.org/10.22161/ijaers/3.11.25
10. Faessel, M., Delisée, C., Bos, F., Castéra, P.: 3D modelling of random cellulosic fibrous networks based on x-ray tomography and image analysis. Composites Science and Technology **65**(13), 1931–1940 (2005). https://doi.org/10.1016/j.compscitech.2004.12.038
11. Fan, J., Hunter, L.: A worsted fabric expert system: Part II: An artificial neural network model for predicting the properties of worsted fabrics. Textile Research Journal **68**(10), 763–771 (1998). https://doi.org/10.1177/004051759806801010
12. Farukh, F., Demirci, E., Sabuncuoglu, B., Acar, M., Pourdeyhimi, B., Silberschmidt, V.: Mechanical analysis of bi-component-fibre nonwovens: Finite-element strategy. Composites Part B: Engineering **68**, 327–335 (2015). https://doi.org/10.1016/j.compositesb.2014.09.003
13. Friedman, J.H.: Multivariate adaptive regression splines. The Annals of Statistics **19**(1), 1–67 (1991). https://doi.org/10.1214/aos/1176347963
14. Gramsch, S., Klar, A., Leugering, G., Marheineke, N., Nessler, C., Strohmeyer, C., Wegener, R.: Aerodynamic web forming: process simulation and material properties. Journal of Mathematics in Industry **6**(1), 1–13 (2016). https://doi.org/10.1186/s13362-016-0034-4
15. Harmening, M., Marheineke, N., Wegener, R.: Efficient graph-based tensile strength simulations of random fiber structures. ZAMM Journal of Applied Mathematics and Mechanics **101**(13) (2021). https://doi.org/10.1002/zamm.202000287
16. van Huffel, S., Vandewalle, J.: The Total Least Squares Problem. Society for Industrial and Applied Mathematics (1991). https://doi.org/10.1137/1.9781611971002
17. Johnson, R.A., Wichern, D.W.: Applied multivariate statistical analysis, 6 edn. Pearson, Upper Saddle River, NJ (2007)
18. Karpatne, A., Watkins, W., Read, J., Kumar, V.: Physics-guided neural networks (PGNN): An application in lake temperature modeling. arXiv e-prints 1710.11431 (2017)
19. Kufner, T., Leugering, G., Semmler, J., Stingl, M., Strohmeyer, C.: Simulation and structural optimization of 3D Timoshenko beam networks. ESAIM: Mathematical Modelling and Numerical Analysis **52**(6), 2409–2431 (2018). https://doi.org/10.1051/m2an/2018065
20. Le Bris, C.: Some numerical approaches for weakly random homogenization. In: Numerical Mathematics and Advanced Applications 2009, pp. 29–45. Springer (2010). https://doi.org/10.1007/978-3-642-11795-4_3
21. Lu, Y., Rajora, M., Zou, P., Liang, S.Y.: Physics-embedded machine learning: Case study with electrochemical micro-machining. Machines **5**(1), 4 (2017). https://doi.org/10.3390/machines5010004
22. Nizam, A., Rosenberg, E., Kleinbaum, D.G., Kupper, L.L.: Applied regression analysis and other multivariable methods. Brooks/Cole (2013)
23. Ohser, J., Mücklich, F.: Statistical Analysis of Microstructures in Materials Science. John Wiley, Weinheim (2000)
24. Ohser, J., Schladitz, K.: 3D Images of Materials Structures: Processing and Analysis. Wiley-VCH, Weinheim (2009)
25. Ostertagová, E.: Modelling using polynomial regression. Procedia Engineering **48**, 500–506 (2012). https://doi.org/10.1016/j.proeng.2012.09.545
26. Rahnama, M., Semnani, D., Zarrebini, M.: Measurement of the moisture and heat transfer rate in light-weight nonwoven fabrics using an intelligent model. Fibres and Textiles in Eastern Europe **21**, 89–94 (2013)
27. Raina, A., Linder, C.: A homogenization approach for nonwoven materials based on fiber undulations and reorientation. Journal of the Mechanics and Physics of Solids **65**, 12–34 (2014). https://doi.org/10.1016/j.jmps.2013.12.011

28. Ribeiro, R., Pilastri, A., Moura, C., Rodrigues, F., Rocha, R., Morgado, J., Cortez, P.: Predicting physical properties of woven fabrics via automated machine learning and textile design and finishing features. In: Artificial Intelligence Applications and Innovations, pp. 244–255. Springer (2020)
29. von Rüden, L., Mayer, S., Beckh, K., Georgiev, B., Giesselbach, S., Heese, R., Kirsch, B., Walczak, M., Pfrommer, J., Pick, A., Ramamurthy, R., Garcke, J., Bauckhage, C., Schuecker, J.: Informed machine learning – a taxonomy and survey of integrating prior knowledge into learning systems. IEEE Transactions on Knowledge and Data Engineering (2021). https://doi.org/10.1109/TKDE.2021.3079836
30. von Rüden, L., Mayer, S., Sifa, R., Bauckhage, C., Garcke, J.: Combining machine learning and simulation to a hybrid modelling approach: Current and future directions. In: Advances in Intelligent Data Analysis, *Lecture Notes in Computer Science*, vol. 12080, pp. 548–560. Springer (2020). https://doi.org/10.1007/978-3-030-44584-3_43
31. Schladitz, K., Peters, S., Reinel-Bitzer, D., Wiegmann, A., Ohser, J.: Design of acoustic trim based on geometric modeling and flow simulation for non-woven. Computational Materials Science **38**(1), 56–66 (2006). https://doi.org/10.1016/j.commatsci.2006.01.018
32. Walczak, B., Massart, D.: The radial basis functions — partial least squares approach as a flexible non-linear regression technique. Analytica Chimica Acta **331**(3), 177–185 (1996). https://doi.org/10.1016/0003-2670(96)00202-4
33. Wegener, R., Marheineke, N., Hietel, D.: Virtual production of filaments and fleeces. In: Currents in Industrial Mathematics, pp. 103–162. Springer (2015)

# Chapter 5
# Machine Learning for Optimizing the Homogeneity of Spunbond Nonwovens

Viny Saajan Victor, Andre Schmeißer, Heike Leitte, and Simone Gramsch

**Abstract** According to the Global Nonwoven Markets Report 2020–2025, published in 2021 by the two leading trading organisations representing nonwovens and related industries INDA and EDANA, the average annual growth rate of nonwoven production was 6.2% (INDA and EDANA Jointly Publish the Global Nonwoven Markets Report, A Comprehensive Survey and Outlook Assessing Growth Post-Pandemic, edana, 2021, Published September 29, 2021, from https://www.edana.org/about-us/news/global-nonwoven-markets-report) during the period from 2010 to 2020. In 2020 and 2021, nonwoven production has increased even further due to the huge demand for nonwoven products needed for protectiedanave clothing such as FFP2 masks to combat the COVID19 pandemic. Optimizing the production process is still a challenge due to its high nonlinearity. In this chapter, we present a machine learning-based optimization workflow aimed at improving the homogeneity of spunbond nonwovens. The optimization workflow is based on a mathematical model that simulates the microstructures of nonwovens. Based on training data coming from this simulator, different machine learning algorithms are trained in order to find a surrogate model for the time-consuming simulator. Human validation is employed to verify the outputs of machine learning algorithms by assessing the aesthetics of the nonwovens. We include scientific and expert knowledge into the training data to reduce the computational costs involved in the optimization process. We demonstrate the necessity and effectiveness of our workflow in optimizing the homogeneity of nonwovens.

V. S. Victor (✉) · A. Schmeißer · S. Gramsch
Fraunhofer ITWM, Kaiserslautern, Germany
e-mail: viny.saajan.victor@itwm.fraunhofer.de; andre.schmeisser@itwm.fraunhofer.de;
simone.gramsch@itwm.fraunhofer.de

H. Leitte
RPTU Kaiserslautern-Landau, Kaiserslautern, Germany
e-mail: leitte@cs.uni-kl.de

© The Author(s) 2025
D. Schulz, C. Bauckhage (eds.), *Informed Machine Learning*,
Cognitive Technologies, https://doi.org/10.1007/978-3-031-83097-6_5

## 5.1   Introduction

Spunbond processes are highly effective and cost-efficient methods of producing industrial nonwovens with desirable properties. The resulting nonwovens possess excellent tensile strength and tear resistance, which makes them ideal for applications requiring durability. Spunbond fabrics also have a consistent structure and thickness, thus rendering them highly desirable for applications that demand uniformity. Due to their ability to allow air and moisture to pass through, these materials are well-suited for use in filtration. Furthermore, spunbond fabrics are an economical alternative to woven or knitted fabrics, making them appropriate for a variety of specialized applications. These applications include liquid and gas filtration (such as vacuum cleaner bags and water filtration systems), insulation (for roofs, floors, and walls), automotive applications (such as seat covers, door panels, and headliners), medical applications (such as surgical gowns, masks, and drapes), hygiene products (such as diapers, sanitary pads, and wipes), as well as in batteries, fuel cells, and numerous other areas. It is projected that the market for nonwoven fabrics on a global scale will experience a compound annual growth rate of 7% [16] for the period from 2024 to 2032.

Using a spunbond process a nonwoven fabric is produced in several steps (cf. Fig. 5.1). First, a polymer is melted and pressed through hundreds of nozzles that are positioned in a spinneret. Thereby, filaments are formed from the molten polymer. Air coming from the side cools the polymer fibers. A second air stream stretches the fibers to their final diameter by drawing. After the fibers leave the drawing unit they are twirled by air until they lay down and form the nonwoven fabric. Due to the stochastic nature of the spunbond production process, achieving the desired quality of nonwoven products is a significant challenge for the industry that requires effective control measures. The complex production lines (>300 influencing variables) are typically adjusted by experienced line operators through a trial-and-error process. However, this process fails to reveal the true quality of the line's settings, thus preventing the full utilization of its potential for efficient and sustainable production. As a result, the level of performance, product quality, and line availability remain low. Additionally, the knowledge of how to operate production equipment is usually ingrained in the minds of the operators. This makes it difficult to increase production capacity when demand for certain products suddenly increases. This problem was exemplified in 2020, when the demand for FFP2 masks suddenly led to a strong demand for nonwoven products. Despite the surge in demand, production capacities could not be scaled up quickly enough.

The first approach to overcome this problem is to simulate the nonwoven production process. Using the simulation, optimal parameter settings for an altered production process can be searched offline and directly applied when changing the process. Depending on which production process is to be simulated, we need different numerical models. Figure 5.1 shows the process steps and lists the corresponding physical models that are necessary in order to simulate the spunbond process. These simulations are very accurate and independent of the amount of

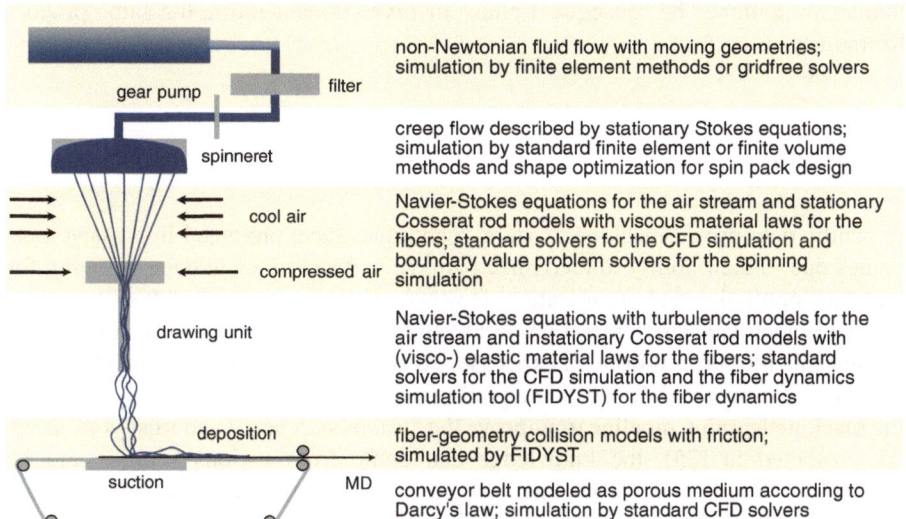

| | non-Newtonian fluid flow with moving geometries; simulation by finite element methods or gridfree solvers |
| --- | --- |
| gear pump   filter | |
| spinneret | creep flow described by stationary Stokes equations; simulation by standard finite element or finite volume methods and shape optimization for spin pack design |
| cool air   compressed air | Navier-Stokes equations for the air stream and stationary Cosserat rod models with viscous material laws for the fibers; standard solvers for the CFD simulation and boundary value problem solvers for the spinning simulation |
| drawing unit | Navier-Stokes equations with turbulence models for the air stream and instationary Cosserat rod models with (visco-) elastic material laws for the fibers; standard solvers for the CFD simulation and the fiber dynamics simulation tool (FIDYST) for the fiber dynamics |
| deposition | fiber-geometry collision models with friction; simulated by FIDYST |
| suction   MD | conveyor belt modeled as porous medium according to Darcy's law; simulation by standard CFD solvers |

**Fig. 5.1**  Principle of a spunbond process and necessary simulation models (according to [10])

data available. But the long simulation times prevent a timely prediction of quality when process conditions change as these tools are computationally expensive. Additionally, the simulation tools cannot easily incorporate human knowledge and weakly measurable criteria such as aesthetics.

Another approach is using data-driven machine learning (ML) models as they have gained immense importance in the last two decades. According to a 2018 analysis [7], however, not all potential application industries are developing at the same pace. Regarding the turnover potential of Machine Learning, the experts rank "Production and Industry 4.0" in 9th place after marketing, consumer electronics, banking, mobility, services, agriculture, and telecommunication. The reason for this is that especially in mechanical engineering often only small amounts of data are collected. Usually, production operates around the clock, 24 hours a day and 7 days a week, which restricts the duration of experimental series. Generating target data at industrial production facilities tends to be expensive and any downtime can result in production losses, leading to reduced revenue.

The goal of this chapter is therefore to combine simulation models with data-driven machine learning models along with human validation to improve the optimization of nonwoven production processes. We design a machine learning model to accelerate the mapping of process parameters to nonwoven quality. To address the issue of missing training data, we employed simulation tools. We further incorporate scientific and expert knowledge into this training data making our ML model "Informed". A visualization tool based on the proposed informed ML model is presented in our work [24]. The tool has been designed to cater to users who are domain experts, material scientists, and textile engineers. The proposed workflow is currently being used and tested by academic simulation experts with offline human validation. The subsequent phase involves implementing the same process for industry use.

## 5.2  Related Work

Recently, data-driven machine learning techniques have produced impressive outcomes due to their ability to recognize patterns and structures in data, allowing for real-time prediction and optimization. However, when confronted with systems that lack sufficient data and demand physical validity, these models are constrained due to their inherent lack of domain expertise. To address this issue, Informed Machine Learning is used which involves integrating prior, problem-specific knowledge into the machine learning pipeline to improve the system's accuracy and trustworthiness. As presented in [25], the knowledge can come from various sources and be represented in different forms and injected at various stages of the ML pipeline. Our approach involves incorporating the knowledge derived from simulation results into the training phase of the pipeline. Many previous methods include simulation-based knowledge integration in the training data by transforming or supplementing input and output features [5, 6, 13, 14, 18, 19, 22]. Our proposed workflow first incorporates expert knowledge to select relevant features and establish their acceptable ranges for the creation of training data. Following this, scientific knowledge is utilized with the aid of simulators to select and validate the input and output features. The presented Fig. 5.2 depicts how the pieces of prior knowledge is represented and integrated into the machine learning workflow.

In the textile industry, ML methods are commonly utilized as a substitute model to expedite the optimization process. These models forecast the physical characteristics of the product based on process parameters, allowing for opti-

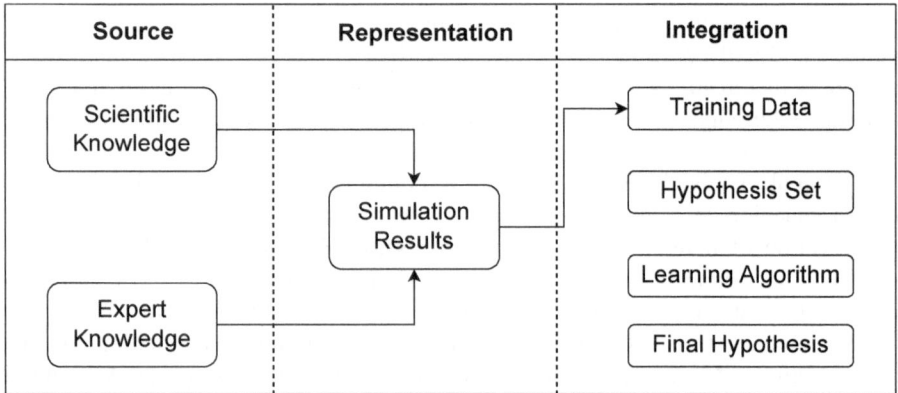

**Fig. 5.2** Information flow of the knowledge integration in the proposed machine learning method. Diagram adapted from [25]

mization. Various machine learning algorithms are employed to successfully solve classification and regression tasks such as defect detection and quality estimation [1, 2, 8, 9, 20, 21, 26]. These studies have assessed different ML algorithms to select the most efficient surrogate model for a particular dataset and application. Our study involves training a surrogate ML model to predict product quality from process parameters, which is then utilized to create a visualization tool to assist textile engineers in optimizing nonwoven quality. Additionally, our approach incorporates offline human validation of machine learning model results based on product-specific aesthetics. To the best of our knowledge, no previous study has provided a comprehensive workflow that covers dataset generation to visual application in the context of parameter space exploration to optimize nonwoven quality. Our approach minimizes the time required for optimization, through the utilization of ML, and reduces the need for domain expertise by providing a visual aid to navigate the parameter space. This workflow can be generalized to other applications that seek to optimize product quality by identifying the optimal combination of process parameters.

## 5.3   Machine Learning-Based Optimization Workflow Using Simulation Models

In this section, we propose a workflow for optimizing the quality of spunbond nonwovens based on Machine Learning. The workflow relies on a numerical tool that simulates the microstructures of nonwoven products using input parameters. However, due to the tool's high computational cost, it is not feasible to utilize it for real-time analysis of nonwoven product quality. Therefore, we use a machine learning model as a substitute for the tool. The task of the ML model is to predict the quality of the product for varying process conditions. We formulated this ML problem as multi-output regression based on the type of parameters involved in the production process and product quality. The dataset for the ML model is created using the numerical simulation tool as seen in Fig. 5.3. We integrate scientific and expert knowledge into this dataset. Based on the collected dataset, different regression models are trained and evaluated. The outcomes of the selected ML model are further verified through human validation. Below is a detailed discussion of the workflow that consists of five stages: parameter selection (Sect. 5.3.1), data collection with knowledge integration (Sect. 5.3.2), model selection (Sect. 5.3.3), training and testing (Sect. 5.3.4), homogeneity optimization with human validation (Sect. 5.3.5).

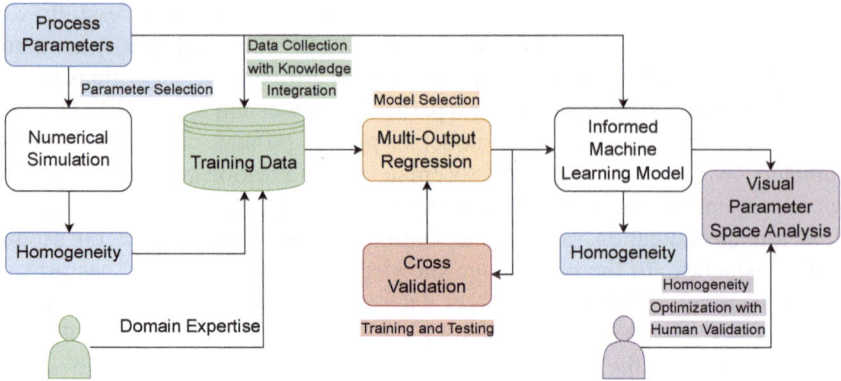

**Fig. 5.3** Workflow of machine learning-based optimization of spunbond nonwovens. Different stages are colored corresponding to the sub-section headings in this chapter

## 5.3.1   Parameter Selection

We identified six process parameters that control the quality of spunbond nonwovens. A specific combination of these parameters represents a particular condition involved in the production process. The ranges for the parameters are defined by the domain experts based on typical application scenarios of the final products (for example production of manufacturing material, medical protective masks, etc.).

### 5.3.1.1   Process Parameters

The process parameters correspond to the inputs taken by the numerical simulation tool. They capture the deterministic and stochastic properties of the production process. Each parameter corresponds to a characteristic property. They are chosen such that the characteristic qualities of a real nonwoven and the nonwoven produced by the tool are identical. The selected process parameters are $\sigma_1$, $\sigma_2$, $A$, $v$, $n$, and $d_s$:

1. $\sigma_1$ is the standard deviation of the 2D normal distribution of the fibers around the spinning outlet in the direction which is parallel to the running conveyor belt (machine direction) without belt movement. The feature values vary from 1 mm to 50 mm.
2. $\sigma_2$ is the standard deviation of the 2D normal distribution of the fibers around the spinning outlet in the direction which is perpendicular to the running conveyor belt (cross direction) without belt movement. The feature values vary from 1 mm to 50 mm.
3. $A$ is the noise amplitude of the stochastic process. It is the feature that contains all random effects of the production process, e.g., the influence of the turbulent flow during the fiber spinning, the contacts between fibers, and laydown. This

feature specifies whether the simulated fiber lays down in a deterministic (value 0) or stochastic (value $\infty$) manner. The values vary from 1 to 50.
4. $v$ is the ratio of spinning speed and belt speed. The feature values vary from 0.01 to 0.25.
5. $n$ is the number of spin positions per meter, which can be altered when designing the machine but is fixed during production. The feature values vary from 200 to 10,000.
6. $d_s$ is the discretization step size which is the distance of discrete points along the simulated fiber curves. The feature values varies from $2.5 \times 10^{-5}$ to $5 \times 10^{-5}$ mm.

The selected parameters, apart from $v$ and $n$, do not correspond directly to machine settings, but rather reflect the stochastic properties of the nonwoven produced, which are influenced by various factors such as machine geometry, process parameters (e.g., airflow), process conditions (e.g., air pressure, temperature), and complex interactions (e.g., fiber-fiber, fiber-flow interactions). These parameters are not constant for a machine, except for $n$, and are influenced by the nonwoven material itself. For instance, the density and stiffness of the material affect the curvature of the fibers, which in turn affects the standard deviations $\sigma_1$ and $\sigma_2$. Moreover, these parameters may change depending on the desired application; for instance, producing nonwovens for FFP2 masks requires different operating conditions compared to those for roof materials, as the former must meet strict quality standards. However, changing process conditions in real-time can be challenging because of the down times, such as the time required to clear the old polymer material, refill with new material, and adjust suction position and velocity. To address this issue, the fibers are simulated to depict real-world scenarios, and stochastic properties are assigned to each of them to create virtual nonwoven materials for quality inspection.

### 5.3.1.2 Product Quality: Homogeneity

The quality of the virtual nonwoven produced from the simulation tools is measured using the coefficient of variation (CV). The coefficient of variation is a statistical measure of the relative dispersion of data points in a data series around the mean $\mu$. With the standard deviation $\sigma$ it is defined as

$$CV = \frac{\sigma}{\mu}. \tag{5.1}$$

It establishes the nonwoven web's homogeneity. A more homogeneous nonwoven typically has a lower CV value, which can have an impact on characteristics like filter quality and tensile strength. We compute the CV value at multiple grid resolutions, resulting in an output feature vector (one entry per resolution), to take into consideration homogeneity at various levels of resolution. We regularly discretize the data and compute the fiber mass per bin to obtain the different resolutions. Seven levels of resolution, according to our experiments, had the best agreement with the results of manual inspection.

## 5.3.2 Data Collection with Knowledge Integration

This section explains the process of collecting data points that are used to train the ML models. We incorporate scientific knowledge and expert knowledge in the data collection process. This reduces the computational time and memory required in the process and facilitates feature selection. The knowledge integration into the data is validated by experimental results. Training on the 'informed data' obtained by this approach makes our ML model 'informed'. In the following sections, we discuss knowledge integration at various stages of the data collection process.

### 5.3.2.1 Sample Size Estimation for Simulation Model Setup

The computation time and memory required by the numerical simulation tool to simulate virtual nonwoven material increases with the quantity of the material. This makes it not practical to simulate the entire material during data collection. Hence, we decided to simulate a smaller sample of the material that is representative of the whole material in terms of homogeneity. In order to achieve this, we identified two steps: estimating the size of the sample and simulating only the nonwovens that overlap with the material within this sample size.

The simulation tool is non-deterministic in nature as it produces slightly different results each time when run with the same process parameter setting. Hence, we need to ensure that the selected sample size should have the least statistical uncertainty. Therefore, we created a dataset with 3125 combinations of process parameters using uniform sampling. We simulated this dataset five times with a sample material size of 5 cm × 5 cm. The statistical uncertainty was quantified as the coefficient of variation of the simulation results for each process parameter setting across five simulation runs. The parameter setting with the maximum uncertainty was further simulated 100 times with three sample sizes: 5 cm × 5 cm, 15 cm × 50 cm and 25 cm × 50 cm for detailed analysis. Table 5.1 shows the coefficient of variation for the three sample sizes. We can observe from the table that uncertainty reduces with an increase in the sample size. Hence, we decided to choose the sample size of 25 cm × 50 cm and further reduced the uncertainty by sampling each process parameter five times in the data. This averaging of samples reduces the uncertainty by $\sqrt{5}$ according to the central limit theorem.

After determining the size of the sample material, we simulate the nonwovens that intersect with this sample material. The laydown of filaments on the conveyor belt is modeled as a 2D normal distribution with standard deviations $\sigma_1$ and $\sigma_2$. According to the empirical rule in statistics, the nonwovens that are a bit more than $3\sigma_2$ away from the sample in the cross direction and a bit more than $3\sigma_1$ away

**Table 5.1** Table showing coefficient of variation for three sample sizes across seven grid resolutions

| Grid resolutions | | | | | | | |
|---|---|---|---|---|---|---|---|
| Sample size | 0.5 mm | 1 mm | 2 mm | 5 mm | 10 mm | 20 mm | 50 mm |
| 5 cm × 5 cm | 0.04 | 0.05 | 0.07 | 0.12 | 0.21 | 0.51 | 0.72 |
| 15 cm × 50 cm | 0.01 | 0.01 | 0.02 | 0.03 | 0.05 | 0.12 | 0.30 |
| 25 cm × 50 cm | 0.01 | 0.01 | 0.01 | 0.02 | 0.04 | 0.09 | 0.20 |

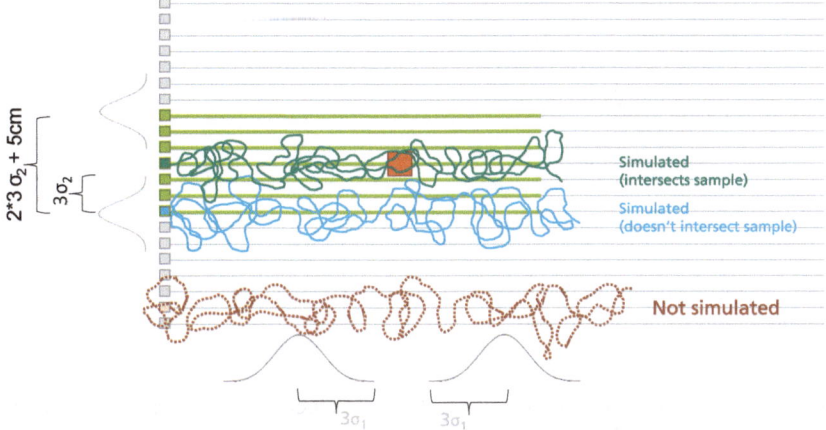

**Fig. 5.4** Simulation construction based on $\sigma_1$ and $\sigma_2$ values for sample size 5 cm × 5 cm: Orange area shows the sample extracted for analysis. Green and cyan filaments are simulated in order to compute the sample, brown filament is part of the nonwoven but not simulated

from the sample in the machine direction do not have the chance of overlapping as depicted in Fig. 5.4. Hence we do not simulate these filaments.

### 5.3.2.2   Influence of Discretization Step Size ($d_s$)

One of the potential parameters that affect the product quality is the discretization step size of the simulated filaments. Based on the domain expertise, it is expected that chosen $d_s$ is small enough such that the simulated nonwoven accurately represents a corresponding real nonwoven and thus this parameter does not affect the simulated product quality. We wanted to investigate whether this assumption is correct. Otherwise, the discretization step size would have to be included as an input parameter in the ML models. For this purpose, we created two small datasets with two different $d_s$ values and analyzed the effect of the two different $d_s$ values on the product quality. Figure 5.5 shows the relative deviation in the CV values between two $d_s$ values. The red color denotes the deviation due to the non deterministic statistical uncertainty of the simulation tools and the blue color denotes the deviation due to $d_s$. We defined a threshold to extract the latter alone (green line). We retrieved

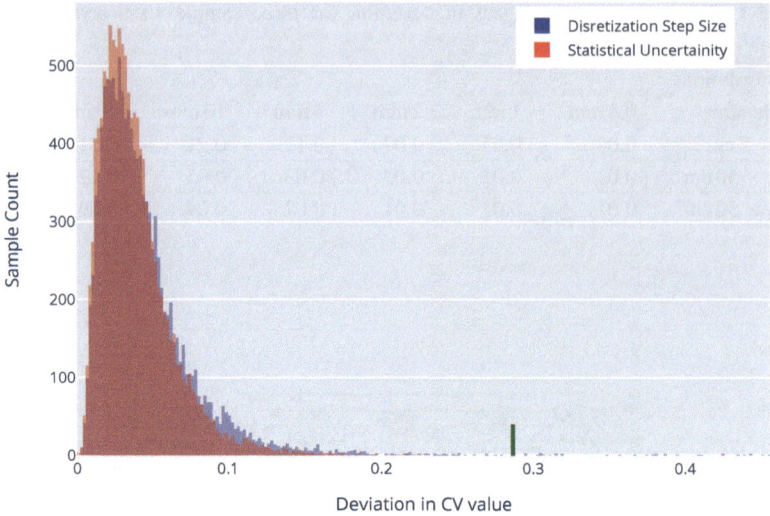

**Fig. 5.5** Histogram depicting the relative deviation in CV values due to discretization step size (blue) and statistical uncertainty (red)

only 0.0025% of the parameter settings that exceeded the threshold. Hence, we concluded that the influence of discretization step size on the product quality is not statistically significant and eliminated this parameter from the input feature set.

### 5.3.2.3  Input Data Sampling

Once the simulation setup is completed, determining which parameter values to investigate is a crucial question to handle during the data collection. We chose a sparse sampling strategy since the parameter space is very large. Specifically, we used two data sampling techniques to effectively capture the behaviour of the production process for the desired output. The first technique involves the determination of optimal parameter ranges over various small subsets of the parameter space based on the specific nonwoven product. This technique uses expert knowledge to choose the precise parameter ranges. The second technique involves Latin hypercube sampling, which selects parameter values uniformly random from the chosen ranges of values. This technique provides unbiased observations from the numerical simulator.

### 5.3.3   Model Selection

We chose multiple-input, multiple-output regression for the machine learning technique since the input data comprises five continuous features ($\sigma_1$, $\sigma_2$, $A$, $v$, $n$) and the output has seven continuous values (CV values at seven grid-sizes). We assessed common regression algorithms such as linear regression, polynomial regression, Bayesian regression, random forests, and neural networks. A brief description of these models is listed as follows. In order to simplify the optimization formulas described in the following sections, we have made the assumption that the regression problem involves a single output variable.

#### 5.3.3.1   Linear Regression (LR)

Linear Regression [15] is a supervised machine learning algorithm. It is used to determine the linear relationships between the input parameters and output values. We evaluated four different flavors of the linear regression model: Vanilla, Ridge, Lasso, and ElasticNet. Each of the flavors differs in the type of regularization used in the optimization function. Vanilla regression uses no regularization. Lasso regression and Ridge regression use $L_1$ and $L_2$ regularization respectively. ElasticNet regression uses both $L_1$ and $L_2$ regularization. The minimization problem of the linear regression algorithm with $n$ data points is defined below

$$\min_{\boldsymbol{w}} \sum_{i=1}^{n} \left\| \boldsymbol{w}^T \boldsymbol{x_i} - y_i \right\|^2 , \tag{5.2}$$

where $\boldsymbol{w} = \{w_1, w_2, ..., w_p\}$ is the coefficient vector and $p$ is the number of input features. The target value $y_i$ is expected to be the linear combination of the ith input feature vector $\boldsymbol{x_i}$. We observed that the linear models are not adequate to fit our data as their error rate is very high. This implies that the relationship between the input parameters and the output values is not linear.

#### 5.3.3.2   Support Vector Regression (SVR)

Support Vector Regression [23] is a supervised learning algorithm that is used to predict continuous values. SVR can efficiently perform a non-linear regression using the kernel trick by implicitly mapping the input data into high-dimensional feature spaces. We use the radial basis function as the kernel for our model. The simplest form of the minimization problem for support vector regression is

$$\min \frac{1}{2} \|\boldsymbol{w}\|^2 \tag{5.3}$$

subject to the constraints

$$\begin{cases} y_i - \langle\, w, x_i\,\rangle - b \leq \epsilon \\ \langle\, w, x_i\,\rangle + b - y_i \leq \epsilon\,, \end{cases}$$

where $y_i$ is the target value for the ith input feature $x_i$ and $\langle\,,\,\rangle$ denotes the dot product. The goal is to find a hyperplane with optimal values for the weight vector $w$, and the bias $b$ that maximizes the width of the margin $\epsilon$ between the predicted outputs and the actual outputs of the training data. SVR uses relatively less memory compared to random forests (with a large number of trees) and artificial neural networks (with a complex architecture) and performs better than linear regression models in high-dimensional spaces. However, it does not scale well with data and is prone to noisy data.

### 5.3.3.3 Polynomial Regression (PR)

Polynomial regression [17] is a supervised learning algorithm in which the relationship between the input parameters and the output values is modeled as a polynomial of degree $n$. The minimization problem for polynomial regression with $n$ data points is

$$\min \sum_{i=1}^{n} (y_i - f(x_i))^2\,, \tag{5.4}$$

where $y_i$ and $f(x_i)$ are actual and predicted values respectively for the ith input feature $x_i$. The polynomial function $f(x)$ can be represented as

$$f(x) = \beta_0 + \beta_1 x + \beta_2 x^2 + \ldots + \beta_m x^m\,, \tag{5.5}$$

where $\beta_0$, $\beta_1$, $\beta_2$, ..., $\beta_n$ are the coefficients of the polynomial function to be estimated. The degree $m$ of the polynomial that is chosen is crucial in polynomial regression. A very small degree would under-fit the model. As we pursue higher degrees, the training and validation error initially decreases, as seen in Fig. 5.6. After degree 11, the training error is still falling while the validation error starts to rise, indicating that the model is beginning to overfit the data. Therefore, we decided that degree 11 would be a good fit for our model. Given that it offers a large variety of functions for data fitting, polynomial regression is well suited to model non-linear relationships between the data. Nevertheless, they are susceptible to overfitting and sensitive to outliers, which has an impact on the generalizability of the models.

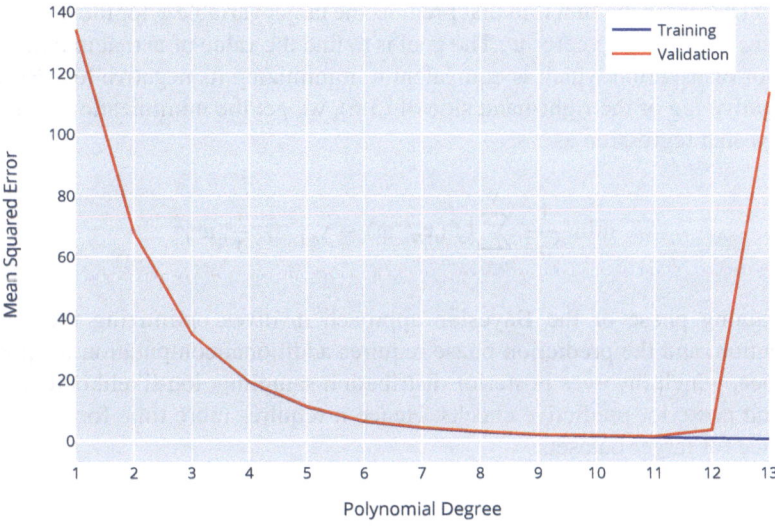

**Fig. 5.6** Plot showing the mean squared error for different degrees of the polynomial regression

### 5.3.3.4 Bayesian Regression (BR)

In Bayesian regression [3], problems are formulated using probability distributions rather than point estimates. This enables us to assess the level of uncertainty and confidence in model predictions. The goal of Bayesian regression is to ascertain the posterior distribution for the weight vector $w$ rather than to identify the one "best" value. The posterior distribution is the conditional distribution of weight vector $w$ given target variable $y$, hyper-parameter $\alpha$, and model noise variance $\sigma^2$ and it is calculated using Bayesian theorem as given below:

$$p(w|y, \alpha, \sigma^2) \propto p(w|\alpha)L(w) , \qquad (5.6)$$

where $p(w|\alpha)$ is the prior distribution of the weight vector which is assumed to be drawn from Gaussian distribution and is given by

$$p(w|\alpha) = \left(\frac{\alpha}{2\pi}\right)^{1/2} \exp\left\{-\frac{\alpha}{2}\|w\|^2\right\} \qquad (5.7)$$

and $L(w)$ is the likelihood function which is the conditional distribution of target variable $y$ given weight vector and model noise distribution with mean 0 and variance $\sigma^2$. It is calculated as

$$L(w) = p(y|w, \sigma^2) = \left(\frac{1}{2\pi\sigma^2}\right)^{N/2} \exp\left\{-\frac{1}{2\sigma^2}\sum_{n=1}^{N}|f(x_n; w) - y_n|^2\right\} ,$$

$$(5.8)$$

where $f(x_n; w)$ is the function that predicts the target variable $y$ for the input feature vector $x_n$ with weight vector $w$. The goal is to find the value of $w$ that maximizes the posterior distribution which is equivalent to minimizing its negative log. By taking the negative log of the right-hand side of (5.6), we get the minimization problem of the Bayesian regression as

$$\min \frac{1}{2\sigma^2} \sum_{n=1}^{N} \left| f(x_n; w) - y_n \right|^2 + \frac{\alpha}{2} \|w\|^2 . \tag{5.9}$$

The training phase of the Bayesian approach involves optimizing the posterior distribution, and the prediction phase requires additional computation for posterior inference, which involves posterior distribution sampling, log-likelihood computation, and posterior predictive checks. Hence it requires more time for training and inference for larger datasets.

### 5.3.3.5  Random Forests (RF)

Random Forest Regression [4] is a supervised learning algorithm that leverages the ensemble learning method for regression. This type of learning method combines predictions from various machine learning algorithms to provide predictions that are more accurate than those from a single model. Given the set of input-output pairs ([$x$, $y$]), the goal of random forest regression is to obtain the function $f(x)$ the accurately predict the output for unseen input. The function $f(x)$ is formulated as an ensemble of $T$ decision trees, where each decision tree $t$ is trained on a subset of training data. The predictions of the decision trees are then combined to obtain final prediction. Each decision tree is built by choosing the feature and threshold that provides the optimal split as determined by a certain criterion (e.g., mean squared error). The data is then divided into subsets based on the chosen feature and threshold. The same steps are repeated for each subset until a stopping criteria is met (e.g., the maximum depth of the tree has been reached). Finally, a constant value (e.g., mean or median) is assigned to each leaf node as the predicted output value.

Random Forests typically perform well on problems that include features with non-linear correlations. This is supported by their capacity for efficient feature subset selection and rapid decision tree construction, which facilitates faster training and prediction process. However, random forests can be prone to over-fitting, lack interpretability, and suffer from imbalanced datasets.

### 5.3.3.6  Artificial Neural Networks (ANN)

An artificial neural network [12] is a computational model that uses a network of functions to comprehend and translate a data input of one form into the desired

**Table 5.2** Neural network architecture chosen from hyper-parameter tuning

| Layer type | Number of nodes | Activation function |
|---|---|---|
| Input layer | 5 | Linear |
| Hidden layer | 256 | Relu |
| Hidden layer | 512 | Relu |
| Hidden layer | 512 | Relu |
| Hidden layer | 256 | Relu |
| Hidden layer | 768 | Relu |
| Output layer | 7 | Linear |

output. The basic building blocks of a conventional neural network are nodes, which are organized into layers. The input features are passed through these layers (input, hidden, and output) with a sequence of non-linear operations to obtain the final prediction. The minimization problem of neural network for regression with $n$ data points can be formulated as

$$\min_{W,b} \sum_{i=1}^{n} L\big(y_i, f(x_i; W, b)\big) , \tag{5.10}$$

where $W$ represents the weights applied to the inputs of a node, while the bias $b$ represents the value added to the weighted sum of the inputs of the same node. and $L$ is the loss function. We used Mean Squared Error (MSE) shown in (5.13) as the loss function for our network. The goal is to find the optimal values for $W$ and $b$ that minimize the loss function. The flexibility of neural networks allows them to learn complex non-linear correlations between inputs and outputs. They can learn to smooth out noise and capture underlying patterns in the data, making them robust to noisy data. However, training neural networks can take a long time, especially if the dataset is large or the model is complex.

We design a Neural Network to learn the non-linear dependency of our output values with respect to the input features. The accuracy of the neural network is determined by the optimal choice of hyper-parameters that decide its architecture. We performed network parameter tuning to find the optimal hyper-parameters while keeping the desirable accuracy. The hyper parameters we used for optimization are the number of hidden layers (from one to five), number of nodes in each layer (from 8 to 1024 with an increment of 8) and the activation functions ('relu', 'sigmoid' and 'tanh'). We used a randomized search (1000 samples) over the ranges of hyper parameters and selected the parameters based on the error on validation data. Adam optimizer and a learning rate of $1 \times 10^{-3}$ was used for the analysis. We used early stopping method to avoid over-fitting of the data. Table 5.2 shows the chosen network architecture based on the evaluation metrics.

### 5.3.4  Training and Testing

We divided the dataset into an 80% training data and a 20% testing set. The training data is further divided into an 80% training set and a 20% validation set. As discussed in Sect. 5.3.2.1, we sample each data point five times to account for the non-deterministic behavior of the simulation tool. Hence, we divided the dataset into groups of five identical data points and assigned indices to these groups. The indices are then randomly shuffled and split into training, validation, and testing sets based on the proportions described above. This makes sure that the identical data points are assigned entirely to one of the three data sets and the trained ML models can be evaluated for unseen data. The training set is then used to train the regression models and the validation set is used to tune the model hyper-parameters. The testing set is used for unbiased evaluation of the model. Since the input features are measurements of different units, we also performed feature scaling to tailor the data for the machine learning models. For each data point $x_i$ of the individual input feature distribution $\mathbf{x}$ having $n$ data points, mean $\mu$, and standard deviation $\sigma$, we calculate the re-scaled feature value $z_i$ as

$$z_i = \frac{x_i - \mu}{\sigma}, i \in \{1, 2, \ldots, n\}. \tag{5.11}$$

The models are evaluated using the metrics defined below. For the equations used in the following section, we define $\mathbf{Y}$ and $\hat{\mathbf{Y}}$ as the matrices with $m$ rows and $n$ columns, where $m$ is the number of test data points and $n$ is the number of indices corresponding to the seven grid sizes. The individual elements $\mathbf{Y_{i,j}}$ and $\hat{\mathbf{Y}}_{\mathbf{i,j}}$ represent the actual and predicted values respectively for the data point $i$ and the grid size corresponding to the index $j$. $\bar{\mathbf{Y}}_{:,\mathbf{j}}$ represents the mean value of the data points corresponding to the grid size with index $j$.

1. Mean absolute percentage error (MAPE) is a statistical measure to evaluate the accuracy of a regression model. The error is independent of the scale of the output as it measures the accuracy as a percentage. MAPE is calculated as

$$MAPE = \frac{100}{n} \sum_{i=1}^{m-1} \sum_{j=0}^{n-1} \left| \frac{\mathbf{Y_{i,j}} - \hat{\mathbf{Y}}_{\mathbf{i,j}}}{\mathbf{Y_{i,j}}} \right|. \tag{5.12}$$

2. Mean squared error (MSE) measures the average of squares of the errors. The MSE is a good estimate for ensuring that the ML model has no outlier predictions with huge errors since it puts larger weight on these errors due to the squaring. MSE is calculated as

$$MSE = \frac{1}{n} \sum_{i=1}^{m-1} \sum_{j=0}^{n-1} (\mathbf{Y_{i,j}} - \hat{\mathbf{Y}}_{\mathbf{i,j}})^2. \tag{5.13}$$

**Fig. 5.7** Nonwoven materials with same base weight (2.20 g m$^{-2}$) and similar average homogeneity (13.82 and 14.11), but with different aesthetics

3. Coefficient of determination ($R^2$ Score) is the measure of how close the data points are to the fitted regression line. It explains how much of the variance of actual data is explained by the predicted values. $R^2$ Score is calculated as

$$R^2(y, \hat{y}) = 1 - \frac{\sum_{i=1}^{m-1} \sum_{j=0}^{n-1} (\mathbf{Y_{i,j}} - \hat{\mathbf{Y}}_{\mathbf{i,j}})^2}{\sum_{i=1}^{m-1} \sum_{j=0}^{n-1} (\mathbf{Y_{i,j}} - \bar{\mathbf{Y}}_{:,\mathbf{j}})^2}. \tag{5.14}$$

### 5.3.5 Homogeneity Optimization with Human Validation

After selecting the best ML model based on the evaluation metrics, it is used to forecast the homogeneity of the spunbond nonwoven given a set of process parameters. However, due to the size of the parameter search space, we are unable to scan the entire parameter space. In order to address this issue, we developed a visualization tool [24] built on the best ML model that aids textile engineers in parameter space exploration. This tool provides real-time navigation through the parameter space. It also supports the identification of promising regions in the parameter space and the sensitivity of the individual parameter settings. The tool reduces the domain expertise required in the optimization by visually guiding the engineers toward local and global minima. The tool is utilized to identify and choose $n$ potential parameter settings (e.g. $n = 10$) based on the optimal CV values. These settings are subsequently used to generate corresponding nonwoven images from the simulator, which is an offline process as it demands a considerable amount of time. Once generated, textile engineers validate the simulated images, discarding any that do not meet the product requirements in order to determine the best parameter setting. One such requirement involves applications where specific aesthetic features (e.g. seat covers for cars, furniture covers) are required. In this case, the best parameter setting can be chosen based on the generated nonwoven's aesthetics. Figure 5.7 shows the images of two virtual nonwoven materials with the same base weight (2.20 g m$^{-2}$) and similar average homogeneity (13.82 and 14.11) with different aesthetics.

**Table 5.3** Performance of machine learning models on the test dataset

|                              | MAPE    |          | MSE     |          | R^2 score |                        |
| ---------------------------- | ------- | -------- | ------- | -------- | --------- | ---------------------- |
| ML algorithm                 | Mean    | Variance | Mean    | Variance | Mean      | Variance               |
| Linear regression            | 95.2983 | 1.9254   | 93.2949 | 2.0498   | 0.595     | $2.77 \times 10^{-5}$  |
| Support vector regression    | 11.0702 | 0.0038   | 22.9310 | 0.4066   | 0.92      | 0.00                   |
| Polynomial regression        | 5.9486  | 0.0093   | 1.767   | 1.1715   | 0.9890    | $1 \times 10^{-5}$     |
| Bayesian regression          | 7.7343  | 0.0058   | 1.5080  | 0.0026   | **0.99**  | 0.00                   |
| Random forests               | 10.0065 | 0.0280   | 1.8280  | 0.0137   | 0.98      | 0.00                   |
| Artificial neural networks   | **3.8214** | 0.0126 | **0.358** | 0.0207 | **0.99**  | 0.00                   |

The bold values indicate the best values for the corresponding error metric

## 5.4 Experiments

In this section, we discuss the necessity and effectiveness of our workflow in optimizing the homogeneity of spunbond nonwovens. For the experiments, we sampled an input database with 311,740 data points. The dataset included 12,348 discrete and 50,000 Latin hypercube data points, each of which was sampled five times as discussed in Sect. 5.3.2.1. For each input data point, we simulated the digital nonwoven image using the numerical tool. This image with a selected sample size of 25 cm × 50 cm is then used to calculate the CV values at seven different grid-sizes. For machine learning analysis, we used 199,514 data points for training, 49,878 data points for validation, and 62,348 data points for testing.

We evaluated execution times on a workstation with a 40 core Intel® Xeon® E5-2680 v2 (2.80 GHz) CPU. The execution time to produce 311,740 virtual spunbond nonwoven samples on this machine is approximately $6,479,983 \times 10^3$ ms with an average time of 20,786.5 ms per sample.

### 5.4.1 Models Evaluation Based on the Accuracy

For statistical evaluation of the ML models, the training and testing sets are randomly selected ten times and for each pair, the models are trained on the training set and evaluated on the testing set. Table 5.3 shows the mean and variance of the ten testing set errors for different regression models. We can see from the table that ANNs have the best MAPE, MSE, and $R^2$ Score compared to other models. Figure 5.8 shows the CV value predictions versus the actual CV values using ANNs for grid-sizes 0.5 mm and 50 mm. The grid-size 0.5 mm corresponds to the best case with least number of predictions ($6.4031 \times 10^{-3}\%$ of the testdata) outside the error tolerance range of 5%. And the grid-size 50 mm corresponds to the worst case with most number of predictions (1.6712% of the testdata) outside the error tolerance range of 5%.

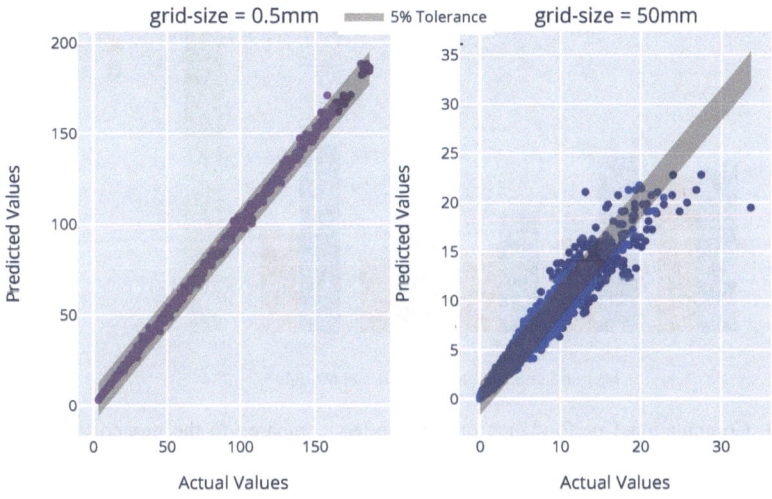

**Fig. 5.8**  Predicted vs actual CV values for grid resolutions 0.5 mm and 50 mm

## 5.4.2  Models Evaluation Based on Computational Performance

The computational efficiencies of the ML models compared to the numerical simulator for 10,000 data samples are displayed in Fig. 5.9. The training time in the figure represents the time required by the model to train on 10,000 samples, the prediction time shows how long the model requires to forecast 10,000 samples. The figure shows that Bayesian regression has the shortest training time, and Random Forests have the shortest prediction time. In general, model training only needs to be done a single time. However, it is necessary to repeat the procedure of employing the models to predict numerous times during optimization. Therefore, Random Forests have a greater advantage relative to other models in terms of computing performance.

For practical applications, the scalability of the ML models plays a significant role. Scalability refers to the ability of ML models to handle large amounts of data and carry out a large number of computations efficiently and quickly. In order to determine whether the models scale well with the data, we computed the computational time required by the model for a sizable data set. Figure 5.10 shows the time taken by the ML models to train and test the entire dataset (311,740 samples) in comparison with the numerical simulator. We mainly focus on the prediction time as the training is done only once. The table shows that, with the exception of the SVR, all the models scale reasonably well with the data. The results reveal the effectiveness of using the ML models as a surrogate to the numerical simulation that significantly reduces the time involved in the optimization process (from 20,786.5 ms to 0.0588 ms per sample with ANN as a surrogate). We save

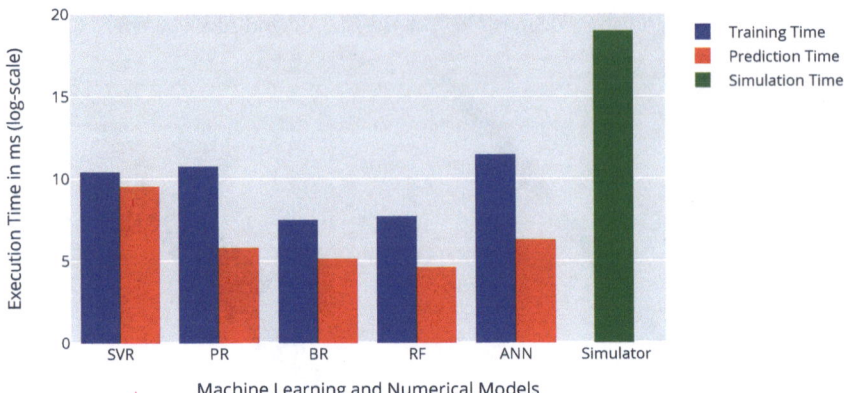

**Fig. 5.9** Computational performance of ML models compared to the numerical simulator for 10,000 data samples

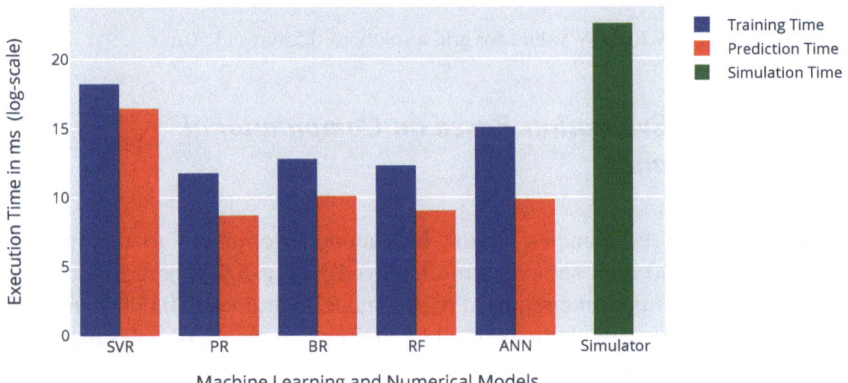

**Fig. 5.10** Computational performance of ML models compared to the numerical simulator for 311,740 data samples

approximately 74 days for the entire dataset using the ANN model (including the training time).

Based on the evaluation, we chose the artificial neural network as the best surrogate model for the simulation tool as it provided the best accuracy with comparable scalability. This chosen model can be used in real-time optimization of homogeneity of the nonwovens. One successful application of this model is presented in [24] as discussed before. The proposed visualization tool uses the ANN model for exploring the space of process parameters in real-time to optimize the quality of the nonwovens. The tool is currently tested by academic simulation experts for its efficacy in the optimization process.

## 5.5 Conclusion

In this chapter, an ML-based workflow for optimizing the homogeneity of spunbond nonwovens is proposed and a model based on multi-output regression is established. During the data collection phase of the training process, we showcased the successful integration of scientific and expert knowledge, leading to the establishment of an Informed ML model. Furthermore, several machine learning algorithms for process parameter tuning are explored based on the model that is verified by human validation. Experimental results show that Artificial Neural Networks have good accuracy and Random Forests have good computational performance across different sizes of training and testing data. Additionally, experimental findings demonstrate the efficacy of our strategy for real-time optimization.

**Acknowledgments** This contribution was supported by the Fraunhofer Cluster of Excellence "Cognitive Internet Technologies".

## References

1. Abou-Nassif, G. A.: Predicting the tensile and air permeability properties of woven fabrics using artificial neural network and linear regression models. Textile Sci Eng. **5**, 5–209 (2015)
2. Beltran, R. and Wang, L. and Wang, X.: Predicting the pilling propensity of fabrics through artificial neural network modeling. Textile research journal. **75**, 557–561 (2005)
3. Bishop, C. M. and Tipping, M. E. and others: Bayesian regression and classification. Nato Science Series sub Series III Computer And Systems Sciences. IOS PRESS. **190**, 267–288 (2003)
4. Breiman, L.: Random forests. Machine learning. Springer. **45**, 5–32 (2001)
5. Daw, A. and Karpatne, A. and Watkins, W. and Read, J. and Kumar, V.: Physics-guided neural networks (PGNN): An application in lake temperature modeling. arXiv preprint arXiv:1710.11431. (2017)
6. Deist, T. and Patti, A. and Wang, Z. and Krane, D. and Sorenson, T. and Craft, D.: Simulation assisted machine learning. Bioinformatics. Oxford, England. (2019)
7. Döbel, I. and others: Maschinelles Lernen. Eine Analyse zu Kompetenzen, Forschung und Anwendung. https://www.bigdata-ai.fraunhofer.de/content/dam/bigdata/de/documents/Publikationen/Fraunhofer_Studie_ML_201809.pdf last accessed: 09/05/2022
8. Eltayib, H. E. and Ali, A. HM and Ishag, I. A.: The prediction of tear strength of plain weave fabric using linear regression models. International Journal of Advanced Engineering Research and Science. **3**, 151–154 (2016)
9. Fan, J. and Hunter, L.: A worsted fabric expert system: Part II: An artificial neural network model for predicting the properties of worsted fabrics. Textile Research Journal. **68**, 763–771 (1998)
10. Gramsch, S. and Sarishvili, A. and Schmeißer, A.: Analysis of the fiber laydown quality in spunbond processes with simulation experiments evaluated by blocked neural networks. Advances in Polymer Technology, 2020.
11. INDA and EDANA Jointly Publish the Global Nonwoven Markets Report. A Comprehensive Survey and Outlook Assessing Growth Post-Pandemic (2021). https://www.edana.org/about-us/news/global-nonwoven-markets-report
12. Jain, A. K. and Mao, J. and Mohiuddin, K. M.: Artificial neural networks: A tutorial. Computer. IEEE. **29(3)**, 31–44 (1996)

13. Lee, K. and Ros, G. and Li, J. and Gaidon, A. Spigan: Privileged adversarial learning from simulation. In: Int. Conf. Learning Representations. ICLR. (2018)
14. Lerer, A. and Gross, S. and Fergus, R.: Learning physical intuition of block towers by example. International conference on machine learning. PMLR. 430–438 (2016)
15. Montgomery, D. C. and Peck, E. A. and Vining, G. G.: Introduction to linear regression analysis. John Wiley & Sons (2021)
16. Nagrale, P.: Global Non Woven Fabric Market Overview. In: Market Research Future. 2019, from https://www.marketresearchfuture.com/reports/nonwoven-fabrics-market-7459
17. Ostertagová: Modelling using polynomial regression. Procedia Engineering. Elsevier. **48**, 500–506 (2012)
18. Pfrommer, J. and Zimmerling, C. and Liu, J. and Kärger, L. and Henning, F. and Beyerer, J.: Optimisation of manufacturing process parameters using deep neural networks as surrogate models. Procedia CiRP. Elsevier. **72**, 426–431 (2018)
19. Rai, A. and Antonova, R. and Meier, F. and Atkeson, C. G.: Using simulation to improve sample-efficiency of Bayesian optimization for bipedal robots. The Journal of Machine Learning Research. **20(1)**, 1844–1867 (2019)
20. Ribeiro, R. and Pilastri, A. and Moura, C. and Rodrigues, F. and Rocha, R. and Cortez, P.: Predicting the tear strength of woven fabrics via automated machine learning: an application of the CRISP-DM methodology. Proceedings of the 22th International Conference on Enterprise Information Systems – ICEIS (2020)
21. Ribeiro, R. and Pilastri, A. and Moura, C. and Rodrigues, F. and Rocha, R. and Morgado, J. and Cortez, P.: Predicting physical properties of woven fabrics via automated machine learning and textile design and finishing features. IFIP International Conference on Artificial Intelligence Applications and Innovations, 244–255 (2020)
22. Shrivastava, A. and Pfister, T. and Tuzel, O. and Susskind, J. and Wang, W. and Webb, R.: Learning from simulated and unsupervised images through adversarial training. Proceedings of the IEEE conference on computer vision and pattern recognition. 2107–2116 (2017)
23. Smola, A. J. and Schölkopf, B.: A tutorial on support vector regression. Statistics and computing. Springer. **14**, 199–222 (2004)
24. Victor, V. S. and Schmeißer, A. and Leitte, H. and Gramsch, S.: Visual Parameter Space Analysis for Optimizing the Quality of Industrial Nonwovens. IEEE Computer Graphics and Applications. **42(2)**, 56–67 (2022)
25. Von Rueden, L. and Mayer, S. and Beckh, K. and Georgiev, B. and Giesselbach, S. and Heese, R. and Kirsch, B. and Pfrommer, J. and Pick, A. and Ramamurthy, R. and others: Informed Machine Learning–A Taxonomy and Survey of Integrating Knowledge into Learning Systems. IEEE Transactions on Knowledge and Data Engineering (2023)
26. Yap, P. H. and Wang, X. and Wang, L. and Ong, K.: Prediction of wool knitwear pilling propensity using support vector machines. Textile research journal. **80**, 77–83 (2010)

# Chapter 6
# Bayesian Inference for Fatigue Strength Estimation

Dorina Weichert, Elena Haedecke, Gunar Ernis, Sebastian Houben, Alexander Kister, and Stefan Wrobel

**Abstract** A vital material property of metals is long life fatigue strength. It describes the maximum load that can be cyclically applied to a defined specimen for a number of cycles that is thought to represent an infinite lifetime. The experimental measurement of long life fatigue strength is costly, justifying the need to create a precise estimate with as few experiments as possible. We propose a new approach for estimating long life fatigue strength that defines a ready-to-use experimental and analysis procedure. It relies on probabilistic machine learning methods, efficiently connecting expert knowledge about the material behavior and the test setup with historical and newly generated data. A comparison to state-of-the-art standard experimental procedures shows that our approach requires fewer experiments to produce an estimate at the same precision—massively reducing experimental costs.

D. Weichert (✉) · E. Haedecke · G. Ernis
Fraunhofer IAIS, Sankt Augustin, Germany
e-mail: dorina.weichert@iais.fraunhofer.de; elena.haedecke@iais.fraunhofer.de;
gunar.ernis@iais.fraunhofer.de

S. Wrobel
Fraunhofer IAIS, Sankt Augustin, Germany
University of Bonn, Bonn, Germany
e-mail: stefan.wrobel@iais.fraunhofer.de

S. Houben
Hochschule Bonn-Rhein-Sieg, Sankt Augustin, Germany
e-mail: sebastian.houben@h-brs.de

A. Kister
Federal Institute for Materials Research and Testing, Berlin, Germany
e-mail: Alexander.Kister@bam.de

© The Author(s) 2025
D. Schulz, C. Bauckhage (eds.), *Informed Machine Learning*,
Cognitive Technologies, https://doi.org/10.1007/978-3-031-83097-6_6

**Table 6.1** List of symbols, grouped by application type

| | |
|---|---|
| | **Long life fatigue strength** |
| $L$ | Long life fatigue strength, assumed to follow a log-normal distribution, therefore $\log L \sim \mathcal{N}(\mu_L, \sigma_L^2)$ |
| $L_+$ | Fatigue strength of a single specimen |
| $\mu_L$ | Mean of log-normally distributed long life fatigue strength |
| $\sigma_L$ | Standard deviation of log-normally distributed long life fatigue strength |
| | **Loads in experiments** |
| $n_k$ | Number a load level $k$ was reached (staircase method) |
| $l_i$ | Load in failure experiment $i$ (BI method) |
| $l_j$ | Load in runout experiment $j$ (BI method) |
| $l^\star$ | Recommended load for next experiment (BI method) |
| | **Quantities in Bayesian Inference method** |
| $x$ | Material parameters |
| $m(x)$ | Mean function of Gaussian Process |
| $k(x, x')$ | Covariance function of Gaussian Process |
| $\hat{\mu}_{L_{GP}}(x)$ | Gaussian Process prediction for some material $x$, $\hat{\mu}_{L_{GP}}(x) \approx \mathcal{N}\left(\mu_{GP}(x), \sigma_{GP}^2(x)\right)$ |
| $\mu_{GP}$ | Estimated mean of the mean of the long life fatigue strength by Gaussian Process |
| $\sigma_{GP}^2$ | Estimated variance of the mean of the long life fatigue strength by Gaussian Process |
| $g(\mu_L, \sigma_L)$ | Posterior estimate of long life fatigue strength |
| $\Phi_{\mu_L, \sigma_L}(l)$ | Value of cumulative fatigue strength distribution at experiment with load $l$ |
| $H(g)$ | Entropy of a distribution $g$ |
| $H(g\vert l)$ | Predictive entropy of posterior estimate of long life fatigue strength when adding an experiment at load $l$ |
| $\hat{\mu}_{L_{MAP}}$ | Maximum a posteriori estimate of the mean fatigue strength $\mu_L$, based on actual posterior distribution $g(\mu_L, \sigma_L)$ |
| | **Quantities in staircase method** |
| $\hat{\mu}_{L_0}$ | Experimenter's estimate of the mean fatigue strength |
| $\hat{\mu}_{L_{stair}}$ | Estimated mean fatigue strength via the staircase method |
| $\hat{\sigma}_{L_{stair}}$ | Estimated standard deviation of the fatigue strength via the staircase method |
| $d_{\log}$ | Load increment in staircase method |
| $F_T$ | Sum of all loads $l_k$ in staircase method |
| $A_T$ | Sum of all $k \cdot l_k$ in staircase method |
| $B_T$ | Sum of all $k^2 \cdot l_k$ in staircase method |
| $D_T$ | Variance for estimating the standard deviation in staircase method |
| $L_0$ | Lowest analysable load in staircase method |

## 6.1 Introduction

Long life fatigue strength is an important metric of metallic materials. It describes the maximum load a product can tolerate without breaking for its lifetime, e.g., a spring or a gear tooth. To generate a material-specific estimate, defined (fixed) loads are applied cyclically to specimens of a defined geometry. The long life fatigue strength (in the following fatigue strength) refers to the maximum load that can be applied for a defined maximum number of load cycles that is thought to refer to an infinite lifetime. The fatigue strength of a material is estimated via testing a collection of specimens of identical characteristics (e.g., geometry). The fatigue strength of each specimen is unique, so the values of the collection follow a distribution. To model this distribution, literature assumes that the fatigue strength $L$ of a specific material follows a log-normal distribution with parameters $(\mu_L, \sigma_L)$ [4]. Fatigue strength estimation is difficult, as it aims to determine this not directly observable random variable $L$. In practice, this means that for a specimen it is only possible to observe if it breaks at the applied load, indicating that the load was larger than the specimen's fatigue strength $L_+$ (a so-called failure); or if it survives the procedure, indicating that the load was smaller than the specimen's fatigue strength $L_+$ (a so-called runout). Additionally, the evaluations are specimen-specific and testing two specimens of the same material at the same load does not necessarily lead to the same outcome: one may fail, while the other might survive. Hence the idea of fatigue testing is to create a valid statistic of failures and runouts at different loads to estimate the mean fatigue strength $\mu_L$ and the standard deviation of the fatigue strength $\sigma_L$ with sufficient confidence.

Traditionally, fatigue strength estimation is performed following a standard procedure [4]. This standard defines multiple analysis methods for test results (i.e., the probit method, the maximum likelihood method, the staircase method) and one experimental procedure specifying which loads to apply to the specimens (e.g., the staircase method). Disadvantageous about these methods is the high dependency on the experience of the process engineers, as either no experimental procedure is defined at all (the experiments fully rely on the experts) or the procedure requires hyperparameters to be defined by the process engineers.

In this chapter, we define a new hyperparameter-free modular approach for fatigue strength estimation that connects the expert knowledge of the process engineers with data. Our approach consists of two modules: in the first module, a Gaussian Process (GP) Regression Model is used to create a prior estimate of the mean fatigue strength $\hat{\mu}_{L_{GP}}$ based on the similarity of different materials. The model incorporates available historical fatigue data and expert knowledge defined in the covariance function. The second module builds on traditional Bayesian Inference (BI) to create a posterior distribution over the parameters $(\mu_L, \sigma_L)$, using actual experimental data and a knowledge-based likelihood. This posterior offers three independent options for the experimenter:

1. the derivation of a maximum a posteriori (MAP) estimate for the distribution parameters

2. the variance of the posterior and therefore an approximation of the confidence of the parameters found by MAP estimation
3. an acquisition function in line with traditional Active Learning, specifying an experimental procedure.

We design an acquisition function that trades off the confidence of the experimental outcome with the predictive entropy of the posterior based on the acquired experiment. We verify our approach by comparison with the staircase method, as it is the only fatigue strength estimation method that defines an experimental procedure. The comparison shows the superiority of our modular approach as it yields similar results to the staircase method at a shorter number of test iterations. This is due to the expert knowledge being abstractly integrated into the covariance function and the likelihood, making it robust against misspecification as there are no material-specific procedural parameters. The approach corresponds to a type of Informed Machine Learning in which expert knowledge is integrated into the hypothesis set in the form of probabilistic relations [29].

We structure this chapter as follows: in the first section, we describe the necessary background on fatigue testing, the state-of-the-art experimental procedures, and the related work. In the second section, we introduce our modular approach for fatigue strength estimation and validate it in the third section. The fourth section summarizes the results and concludes with potential future work. Table 6.1 lists the symbols used in this chapter.

## 6.2   Background

The intuition of this section is to give the reader the necessary background of fatigue strength estimation. Here, we portray standard fatigue testing methods, especially the staircase method, and give an overview of the related work.

### 6.2.1   Fatigue Testing

This paragraph gives a short overview of fatigue testing, closely following the latest valid standard [4]. Fatigue testing determines the tolerable load amplitude at a predefined number of cycles for a material. Therefore, a defined load is applied cyclically to a specimen, which will either be fractured or become a so-called runout, where the (predefined) ultimate number of cycles is reached without breaking. Fatigue strength is a material property that depends on the applied load type (e.g., tension, pressure, torsion, bending), the load amplitude, the applied mean load, and the number of cycles used to define a long life. We interpret fatigue strength as a log-normally distributed random variable. At test time, this means that at exactly the same conditions, one specimen can be a runout while another is fractured. To estimate the fatigue strength, the parameters of the log-normal distribution $(\mu_L, \sigma_L)$ are determined up to a predefined uncertainty. Using hat-notation for estimated values, we find $(\hat{\mu}_L, \hat{\sigma}_L)$ based on the experiments. Nevertheless, we not only have to analyze the parameters of the distribution themselves but also quantities like the standard deviation of the estimated mean $\mathrm{Std}(\hat{\mu}_L)$.

Following the standard, the fatigue strength properties of different materials are assumed to be independent [4]. Therefore, the required number of tests is defined by the chosen uncertainty of the distribution parameters. Different methods are valid for analyzing test results: the probit method, boundary techniques, the staircase method, combinations of the probit and the staircase method, and the maximum likelihood method [4]. In the following, we concentrate on the staircase method, as it not only defines an analysis method but also an experimental procedure similar to our approach.

## 6.2.2 Experimental Procedure and Analysis of the Staircase Method

The term "staircase method" defines a standardized experimental procedure and a standardized analysis method for the gained data. In the following, we describe the method in detail. Afterward, we analyze its drawbacks and derive requirements for a new experimental procedure and analysis method.

### 6.2.2.1 Experimental Procedure

The experiments must follow a defined iterative procedure for determining the fatigue strength by the staircase method.

Before the first test, the experimenter has to define a first estimate for the mean fatigue strength $\hat{\mu}_{L_0}$ and a logarithmic load increment $d_{\log}$. It is recommended to choose this value based on an estimate for the logarithmic standard deviation of the fatigue strength $s_{\log} = \log_{10} \sigma_L$, then $d_{\log} = 10^{s_{\log}}$. Valid loads in the test procedure are given by $\hat{\mu}_{L_0} \cdot d_{\log}^i$, where $i \in \mathbb{Z}$. Given an initial test at a freely chosen load level, the iterative procedure works as follows:

- after a failure, the load level is reduced by the equidistant load increment;
- after a runout, the load level is raised by the load increment.

This procedure is followed until the required number of tests for the estimation of the distribution parameters is reached, or it is evident that the test parameters, especially the load increment $d_{\log}$, have to be changed. The number of tests to take out to reach a given confidence interval of the estimated mean fatigue strength $\hat{\mu}_{L_{stair}}$ is given in a table in [4]; this table relies on simulations of the staircase procedure.

Figure 6.1a shows an exemplary run by the staircase method. While the load levels in the first two runs were too high to be reached again, all other runs show the typical staircase pattern and additionally illustrate the noisiness of fatigue tests: tests at the same load level can both fail or survive the applied load.

### 6.2.2.2 Analysis of Test Results by the Staircase Method

For estimating the distribution parameters of the fatigue strength, the test results have to fulfill several conditions:

1. no interruptions of the load sequence, tests have to follow the test procedure strictly,
2. test results of the first iterations are only taken into account if the corresponding load level is reached again within the test series,
3. test series must contain at least two reversal points,
4. test series have to contain load levels where specimens are both failures and runouts.

If the series is valid, the parameters $(\mu_L, \sigma_L)$ of the fatigue strength are estimated as follows: Firstly, a fictitious experiment with an unknown outcome is added to the current test series to statistically enhance the data basis for analysis. Therefore, the rules of the iterative experimentation procedure are followed: if the last experiment was a failure, the fictitious experiment is at a lower load level; if it was a runout, the fictitious experiment is at a higher load level (see Fig. 6.1a). Afterward, the lowest valid load level $L_0$ is determined and is indexed by $k = 0$. Then, the number $l_k$, how often each load level at index $k$ was reached, is counted. Based on these figures, additional variables are calculated, namely

$$F_T = \sum_k l_k$$

$$A_T = \sum_k k \cdot l_k$$

$$B_T = \sum_k k^2 \cdot l_k$$

$$D_T = \frac{F_T \cdot B_T - A_T^2}{F_T^2},$$

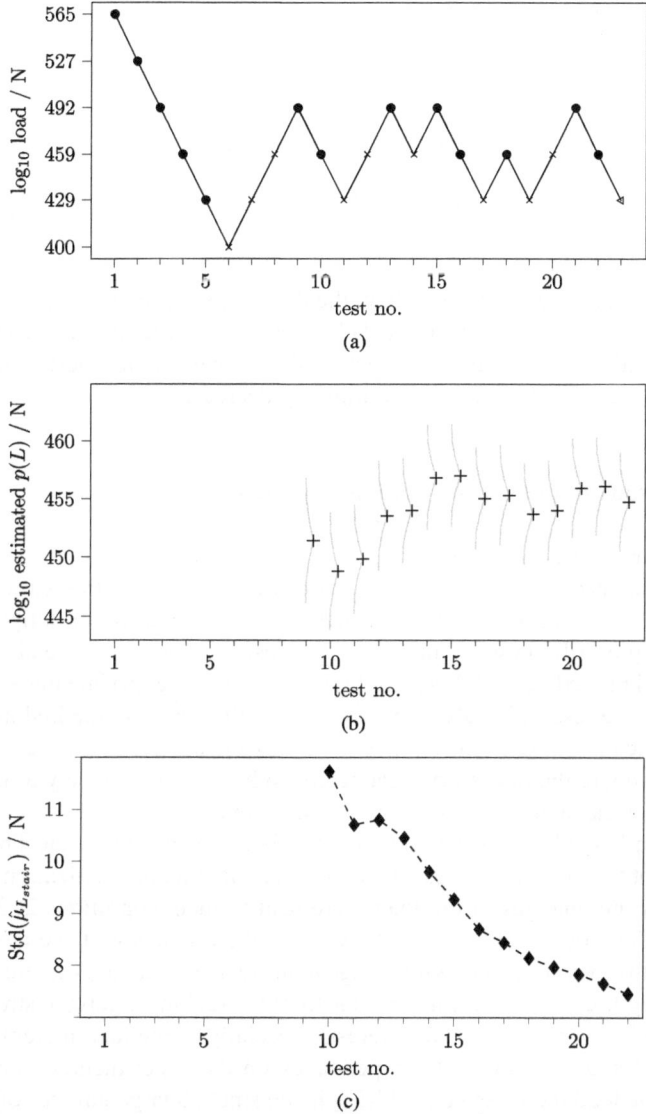

**Fig. 6.1** Estimation of fatigue strength by the staircase method. For the analysis, only series parts that fulfill the analysis conditions (see Sect. 6.2.2.2) are taken into account. (**a**) Test series by the staircase method, the logarithmized y-axis shows the non-equidistant load increments. Filled circle: failure, multiplication symbol: runout, open triangle: fictitious run at end of series. (**b**) Estimated fatigue strength distribution $p(L) \approx \log \mathcal{N}(\hat{\mu}_{L_{stair}}, \hat{\sigma}_{L_{stair}})$ by the staircase method. Plus symbol: estimated mean fatigue strength $\hat{\mu}_{L_{stair}}$. (**c**) Estimated standard deviations of the mean of the fatigue strength

which are then used to find

$$\hat{\mu}_{L_{stair}} = L_0 \cdot d_{\log}^{\frac{A_T}{F_T}}$$

if $D_T < 0.5 : \hat{\sigma}_{L_{stair}} = 0.5 \log d_{\log}$

if $D_T \geq 0.5 : \hat{\sigma}_{L_{stair}} = \log d_{\log} \cdot 10^{4.579494 \cdot (F_T)^{-0.889521}} \cdot D_T^{7.235548 \cdot (F_T)^{-0.405229}}.$

$$(6.1)$$

Figure 6.1b and c exemplarily show the devolution of the these estimated values over the test sequence in Fig. 6.1a. While especially the estimated mean of the fatigue strength $\hat{\mu}_{L_{stair}}$ (+ in Fig. 6.1b) oscillates over the test series, the standard deviation of the mean $\text{Std}(\hat{\mu}_{L_{stair}})$ continually decreases.

### 6.2.2.3 Disadvantages of the Staircase Method

**From Application Perspective**

The staircase method is often used in practice, as it specifies experimentation guidelines. This contrasts with the other analysis methods for fatigue strength, which only prescribe how to analyze given test data but do not elaborate on the experimental procedure and therefore highly rely on the experimenter's knowledge.

However, the test and analysis procedures by the staircase method are problematic for application in real life, as they are inflexible and rely on the experimenter's ability to estimate the necessary parameters. When looking closely at the different conditions for the analysis, this point becomes obvious:

Condition 1 is only met when the tests strictly follow the defined test procedure. It indicates that the order of the tests is relevant and that no parallelization is possible. Additionally, no changes in the load increment to meet conditions 2, 3, and 4 are permitted. Condition 2 exemplifies how much the efficiency of the test procedure depends on the experimenter's knowledge to make a reasonable estimate of the mean fatigue strength $\hat{\mu}_{L_0}$. If his estimate is far from the real mean fatigue strength, many runs are wasted to reach loads of interest, especially as the load increment is fixed. Conditions 3 and 4 illustrate the dependency on the experimenter's knowledge to determine the load increment $d_{\log}$. If $d_{\log}$ is too small, a large number of tests has to be taken out to generate two reversal points, especially if $\hat{\mu}_{L_0}$ is misspecified. A too large $d_{\log}$ causes the series not to meet condition 4.

**From a Mathematical Perspective**

From a mathematical perspective, the analysis by the staircase method has several drawbacks: first, the fixed step size hinders an exact determination of the distribution parameters. A flexible step size facilitates experimentation and allows the experimenter to do more experiments at load levels with a high impact on the parameter estimates.

Furthermore, the method only indirectly considers the information if an experiment is a runout or a failure. For a high number of samples, this is plausible. However, for a lower one (that is the case in practice), this additional information is expected to improve the quality of the parameter estimates.

#### 6.2.2.4 Requirements for an Alternative Experimental Approach

From the application side, there exists a high interest in developing a more flexible test and analysis procedure, (ideally) offering the following advantages:

**Use of All Existing Data About the Same Material Type** A main drawback of the staircase method is that the load increment $d_{\log}$ must not be changed during the test procedure. A false estimate of the load increment results in non-valid test results, as they do not fulfill the analysis conditions 3 and 4. An improved test and analysis procedure should consider changes of the load increment and load levels that were reached only once (condition 2) to include additional historical data, if available.

**No Necessity to Estimate the Mean Fatigue Strength by Experts Beforehand** Even for human experts, it is hard to make a good estimate of the mean fatigue strength $\hat{\mu}_{L_0}$ before conducting experiments. Nevertheless, test efficiency (the number of tests taken into account for the analysis) relies on the estimate, as initial experiments might be wasted due to condition 2. Recommending an estimate based on historical data from similar materials is, therefore, a substantial improvement.

**Potential Parallelization of Tests** A single experiment run takes up to 12 weeks, depending on the limiting number of cycles. The staircase method is a purely iterative approach, prohibiting parallelization. Therefore, an experimental procedure that allows for conducting two or more experiments in parallel would significantly reduce experimentation time.

### 6.2.3 Related Work

Several authors show the high potential of using Machine Learning based methods in Material Science [14, 24]. As a vast research field, we focus on publications that either specifically deal with fatigue strength estimation or summarize multiple relevant publications in terms of a review.

In general, there exist three main potentials of applying Machine Learning in Material Science: the prediction of material properties, the discovery of new materials fulfilling specific material properties using Bayesian Optimization [27] and the efficient data acquisition for modeling material properties via Active Learning [26]. Especially the two latter methods are essential for fatigue strength estimation: these sequential methods deeply exploit the information in the available data, which is advantageous as fatigue strength experiments are costly.

For the first time, Agrawal and Choudhary [1] show how to learn a prediction model for fatigue strength based on the NIMS fatigue data set [8], a data set consisting of 400 observations from the Japan National Institute of Materials Science (NIMS). They were quickly followed by other authors, such as [3, 10, 11, 25, 28, 31, 34]. While these models are based on heuristic models, such as a Multilayer Perceptron or a Regression Tree, our work uses a probabilistic model: a GP with an engineered covariance function. By using this method, we are able to not only predict a point estimate of the fatigue strength but a probability distribution revealing the uncertainty of the prediction.

Regarding the second potential, Ling et al. [19] introduce a framework named FUELS for material discovery based on Bayesian Optimization. Their experiments demonstrate its use for finding a new material composition with a high fatigue strength using historical data. Our work differs from this approach as we focus not on material discovery but on designing an efficient test methodology for a given material.

This relates to the third potential: Active Learning, which is used for surrogate modeling, e.g., for finding models of a process [16], or material properties [14, 20]. The idea of Active Learning is ideal for formulating new experimental procedures. Simultaneously to our work, a similar approach aiming for a Bayesian staircase setting was proposed by Magazzeni et al. [21]. It is comparable to our approach regarding the definition of the likelihood based on the experimental setup but differs in the exploitation of the formulated posterior. To be more precise, our contribution is the maximum a posteriori approach for the distribution parameters and the derivation of a stopping criterion. Their acquisition function is similar to ours, but they limit it to fixed step sizes in all their experiments, thus adding an additional hyperparameter for process engineers.

## 6.3 Informed Fatigue Strength Estimation

This section describes our new fatigue testing approach. After a short overview of the basic ideas, we describe the two main modules and their opportunities in detail. We conclude by giving necessary further information for deploying the derived procedures.

### 6.3.1 Overview of Approach

The primary goal of informed fatigue strength estimation is to increase the test efficiency, thus, reducing the number of tests to determine the fatigue strength. Therefore, we consider expert knowledge and historical data about material behavior. Loosely speaking, our approach builds on the similarity of the fatigue strength estimates for similar steels, and of the behavior of the specimens at test time. This

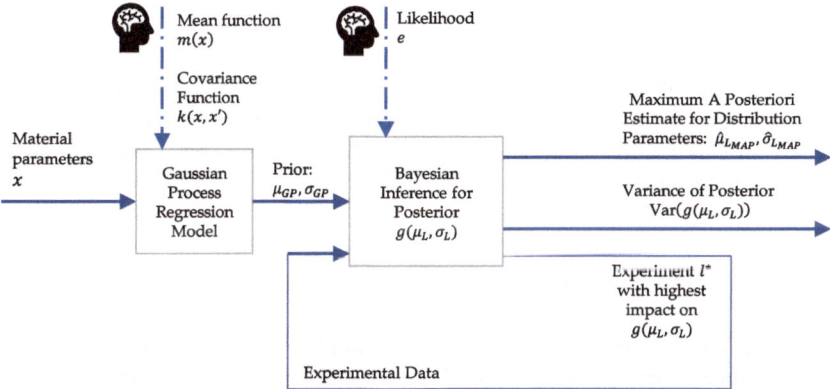

**Fig. 6.2** An overview of the experimental procedure. Expert knowledge is introduced in the covariance function of the Gaussian Process regression model and in the definition of the likelihood for determining the posterior distribution over the fatigue strength $L$

similarity assumption violates the conservative independence assumption made in the standard [4]. However, both the state-of-the-art literature (e.g., [1]) and the statements of the domain experts expect correlations to be plausible.

Our approach consists of two main modules, as shown in Fig. 6.2: a GP regression model and a module building on BI.

The GP is used to express the similarity of the mean fatigue strength $\mu_L$ of different materials. Similar to the expert estimate of $\hat{\mu}_{L_0}$ in the staircase method, it is used to create a prior estimate of the mean fatigue strength; here, in the form of a normal distribution $\hat{\mu}_{L_{GP}} \sim \mathcal{N}(\mu_{GP}, \sigma_{GP})$.

The BI module creates a posterior distribution over the fatigue strength parameters, following Bayes' Rule [2]. Necessary ingredients for this approach are a prior (calculated from the GP's prediction), a likelihood (derived from expert knowledge), and experimental data from fatigue testing.

From the posterior distribution of the parameters $g(\mu_L, \sigma_L)$, three quantities can be derived:

1. the most probable parameters $(\hat{\mu}_{L_{MAP}}, \hat{\sigma}_{L_{MAP}})$ that correspond to both the experimental data and the assumptions made; derived by a maximum a posteriori (MAP) estimate,
2. the confidence about the parameters, expressed as the variance $\text{Var}(g(\mu_L, \sigma_L))$, and
3. the load $l^*$ with the highest impact on the distribution $g(\mu_L, \sigma_L)$, recommended as the next experiment for data acquisition.

### 6.3.2   Machine Learning Model

The first module of our approach consists of a GP to predict an estimate of the mean fatigue strength $\hat{\mu}_{LGP}$. In the next subsection, we describe the necessary technical background and details on the learning procedure of the model, following the well known CRISP-DM [33].

#### 6.3.2.1   Gaussian Processes

Formally, a GP is defined as a collection of random variables, any finite number of which have a joint Gaussian distribution [23]. For full specification, a mean function $m(x)$ and a covariance function $k(x, x')$ are required, so if we approximate a function $f$ by a GP, we write $f(x) \approx GP(m(x), k(x, x'))$.

The goodness of fit of a GP model depends on the available data and on the GP prior that is formulated in the mean and covariance function. The GP prior expresses the expected behavior of the function $f(x)$, where the mean function $m(x)$ is often used to describe the extrapolation behavior far from the data, while the covariance function $k(x, x')$ carries the information about similarity of the function values $f(x)$ depending on the covariance of the inputs $(x, x')$. By definition, the prediction of a GP at a single new location $x^{\star}$ is normally distributed with mean $\mu_{GP}(x^{\star})$ and variance $\sigma^2_{GP}(x^{\star})$

$$
\begin{aligned}
\mu_{GP}(x^{\star}) &= K(x^{\star}, X)(K(X, X) + \sigma_n I)^{-1}(y - m(x^{\star})) \\
\sigma^2_{GP}(x^{\star}) &= K(x^{\star}, x^{\star}) - K(x^{\star}, X)(K(X, X) + \sigma_n I)^{-1}K(X, x^{\star}) ,
\end{aligned}
\tag{6.2}
$$

where $X$ is the matrix of the training inputs, $y$ are the function values at the train inputs, $K(X, x^{\star})$ is a matrix containing the covariance function $k(x, x')$ applied to the pairs of inputs $X$ and $x^{\star}$, and $\sigma_n$ is the noise hyperparameter.

#### 6.3.2.2   Gaussian Process for Estimating Fatigue Strength

We use a GP to express the similarity of the mean fatigue strength $\mu_L$ for different materials: the label of our data is the mean fatigue strength $\mu_L$. Therefore, the prediction of the resulting GP model for a single new material $x^{\star}$ will be a normal distribution $\hat{\mu}_{LGP} \sim \mathcal{N}(\mu_{GP}(x^{\star}), \sigma_{GP}(x^{\star}))$, which means it is an estimation of the mean fatigue strength, with standard deviation $\sigma_{GP}$. Please note that this standard deviation is not the same as the standard deviation of $L$, but the standard deviation of the estimated mean $\hat{\mu}_{LGP}$.

For model definition, we follow CRISP-DM [33]. Our raw data consists of 277 labeled points and four material parameters. The material parameters are the loaded volume $V90/mm^3$, the load type, the edge hardness of the specimen/HV, and the load ratio R, which expresses the ratio between the maximum and minimum applied load.

During preprocessing, we rigidly clean the data by excluding outliers in both input and output space and removing full duplicates resulting in 112 valid data points. Several duplicates exist in the input space (so identical materials with identical loads) but different values for the label, which prohibits learning of the GP. As averaging the labels results in another substantial data reduction, we opted to add small noise to the duplicated feature values and a lower bound to the noise hyperparameter $\sigma_n$ of the GP. Then, we perform a train-test split at a ratio of 80/20. Additionally, we take the logarithm of the label as the data distribution suggests that not only the fatigue strength $L$ for a specific material is log-normally distributed but also the mean fatigue strength $\mu_L$ for different materials. Afterwards, we standardize the train labels to a mean of 0 and a variance of 1.

In our model, we include the following expert knowledge in the GP prior:

- linearity of mean fatigue strength with respect to certain material parameters as a trend,
- a smooth behavior of the fatigue strength (no jumps when changing the input parameters),
- a varying similarity of function values depending on the input location but smooth transitions between these input areas.

The mean and covariance function incorporate this knowledge as follows: We apply a constant mean function $m(x) = 0$, so the estimate for the mean fatigue strength always falls back to the constant mean value at extrapolation regions. The covariance function is constructed as a sum covariance function [5], in which the ingredients are a linear covariance function with automatic relevance determination $k_{lin} = \sum_{d=1}^{D} \sigma_d^2 x_d x_d'$ and rational quadratic covariance function $k_{RQ}(x, x') = \left(1 + \frac{(x-x')^2}{2\alpha l^2}\right)^{-\alpha}$. Here, $k_{lin}$ expresses the linearity with respect to the material parameters, while $k_{RQ}$ induces a smooth but varying behavior that fluctuates around the trend. Overall, this means that we approximate the mean fatigue strength $\mu_L$ assuming $\mu_L \approx GP(m(x), k(x, x'))$ with $m(x) = 0, k(x, x') = k_{lin}(x, x') + k_{RQ}(x, x')$.

We train the model using a 10-fold cross-validation and determine the values of the covariance function's hyperparameters in each fold via maximization of the log marginal likelihood. Afterwards, we apply the hyperparameters of the best fold in the model for all train data.

Our implementation uses the state-of-the-art python packages GPy [9] for the GP and scikit-learn [22] for the cross-validation.

Figure 6.3 shows the prediction quality of our model on the test set in real space. In most cases, the ground truth value is within one standard deviation of the model's predicted mean, confirming that the model captures the overall behavior of the mean fatigue strength. Please note that the unsymmetric error bars are due to the preprocessing of the data. Overall, the model reaches a $R^2 = 0.91$ (metric was chosen by the process engineers) for the mean estimates on the test data. This

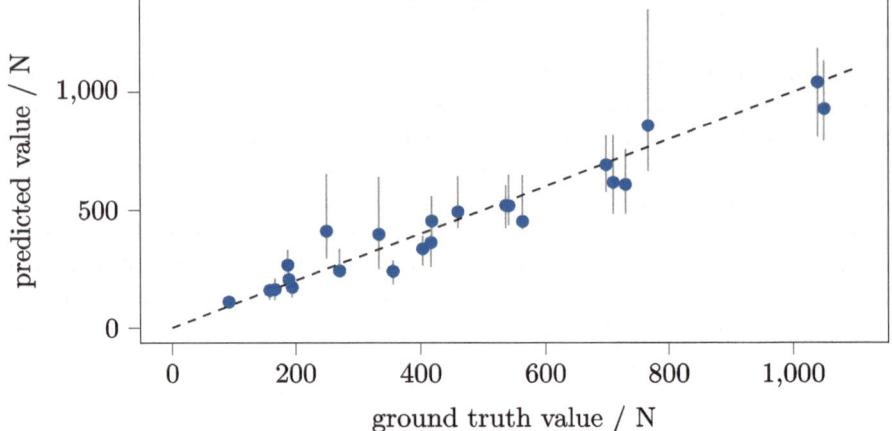

**Fig. 6.3** Prediction quality of the found GP model for $\hat{\mu}_{L_{GP}}$ on the test data. Dashed line: ideal prediction; blue filled circle: predicted mean values $\mu_{GP}$; error bars: one predicted standard deviation $\sigma_{GP}$

performance is slightly lower than the results of recent fatigue strength prediction models, such as the work of Agrawal and Choudhary [1], who reach a value of $R^2 = 0.98$ using an ensemble of tree-based models. For our approach, the GP is nevertheless at an advantage since it provides a suitable and reliable prior distribution for the BI module by design, is justified by expert knowledge, and was built from fewer instances (277 instead of 437).

### 6.3.3 Bayesian Inference on the Distribution Parameters

Traditional Bayesian Inference BI allows us to estimate a distribution over the parameters $(\mu_L, \sigma_L)$ of the log-normally distributed fatigue strength $L$ from expert knowledge and experimental data. As ingredients, it requires a prior estimate of these parameters and a likelihood to generate a posterior distribution over them. Both the accuracy of the prior and the likelihood are critical for the quality of the posterior—they encode the assumptions (i.e., the expert knowledge) about the data and the data generating process. In general,

$$\text{posterior} = \text{prior} \times \text{likelihood} .$$

In our case, the likelihood $e$ is derived from the experimental setup and is based on the following chain of arguments: the fatigue strength $L$ for an infinite number of specimens follows a log-normal distribution with parameters $(\mu_L, \sigma_L)$ and is not directly observable. Each single specimen has an individual fatigue strength $L_+$ and carries information about the fatigue strength distribution of the material. If an experiment at load $l$ is a failure, it means that the (unknown) fatigue strength of the specimen is smaller or equal to this load, so (failure at $l$) $\Longleftrightarrow L_+ \leq l$. This indicates that the probability of a failure at load $l$ for all specimens is

$$p(\text{failure}|l) = \Phi_{\mu_L,\sigma_L}(l) \,, \tag{6.3}$$

while the probability of a runout equals

$$p(\text{runout}|l) = 1 - p(\text{failure}|l) = 1 - \Phi_{\mu_L,\sigma_L}(l) \,, \tag{6.4}$$

where $\Phi_{\mu_L,\sigma_L}(l)$ is the cumulative density function of the log-normal distribution over $L$. For multiple experiments at different loads with different outcomes, we multiply the probabilities to gain the likelihood of the experimental outcomes

$$e(\text{outcome}|\mu_L, \sigma_L, l) = \prod_i \Phi_{\mu_L,\sigma_L}(l_i) \cdot \prod_j \left(1 - \Phi_{\mu_L,\sigma_L}(l_j)\right) \,, \tag{6.5}$$

where the index $i$ refers to the failures while the index $j$ refers to the runouts.

As a prior for the mean of the fatigue strength distribution $\mu_L$, we use the prediction of the GP regression model, assuming that the mean is normally distributed (in log space, as the data was scaled to train the model). For the standard deviation $\sigma_L$, there exists no prior knowledge or valid assumptions except for positivity. In the following, we apply a simple flat prior , which can lead to unwanted artifacts, such as a negative estimate of the standard deviation $\sigma_L$, in unfavorable cases. Alternatively, a (flat) positive prior or a gamma prior could be used to suppress this behavior.

Therefore, we obtain the following equation for the distribution over the parameters

$$g(\mu_L, \sigma_L | \text{outcome}, l)$$

$$\propto p(\mu_L) \cdot p(\sigma_L) \cdot e(\text{outcome}|\mu_L, \sigma_L, l)$$

$$\approx \frac{1}{\sqrt{2\pi \sigma_{GP}^2}} \exp\left(-\frac{1}{2}\left(\frac{\mu_L - \mu_{GP}}{\sigma_{GP}}\right)^2\right) \tag{6.6}$$

$$\cdot \left(\prod_i \Phi_{\mu_L,\sigma_L}(l_i) \cdot \prod_j \left(1 - \Phi_{\mu_L,\sigma_L}(l_j)\right)\right) \,.$$

### 6.3.3.1 Maximum A Posteriori Estimate

In (6.6), we derived an expression for the distribution over the parameters describing the distribution of the fatigue strength $L$. The location of the maximum of the posterior reveals the most probable values for the unknown parameters $\mu_L$ and $\sigma_L$ based on the prior knowledge about the specimen and its similarity to other material probes, the experimental setup, and experimental data about runouts or failures at different loads. The maximum locations of the posterior distribution are the maximum a posteriori (MAP) estimates for these parameters:

$$\hat{\mu}_{L_{MAP}}, \hat{\sigma}_{L_{MAP}} = \underset{\mu_L, \sigma_L}{\arg \max}\, g(\mu_L, \sigma_L)\,. \tag{6.7}$$

### 6.3.3.2 Active Learning-Inspired Acquisition Function

The data analysis procedure via BI also provides a method for efficient sequential experiment planning. As in basic Active Learning [26], and Bayesian Optimization [27], there are multiple possible formulations for the acquisition function—the function that maps each possible load to apply to the potential of improving the posterior estimates of the distribution parameters. Traditionally, the formulations vary from the experiment that mostly change the parameter estimates [6] to information-based acquisition functions [12, 13, 30], improvement-based methods [15, 17] and the traditional uncertainty sampling [18].

In our case, the acquisition has to merge two antagonistic requirements: on the one hand, the predictive entropy, i.e. the entropy $H(g|l)$ of the distribution over $L$ when adding the next experiment $l$, is to be minimized. Intuitively, the distribution is most changed when an unexpected experimental outcome occurs, e.g., a runout at a high load level. On the other hand, the experimental outcome is random. Our expectations about each experiment's results are expressed in (6.3) and (6.4). We actively decide to use this uncertainty as a proxy for the probability-weighted predictive entropy of the future distribution to find the next experiment

$$l^{\star} = \underset{l}{\arg \min}$$

$$H(g, \text{outcome} = \text{failure}|l) \cdot \Phi_{\hat{\mu}_{L_{MAP}}, \hat{\sigma}_{L_{MAP}}}(l) + \tag{6.8}$$

$$H(g, \text{outcome} = \text{runout}|l) \cdot (1 - \Phi_{\hat{\mu}_{L_{MAP}}, \hat{\sigma}_{L_{MAP}}}(l))\,,$$

where we estimate the real values of the distribution parameters $(\mu_L, \sigma_L)$ by the actual MAP-estimates. Unfortunately, this expression cannot be solved analytically. Instead, we discretize the two-dimensional support of the upcoming posterior and approximate numerically via the Shannon entropy.

### 6.3.3.3   Stopping Criterion

The sharpness of the distribution over $(\mu_L, \sigma_L)$ is a measure of the uncertainty about these parameters. If only $\mu_L$ is estimated, then the uncertainty about $\mu_L$ and thus the estimate $\hat{\mu}_{L_{MAP}}$ can be expressed as the variance of the distribution

$$\text{Var}(g(\mu_L, \sigma_L)). \tag{6.9}$$

This variance corresponds to the variance used for calculating the minimum number of experiments in the other analysis approaches like the staircase method. In the experimental procedure, it can be used as a stopping criterion: when the variance is below a specified limit, the experimental sequence is stopped.

## 6.3.4   Details on the Overall Experimental Procedure

Using the trained GP model and the derived expressions for the acquisition function and the posterior's variance, the experimental procedure to estimate the fatigue strength for a new material $x^\star$ works as follows:

First, a prediction $\hat{\mu}_{L_{GP}} \approx \mathcal{N}(\mu_{GP}(x^\star), \sigma_{GP}(x^\star))$ for the mean fatigue strength is created using the GP model. Based on this prediction, the data for one fracture and one runout are created experimentally. The necessary values are easily obtained by using the mean prediction of the GP and adding/subtracting some multiples of the standard deviation $\sigma_{GP}(x^\star)$.

Using the two data points and the GP prediction as prior, the posterior $g(\mu_L, \sigma_L)$ is calculated. Then an initial estimate for $\hat{\mu}_{L_{MAP}}$ is found by maximum a posteriori estimation. As the posterior may have multiple optima and inferring the posterior for a specific load value is computationally cheap, a multistart numerical optimization is performed.

Additionally, the variance of the posterior $\text{Var}(g(\mu_L, \sigma_L))$ is calculated using an equally spaced grid of 100,000 points in the area of interest.

For finding the new experimental load, we apply the acquisition function, as defined in (6.8) and find the load $l^\star$ with maximum impact on the entropy of the posterior distribution $H(g|l)$. The entropy is calculated via 1000 integration points that are distributed corresponding to the current estimate of the posterior.

This procedure of calculating the posterior, finding the MAP estimate, computing the standard deviation, and executing a new experiment at the value of the MAP estimate is repeated until the maximum number of iterations is met. In that case, the found value for $\mu_L$ is added to the train data of the GP, which is then retrained.

## 6.4   Validation of Approach

For validation purposes, we compare the BI-based approach with the staircase method on historical data of a JIS SUS 403 steel; a corrosion-resisting martensitic stainless steel [7]. The dataset is used because a real-world feasibility study is not possible due to the immense experimental cost. It consists of 136 measurements of the steel at different loads.

From the dataset, we derive the fatigue strength parameters via a maximum likelihood approach and find $(\mu_L, \log_{10} \sigma_L) = (389\,\text{N}, 0.038 \log_{10} \text{N})$. Using these values as ground truth, we create a simulator for the experimental outcome for any given load.

To cope with the randomness of the experimental outcomes, we compare 500 runs of each method, each using the fixed ground truth value for $\sigma_L$. The initial prior for BI builds on the prediction of our trained model. The initial load for the staircase method is drawn at random from a fixed interval, $\hat{\mu}_{L_0} \in [389 \pm d]$, where $d \in \{0, 30, 60\}$.

As mentioned in Sect. 6.2.2.2, an analysis of the test results by the staircase method is only valid if it fulfills specified requirements. In the following, we assume the initial value $\hat{\mu}_{L_0}$ to be the fatigue strength estimate of the staircase method $\hat{\mu}_{L_{\text{stair}}}$, as long as the requirements are not fulfilled.

Figure 6.4 visualizes the residual of the mean fatigue strength $|\mu_L - \hat{\mu}_L|$ for the BI-based and the staircase approach. As the residuals are non-normally distributed, we show the median and the 25% and 75% quantiles. We find that after ten iterations, the BI-based method is comparable to the staircase method when starting with no or slight misspecification of $d = 30$ in terms of the residual distribution.

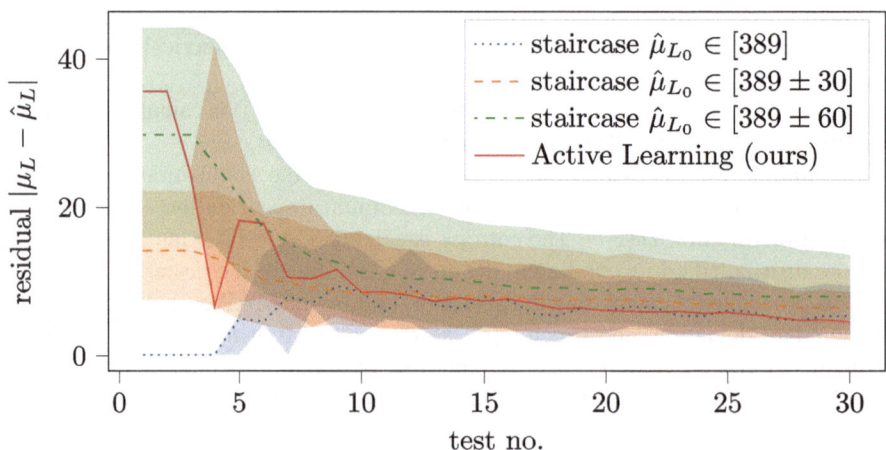

**Fig. 6.4** Residuals for the staircase method and the BI method for 30 iterations. Due to the non-normality of the results at each iteration, we plot the median as line and the 25% and 75% quantiles as a shaded region

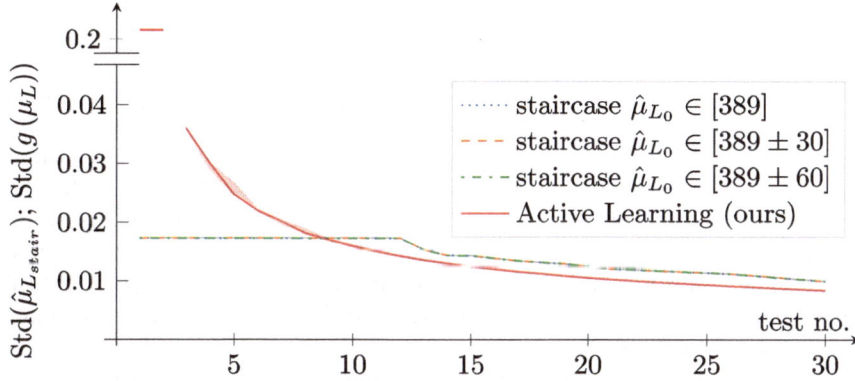

**Fig. 6.5** Development of the standard deviation $\text{Std}(\hat{\mu}_{L_{stair}})$ (staircase method) and the standard deviation of the posterior $\text{Std}(g(\mu_L))$ (BI-based method). Due to the non-normality of the results at each iteration, we plot the median as a line and the 25% and 75% quantiles as shaded regions

Between 10 and 15 iterations, the slight misspecification in the staircase method begins to impact the residuals, becoming worse at iteration 15, where it shows a smaller decrease of the residual and larger quantiles. At the same time, the BI-based method and the staircase method without misspecification become similar in both median and quantiles. Further comparison with the staircase method starting with larger misspecification ($d = 60$) leads to the assumption that the higher the misspecification, the larger the residual.

Figure 6.5 shows the course of the standard deviation of the current mean estimate $\text{Std}(\hat{\mu}_{L_{stair}})$ and the standard deviation of the posterior $\text{Std}(g(\mu_L))$ that is (due to the fixed value of $\sigma_L$) one-dimensional. Again, we show the median and quantiles, as the data are non-normally distributed. The standard deviation of the staircase approaches is very similar, as it depends on a small percentage of the estimated mean fatigue strength $\hat{\mu}_{L_{stair}}$, see (6.1). For the first iterations, it is nearly constant and begins to decrease after 12 iterations. As expected, the BI-based approach shows a smooth decrease of the standard deviation as the mass of the estimated posterior distribution over $\mu_L$ decreases with the evidence of every new data point. After ten iterations, the standard deviation of the posterior is smaller than the one by the staircase approaches.

Overall, both figures substantiate the advantage of the BI-based approach: The quality of the estimated mean fatigue strength $\hat{\mu}_{L_{MAP}}$ is similar to the non-misspecified staircase method after ten iterations, where the confidence of this value (measured by the standard deviation of the posterior) is higher. In general, the BI-based method is favorable to the staircase method in case misspecification of the load is probable, as the correct mean fatigue strength $\mu_L$ is estimated faster with a lower standard deviation.

## 6.5 Conclusion

This chapter presents a modular approach for fatigue strength estimation that reduces the number of necessary experiments based on a combination of expert knowledge and data. It can serve as an example of Informed Machine Learning by means of Bayesian methods for real-life destructive testing, such as life tests and stability tests. In our case of fatigue testing, the expert knowledge is injected in the first module using a tailored covariance function in a GP regression model for estimating a prior distribution over the mean fatigue strength $\mu_L$, while the second (BI) module exploits the knowledge about the experimental procedure in the definition of the likelihood.

The formulation as a BI problem to estimate a posterior distribution over the fatigue strength parameters $(\mu_L, \sigma_L)$ offers the full advantages of Bayesian methods: a MAP estimation of the parameters, an estimation of the confidence of these estimates and a formulation of an Active Learning-inspired acquisition function.

A comparison with experiments by the staircase method shows the superiority of the approach: on the one hand, it estimates the fatigue strength parameters comparable to the staircase method with higher confidence. On the other hand, the experimental procedure does not require hyperparameters and is thus robust against misspecification errors.

For real-life applicability, further studies have to be carried out: In a recent workshop publication, we examine the robustness of the BI module when facing a misspecified prior [32]. Additionally, we plan to investigate the behavior of our approach if the log-normality assumption of the failure probability is not satisfied. As a further highly practice-relevant expansion, we will explore a multi-point acquisition function that allows for acquiring more than one experiment per iteration.

**Acknowledgments** This contribution was supported by the Fraunhofer Cluster of Excellence "Cognitive Internet Technologies".

## References

1. Agrawal, A., Choudhary, A.: An online tool for predicting fatigue strength of steel alloys based on ensemble data mining. International Journal of Fatigue **113**, 389–400 (2018)
2. Bayes, T.: LII. An essay towards solving a problem in the doctrine of chances. By the late Rev. Mr. Bayes, FRS communicated by Mr. Price, in a letter to John Canton, AMFR S. Philosophical transactions of the Royal Society of London (53), 370–418 (1763)
3. Chen, J., Liu, Y.: Fatigue modeling using neural networks: A comprehensive review. Fatigue & Fracture of Engineering Materials & Structures **45**(4), 945–979 (2022)
4. DIN 50100:2016-12: Load controlled fatigue testing – Execution and evaluation of cyclic tests at constant load amplitudes on metallic specimens and components (2016)
5. Duvenaud, D.K.: Automatic model construction with gaussian processes (2014)

6. Frazier, P., Powell, W.B., Dayanik, S.: The knowledge-gradient policy for correlated normal beliefs. INFORMS J. Comput. **21**, 599–613 (2009)
7. Furuya, Y., Nishikawa, H., Hirukawa, H., Nagashima, N.: Data sheets on fatigue properties of SUS403 (12Cr) stainless steel bars for machine structural use (1982)
8. Furuya, Y., Nishikawa, H., Hirukawa, H., Nagashima, N., Takeuchi, E.: Catalogue of NIMS fatigue data sheets. Science and Technology of Advanced Materials **20**(1), 1055–1072 (2019)
9. GPy: GPy: A gaussian process framework in python. http://github.com/SheffieldML/GPy (since 2012)
10. He, L., Wang, Z.L., Akebono, H., Sugeta, A.: Machine learning-based predictions of fatigue life and fatigue limit for steels. Journal of Materials Science & Technology **90**, 9–19 (2021)
11. Ilc, N., Ouyang, R., Qian, Q.: Learning interpretable descriptors for the fatigue strength of steels. AIP Advances **11**(3), 035018 (2021)
12. Hennig, P., Schuler, C.J.: Entropy search for information-efficient global optimization. Journal of Machine Learning Research **13**, 1809–1837 (2012)
13. Hernández-Lobato, J.M., Hoffman, M.W., Ghahramani, Z.: Predictive entropy search for efficient global optimization of black-box functions. Advances in neural information processing systems **27** (2014)
14. Himanen, L., Geurts, A., Foster, A.S., Rinke, P.: Data-driven materials science: status, challenges, and perspectives. Advanced Science **6** (2019)
15. Jones, D.R., Schonlau, M., Welch, W.J.: Efficient global optimization of expensive black-box functions. Journal of Global Optimization **13**, 455–492 (1998)
16. Krause, A., Singh, A.P., Guestrin, C.: Near-optimal sensor placements in gaussian processes: theory, efficient algorithms and empirical studies. Journal of Machine Learning Research **9**, 235–284 (2008)
17. Kushner, H.J.: A new method of locating the maximum point of an arbitrary multipeak curve in the presence of noise. Journal of Basic Engineering **86**, 97–106 (1963)
18. Lewis, D.D., Gale, W.A.: A sequential algorithm for training text classifiers. In: SIGIR '94 (1994)
19. Ling, J., Hutchinson, M., Antono, E., Paradiso, S., Meredig, B.: High-dimensional materials and process optimization using data-driven experimental design with well-calibrated uncertainty estimates. Integrating Materials and Manufacturing Innovation **6**, 207–217 (2017)
20. Lookman, T., Balachandran, P.V., Xue, D., Yuan, R.: Active learning in materials science with emphasis on adaptive sampling using uncertainties for targeted design. npj Computational Materials **5**, 1–17 (2019)
21. Magazzeni, C.M., Rose, R., Gearhart, C., Gong, J., Wilkinson, A.J.: Bayesian optimized collection strategies for fatigue strength testing. Fatigue & Fracture of Engineering Materials & Structures **46**(1), 228–243 (2023)
22. Pedregosa, F., Varoquaux, G., Gramfort, A., Michel, V., Thirion, B., Grisel, O., Blondel, M., Prettenhofer, P., Weiss, R., Dubourg, V., Vanderplas, J., Passos, A., Cournapeau, D., Brucher, M., Perrot, M., Duchesnay, E.: Scikit-learn: Machine learning in Python. Journal of Machine Learning Research **12**, 2825–2830 (2011)
23. Rasmussen, C.E., Williams, C.K.I.: Gaussian processes for machine learning. In: Adaptive Computation and Machine Learning (2009)
24. Schmidt, J., Marques, M.R.G., Botti, S., Marques, M.A.L.: Recent advances and applications of machine learning in solid-state materials science. npj Computational Materials **5**, 1–36 (2019)
25. Schneller, W., Leitner, M., Maier, B., Grün, F., Jantschner, O., Leuders, S., Pfeifer, T.: Artificial intelligence assisted fatigue failure prediction. International Journal of Fatigue **155**, 106580 (2022)
26. Settles, B.: Active learning literature survey (2009)
27. Shahriari, B., Swersky, K., Wang, Z., Adams, R.P., de Freitas, N.: Taking the human out of the loop: a review of bayesian optimization. Proceedings of the IEEE **104**, 148–175 (2016)

28. Shiraiwa, T., Briffod, F., Miyazawa, Y., Enoki, M.: Fatigue performance prediction of structural materials by multi-scale modeling and machine learning. In: Proceedings of the 4th World Congress on Integrated Computational Materials Engineering (ICME 2017), pp. 317–326. Springer International Publishing (2017)
29. Von Rueden, L., Mayer, S., Beckh, K., Georgiev, B., Giesselbach, S., Heese, R., Kirsch, B., Pfrommer, J., Pick, A., Ramamurthy, R., et al.: Informed machine learning–a taxonomy and survey of integrating prior knowledge into learning systems. IEEE Transactions on Knowledge and Data Engineering 35(1), 614–633 (2021)
30. Wang, Z., Jegelka, S.: Max-value entropy search for efficient bayesian optimization. In: International Conference on Machine Learning (2017)
31. Wei, X., Zhang, C., Han, S., Jia, Z., Wang, C., Xu, W.: High cycle fatigue s-n curve prediction of steels based on transfer learning guided long short term memory network. International Journal of Fatigue 163, 107050 (2022)
32. Weichert, D., Kister, A., Houben, S., Ernis, G., Wrobel, S.: Robustness in fatigue strength estimation. In: 2nd Annual AAAI Workshop on AI to Accelerate Science and Engineering (AI2ASE) at the 37th AAAI Conference on Artificial Intelligence (AAAI-23) (2023)
33. Wirth, R., Hipp, J.: CRISP-DM: Towards a standard process model for data mining (2000)
34. Xiong, J., Zhang, T., Shi, S.: Machine learning of mechanical properties of steels. Science China Technological Sciences 63(7), 1247–1255 (2020)

# Chapter 7
# Incorporating Shape Knowledge into Regression Models

Miltiadis Poursanidis, Patrick Link, Jochen Schmid, and Uwe Teicher

**Abstract** Informed learning is an emerging field in Machine Learning that aims at compensating for insufficient data with prior knowledge. Shape knowledge covers many types of prior knowledge concerning the relationship of a function's output with respect to input variables, for example, monotonicity, convexity, etc. This shape knowledge—when formalized into algebraic inequalities (shape constraints)—can then be incorporated into the training of regression models via a constrained optimization problem. The defined shape-constrained regression problem is, mathematically speaking, a semi-infinite program (SIP). Although off-the-shelf algorithms can be used at this point to solve the SIP, we recommend an adaptive feasible-point algorithm that guarantees optimality up to arbitrary precision and strict fulfillment of the shape constraints. We apply this semi-infinite approach for shape-constrained regression (SIASCOR) to three application examples from manufacturing and one artificial example. One application example has not been considered in a shape-constrained regression setting before, so we used a methodology (ISI) to capture the shape knowledge and define corresponding shape constraints. Finally, we compare the SIASCOR method with a purely data-driven automated machine learning method (AutoML) and another approach for shape-constrained regression (SIAMOR) that uses a different solution algorithm.

## 7.1 Introduction

Despite the success of Machine Learning (ML), purely data-driven machine learning models show limited performance when dealing with insufficient data. This is especially problematic in scientific and engineering contexts, where simulation or

M. Poursanidis (✉) · J. Schmid
Fraunhofer ITWM, Kaiserslautern, Germany
e-mail: miltiadis.poursanidis@itwm.fraunhofer.de; jochen.schmid@itwm.fraunhofer.de

P. Link · U. Teicher
Fraunhofer IWU, Chemnitz, Germany
e-mail: patrick.link@iwu.fraunhofer.de; uwe.teicher@iwu.fraunhofer.de

D. Schulz, C. Bauckhage (eds.), *Informed Machine Learning*,
Cognitive Technologies, https://doi.org/10.1007/978-3-031-83097-6_7

experimental data is costly in both time and resources [54]. When the data set is small, machine learning models have difficulties providing reliable models. The main issue is that the models do not behave as expected in regions with sparse or no data. When noise comes into play, this effect is even more severe as the models tend to learn spurious patterns from the data. In addition to that, it is difficult to measure the model's performance at sparse data regions. Also, methods like cross-validation are often misleading because only the data set is considered for measuring the performance.

In many machine learning tasks, by contrast, there is additional prior knowledge available. Informed learning emerged from the need to compensate for shortcomings in the data with supplementary prior knowledge [45]. Machine learning models benefit from prior knowledge in various ways, but we highlight two in particular: interpretability and generalization. Informed machine learning models are interpretable because they behave according to the imposed prior knowledge. For instance, in production, there are usually high costs involved with wrong decision-making. Therefore, practitioners rely more on their knowledge than on the data, especially when the data is sparse and noisy. For this reason, trustworthy prediction models should incorporate prior knowledge to increase acceptance among practitioners. In science, interpretable models are key to the accumulation of scientific knowledge. In contrast to black-box models, theories can be developed based on these interpretable models that have known properties [23]. The other aspect is generalization, that is, the model's ability to achieve low errors on new data. Informed learning is expected to lead to improved generalization. Imposing prior knowledge gives control over regions in the domain with sparse data and makes models less prone to unexpected behavior. This typically leads to models that generalize better outside the data set. Another aspect is extrapolation. The authors in [16] show that, in certain cases, shape constraints can lead to an improvement of the out-of-domain error. However, we do not consider extrapolation here.

In this chapter, we focus on prior knowledge concerning the qualitative shape of the model function. The definition of shape knowledge is very general and captures many properties such as boundedness or monotonicity of the model function. In the terminology of [45], shape knowledge can be categorized as either scientific knowledge (given in explicit formulas) or expert knowledge (common knowledge within a scientific field). Such shape knowledge can often be formulated as algebraic inequalities, so-called shape constraints.

First, we recapitulate the semi-infinite approach to shape-constrained regression (SIASCOR) from [31]. Shape-constrained regression is a constraint problem formulation of a regression problem with the aim to incorporate shape constraints into the model function. Mathematically, this results in a so-called semi-infinite program (SIP) and can be solved, for instance, with adaptive feasible-point methods. In [31], the authors used the core algorithm from [47] as their adaptive feasible-point algorithm to compute an approximate solution to the resulting SIP. In the present chapter, by contrast, we use the simultaneous algorithm from [47] as our adaptive feasible-point algorithm. In contrast to the core algorithm, it computes an approximate solution to the SIP of an arbitrary user-specified precision. According to the taxonomy of [45], SIASCOR integrates shape knowledge—represented as algebraic inequalities—into the training of the regression model. Second, we reconsider the methodology from [31]. This methodology helps practitioners to capture shape knowledge in cooperation with experts in the field, to define shape constraints, and to incorporate the shape constraints into a regression model. From now on, we will refer to this methodology as ISI, which stands for its three steps: inspection, specification, and integration. In terms of the taxonomy of [45], this can be partly viewed as a method that transforms expert knowledge into algebraic inequalities. In addition to that, the ISI methodology uses some shape-constrained regression method, for instance SIASCOR, to integrate these algebraic inequalities into the machine learning model.

We consider three real-world application examples from quality prediction in manufacturing: brushing, press hardening, and milling. On the one hand, all three examples have small and noisy data sets but, on the other hand, they can benefit from shape knowledge provided by experts. The brushing example was already considered in [31] and the press hardening example in [25], thus we reuse the same shape constraints as in these references. The milling case has not been studied in a shape-constrained regression setting before. Therefore, we use the ISI method to ensure that all shape knowledge is captured and, if possible, transformed into shape constraints. After that, we can apply SIASCOR with the obtained shape constraints. Moreover, and in contrast to [31], we compare SIASCOR to more sophisticated machine learning methods. We compare it with an automated machine learning method (AutoML) and another semi-infinite approach to shape-constrained regression but with a different solution algorithm (SIAMOR) [25]. Another difference to [31] is that here we use different settings of SIASCOR, such as another solving algorithm of the SIP and anisotropic polynomial regression functions. We compare the resulting models in terms of shape compliance, training time, and cross-validated test error.

As an extension to the three real-world application examples, we introduce an artificial example to examine the generalization error. The generalization error is the error a model has on data not contained in the training set. Since our real-world examples have small data sets, we cannot analyze the generalization error appropriately, especially in scarce data regions. We compare SIASCOR, for the first time, with AutoML, SIAMOR and with Ridge regression in terms of generalization error.

We organize this chapter as follows. In Sect. 7.2, we give a basic overview of the related work. In Sect. 7.3, we introduce the informed machine learning approach SIASCOR for shape-constrained regression and present the methodology ISI to capture and integrate expert knowledge. Section 7.4 describes our three application examples from manufacturing—namely press hardening, brushing, and milling—and discusses the results of the comparative study. Section 7.5 presents our artificial application example along with the analysis of the generalization error. Finally, we give a conclusion and an outlook in Sect. 7.6.

## 7.2   Related Work

In this section, we present some related work on informed learning, shape-constrained regression, semi-infinite programming, and expert-knowledge-based quality prediction in manufacturing.

Informed learning describes all approaches that incorporate prior knowledge into machine learning models. An overview of the field can be found in [45]. In the present work, we focus on prior knowledge in the form of algebraic inequalities. According to the taxonomy of [45], algebraic inequalities are included in the class of algebraic equations. Algebraic inequalities can be integrated into regression models in four ways: When generating training data [26], by restricting the hypothesis space [3, 34], during the learning algorithm [8, 10, 21, 25, 37, 48] and by modifying the final model [24, 37, 46]. Informed learning methods that integrate algebraic inequalities during training either treat the inequalities as soft constraints by adding a penalty term to the loss function [8, 10, 21, 48] or as hard constraints by adding them as constraints to the loss minimization problem [25, 37]. Among the different constraint types, shape constraints restrict the qualitative shape of the prediction function [15]. One of the most prominent shape constraints is monotonicity and, in the literature, there already exist numerous approaches to enforcing monotonicity constraints during training [1, 6, 14, 27, 44]. Furthermore, the authors of [2] consider various shape constraints in a kernel regression setting. They enforce the shape constraints on a finite set of points but sufficiently tighten the problem to fulfill the constraints on the entire input space. Polynomial shape-constrained regression is considered in [17], where the authors use SDP relaxations to solve the shape-constrained regression problem. The authors of [7] and of [25] approach shape-constrained regression via semi-infinite programming. However, due to new mathematical results from the SIP community [11], more suitable algorithms can be used.

SIPs are optimization problems that have a finite number of decision variables and an infinite number of constraints. For an overview of the theory and how to handle the infinite constraints numerically, we refer to [11, 22, 43]. Popular methods for solving SIPs are discretization methods [4, 33] with the attention shifting to adaptive discretization. Among these, there is also a line of work concerning adaptive feasible-point methods [36, 53], which guarantee termination at a feasible point. In [47], the authors leveraged the convexity—a property inherent in most shape-constrained regression problems—to obtain stronger results such as arbitrary optimality precision under weaker assumptions.

In manufacturing, quality prediction is used to both monitor product quality and optimize processes. Quality prediction models are either data-driven or rely on physical equations. In the context of manufacturing, the use of data-driven models is a challenging task due to data scarcity. As mentioned in [54], complex models are applied to describe complex relationships that have few data available. The problem of data scarcity is also reported in other domains, such as process engineering [39]. In order to handle small data sets, multi-model approaches [5, 29] or polynomial chaos expansion [51] have been used. Another technique is to generate artificial data via bootstrapping [39, 52] or mega-trend-diffusion [28, 30]. Besides these data-driven methods, there are also expert-knowledge-based approaches [17]. In manufacturing, the most common sources of knowledge are scientific and expert knowledge, according to [45]. The authors of [20, 32, 55] integrate probabilistic relationships into the hypothesis sets of Bayesian networks. In addition, [34] and [38] restrict the hypothesis set of neural networks with algebraic equations and knowledge graphs, respectively. The authors of [18] incorporate algebraic equations into the training of Gaussian process models.

## 7.3   Methods

In the first subsection, we describe a methodology (SIASCOR) that integrates shape constraints into regression models via semi-infinite programming. In practice, however, shape knowledge is not always available as algebraic inequalities. Usually, there is merely the expert's intuition that needs to be captured and formalized into shape constraints. Therefore, in the second subsection, we recall a methodology (ISI) that helps to capture shape knowledge and convert it into algebraic inequalities. These algebraic inequalities can then be integrated into the regression model using SIASCOR, for instance.

### 7.3.1   SIASCOR

The goal of classical regression settings usually boils down to finding a model function that fits some data. Assume we have additional prior knowledge about the shape of the input-output relationship to be learned. When shape knowledge is formalized in terms of inequality constraints, we call them shape constraints. Common forms of shape knowledge, for instance, are monotonicity or convexity. The corresponding shape constraints restrict the first or second partial derivative with respect to some input variable of the model function to be positive. Then, the goal of shape-constrained regression is to find a model function that both fits the data and complies with the shape constraints.

Suppose we are given some data set $\mathcal{D} = \{(x^k, y^k) \in X \times \mathbb{R} : k = 1, \ldots, n\}$ consisting of input data points $x^k \in \mathbb{R}^d$ and output data points $y^k \in \mathbb{R}$. We assume that the (unknown) input-output relationship to be learned can be represented by model functions $\widehat{y}_w : X \to \mathbb{R}$ of the form $\widehat{y}_w(x) := w^\top \phi(x)$. In other words, we take our hypothesis space to be the set of functions $\widehat{y}_w$ with $w \in W$. In the above formula, $x \in X$ and $w \in W$ denote the input variables and model parameters, respectively, and we assume the input-variable and model-parameter spaces $X \subset \mathbb{R}^d$ and $W \subset \mathbb{R}^m$ to be compact, convex sets. Also, $\phi : X \to \mathbb{R}^m$ is a feature mapping that is sufficiently often differentiable. We further assume that the shape constraints can be expressed by constraining a function $g_i$ that is given in terms of affine-linear combinations of the partial derivatives of the model function $\widehat{y}_w$. Clearly, boundedness, monotonicity or convexity constraints can be cast in this form; for instance, monotonic increasingness w.r.t. $x_j$ is equivalent to the condition $g(w, x) := -w^T \partial_{x_j} \phi(x) \leq 0$. Consequently, in mathematical terms, shape-constrained regression problems take the form

$$\min_{w \in W} \sum_{k=1}^{n} |y^k - \widehat{y}_w(x^k)|^2 + \lambda \|w\|^2 \tag{7.1}$$

$$\text{s.t. } g_i(w, x) \leq 0 \text{ for all } x \in X \text{ and } i \in I,$$

where $\lambda > 0$ is some regularization parameter, $\|w\|^2$ denotes the squared $\ell^2$-norm of the model parameter $w$ (ridge regularization) and $I$ is a finite index set indexing the shape constraints. Note that we restricted ourselves here to the ridge regression and to model functions that are linear w.r.t. to their model parameters. For more general cases, see [47].

Problem (7.1) is a so-called semi-infinite program (SIP). Assume the problem is feasible, i.e. there exists a $w \in W$ such that $g_i(w, x) \leq 0$ for all $i \in I$ and $x \in X$. Intuitively, this means that there exists a function from our hypothesis space that satisfies all shape constraints. Then, problem (7.1) has a unique solution, by its strict convexity and our continuity assumptions on $\phi$. See [22, 47] for example. There exist many approaches for solving SIPs [11] and in particular convex SIPs [43]. We prefer feasible-point methods because they guarantee termination at a feasible point. Other methods mostly guarantee feasibility only as the iteration number tends to infinity. Hence, we use the simultaneous algorithm from [47], a feasible-point method that leverages the convexity of the problem to provide guarantees for approximate, feasible solutions, while the only assumption is strict feasibility of problem (7.1). There are more algorithmic merits of the approach but we will not detail them here. Note that the algorithm we used in this chapter is different from [31] where we used the core algorithm from [47]. The core algorithm also guarantees feasibility but does not guarantee optimality of arbitrary precision. Besides, it is different from [7] where the authors do not use a feasible-point method in the first place.

After having defined the method more precisely, we can see how it can be embedded into the taxonomy of [45]: knowledge, in our context, is given in the form of shape constraints which, ultimately, are algebraic inequalities. Then, these algebraic inequalities are integrated during the learning algorithm through a constrained optimization problem formulation.

In contrast to soft-constraint methods, SIASCOR imposes hard constraints on the model function. This is suitable for applications where the model function needs to satisfy the constraints strictly, for instance when the model needs to be in accordance with physical laws. Despite that, one can relax the constraints by a small value $\varepsilon > 0$ if the constraints do not need to be fulfilled strictly. This can be done by subtracting the value $\varepsilon$ from all shape constraint functions $g_i$. In this case, the resulting model function complies with the relaxed shape constraints.

### 7.3.2   ISI

In the previous section, we described how SIASCOR incorporates shape constraints into the training of machine learning models of the form $\widehat{y}(x) = w^T \phi(x)$. In practice, these shape constraints need to be developed in collaboration with experts in the field. The ISI methodology supports practitioners in capturing shape expert knowledge, in formalizing it into shape constraints, and in producing a shape-compliant model. In this section, we summarize the ISI methodology that was introduced in [31].

The schematic procedure of ISI is depicted in Fig. 7.1. As its input, the methodology requires an initial model $\widehat{y}^0$ that may be purely data-based. Then, the methodology proceeds in the following three steps:

1. Inspection of the initial model by a process expert
2. Specification of shape expert knowledge by the expert
3. Integration of the specified shape expert knowledge into the training of a new model

These three steps generate a shape-compliant model. If this model does not behave as the expert expects, the three steps can be repeated. Note that in the second iteration one must take the shape-compliant model from the first iteration as the input for the inspection. This can be repeated as often as necessary while always taking the current shape-constrained model as the input model for the next iteration.

The initial model is the starting point for the introduced methodology. The initial model visually supports the expert in analyzing the relationship between the inputs and outputs. In principle, any type of model function can be used as an initial model. However, we recommend choosing the initial model from the same hypothesis space

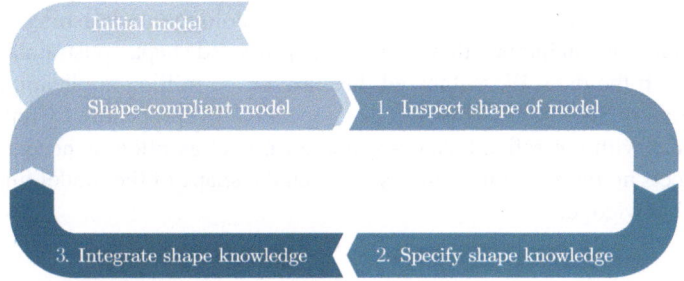

**Fig. 7.1** Schematic of the three-step methodology with an initial purely data-based model as its input and a shape-knowledge-compliant model as its output. Image source: [31]

as the SIASCOR model. In other words, the initial model should be of the form $\hat{y}^0(x) = w_0^T \phi(x)$ for some $w_0 \in W$. This way, the expert can get an intuition of how model functions from this hypothesis space behave when they are not shape-constrained, especially in regions with sparse data. The parameter $w_0 \in W$ can be computed in any purely data-based regression method, such as ridge regression.

After generating an initial model, the first step of the methodology is to provide the expert with one- and two-dimensional graphs at multiple points of high- and low-fidelity. Here, a custom notion of fidelity can be used. For instance, one can consider points $x_{\text{high}}, x_{\text{low}} \in X$ with maximal and minimal distance to all data points, respectively. These one- and two-dimensional graphs at the high- and low-fidelity points help the expert understand the functional relationships behind the data by exploring the initial model along the input space. This way, the expert can find shape behavior that either confirms or contradicts their intuition.

In the second step of the methodology, the process expert specifies the shape expert knowledge, that is, his intuition about the qualitative functional relationship of the output along different dimensions of the input space. Then, shape knowledge is converted into shape constraints. The expert can choose from a variety of common shape knowledge with associated shape constraints such as monotonicity, convexity or concavity or upper and lower bounds; see Fig. 7.2 for a selection of qualitative shape knowledge. Also, multiple shape constraints can be combined. In case where the shape knowledge cannot be composed by common shape constraints, practitioners may consider designing a new shape constraint. This, for instance, was the case in [31] that resulted in defining the rebound constraint.

With the shape constraints at hand, the third step of the procedure is to find a shape-compliant model. At this point, SIASCOR can be used but can be interchanged with any shape-constrained regression method. In comparison to the initial model, the shape-compliant model fits both the data and satisfies the defined shape constraints.

After the first iteration, the ISI methodology produces a model that is compliant with the shape constraints imposed in the first iteration. Yet, we can neither guarantee that the imposed shape constraints are complete nor that they are all correct. It can happen, for instance, that some shape knowledge has been overlooked in the first run and, therefore, necessary shape constraints have not been imposed. We witnessed this when we applied the ISI methodology in a brushing use case in [31]. Similarly, it can happen that some of the proposed shape constraints are not in harmony with the data. We witnessed this case in the milling application example that we detail in Sect. 7.4.3. Therefore, we repeat the ISI procedure by replacing the initial model with the refined shape-compliant model as often as needed until the expert detects no more inconsistencies between the shape of the model function and his shape knowledge.

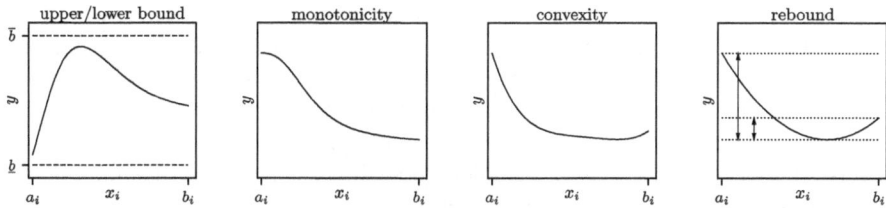

**Fig. 7.2** Sketches of the shape constraints we considered in our applications. The upper- and lower-bound constraints bound the maximal and minimal values of the function, respectively. Monotonicity constraints (either increasing or decreasing) restrict the first derivative of the function to be either positive or negative. Similarly, convexity constraints assume the second derivative to be positive (or negative if concave). The rebound constraint describes, loosely speaking, how much a function can increase again after it has witnessed a decrease. For more details, see [31] and for proofs see [47]

## 7.4   Application Examples

In this section, we present three real-world application examples for shape-constrained regression. We apply the SIASCOR method from Sect. 7.3.1 to incorporate prior knowledge given in the form of shape constraints. For two application examples, the shape constraints have been already captured in previous works, using the ISI methodology—either implicitly in the case of [25] or explicitly in the case of [31]. The other example has not been considered in a shape-constrained regression setting before, so we did not have any shape constraints available. Therefore, we used the ISI methodology to capture the prior knowledge and to define the shape constraints, together with experts in the field. Finally, we compare SIASCOR with both a data-driven machine learning approach and another shape-constrained regression approach (SIAMOR) [25].

Every subsection deals with a distinct application example from quality prediction for manufacturing processes. We consider three manufacturing processes: press hardening, brushing, and milling. Quality prediction for the press hardening case has already been considered in [25] and for the brushing case in [31]. In these two cases, the shape constraints have already been identified and we can directly apply SIASCOR with respect to the shape constraints proposed by the experts. In contrast, the milling case has not been considered for shape-constrained regression so far. Hence, we first apply the ISI procedure in collaboration with experts in the field to capture their shape knowledge and formalize it in terms of shape constraints. With the shape constraints at hand, we finally apply SIASCOR.

First, we compare SIASCOR with a purely data-driven machine learning approach to highlight the necessity of imposing shape constraints. Concretely, we compare it with auto-sklearn [13], an sklearn [41] package for automatic machine learning (AutoML). We choose an AutoML approach, because this has become a preferred choice in many fields of machine learning [19] due to its systematic and automated manner of model and hyperparameter selection. We consider two versions of auto-sklearn in our comparison: one with default settings and another with handpicked regression models, data- and feature preprocessing to avoid unphysical behavior through discontinuities in the shape of the model function. More precisely, we exclude "extra_trees_preproc_for_regression" and "random_trees_embedding" from the feature preprocessing and "adaboost", "decision_tree", "extra_trees", "gradient_boosting", "k_nearest_neighbors" and "random_forest" from the regressor selection.

Second, we compare SIASCOR with its predecessor SIAMOR—an alternative method for shape-constrained regression. It was initially developed to enforce monotonicity of regression models, but it can be easily adapted to general shape constraints. Both methods consider a SIP formulation of shape-constrained regression. In contrast to SIASCOR, however, SIAMOR uses a different algorithm to solve the SIP which does not provide a theoretical guarantee for feasibility. It merely guarantees feasibility on a reference grid and, generally, only after infinitely many iterations. In other words, we can only guarantee shape compliance on a grid of finitely many points in the limit of infinitely many iterations. In all applications, we chose a reference grid of 20 points per dimension. Besides the theoretical differences, we compare these two methods in practice.

We consider three criteria in our comparison: shape compliance, training time and test error. For all three criteria, we conducted a ten-fold cross-validation. First, we counted how many shape constraints the final models violated in each fold and then we averaged these numbers over all folds. For every shape constraint we sampled 10,000 random points in the input space X and tested the shape compliance on 100 equidistant points along the relevant axis. Besides, we visualize some of the violations at selected points. During the training of each fold, we also measured the wall clock time and averaged it across all folds. The training was conducted on a standard office laptop. Lastly, we measured the root-mean-squared errors RMSE of the model on all test sets and, again, averaged along all folds.

### 7.4.1 Press Hardening

Our first application example is press hardening, a forming process for hot sheet metal [40]. During the forming process, the hot sheet metal is formed and subsequently quenched to achieve improved hardness of the parts. The goal in this application is to build a model that predicts the hardness of the metal sheet as a function of four process parameters, namely the furnace temperature $T_f$, the handling time from furnace to press $\Delta t_h$, the press force $F_p$ and the quenching time $t_q$. For the informed learning task, we use the 60 data samples that were generated in an experimental setup and the defined monotonicity constraints, both from [25].

Now we present our parameter settings for SIASCOR. First, we scaled the input data to the unit cube based on the ranges provided in [25]. This way, we can consider $X = [0, 1]^4$ and all the data is contained in that domain $\mathcal{D} \subset X \times \mathbb{R}$. Moreover, we choose $\phi$ to be an anisotropic polynomial feature map. This means that the maximal polynomial degree can be different for each of the four dimensions. Specifically, we chose the maximal degrees to be (4, 1, 3, 4) for $(T_f, \Delta t_h, F_p, t_q)$, as suggested by the process experts. Accordingly, we set the parameter space to be $W = [-10^5, 10^5]^{54}$ and, furthermore, the regularization parameter $\lambda = 0.0001$. The shape constraints from [25] are monotonic increasingness in $T_f$ and $t_q$ and monotonic decreasingness in $\Delta t_h$. The settings above specify the SIP in (7.1). Afterwards, we apply the main algorithm from [47] to solve the SIP with optimality precision $\delta = 0.0001$. We are not going into detail here which parameters values we chose for the adaptive feasible-point algorithm. For more details on the algorithm and its parameter settings, see [47].

For both AutoML approaches, we set the strategy that chooses the best model to be a ten-fold cross-validation and fixed the maximum search time to 600 seconds, as suggested in [12]. Afterwards, the AutoML models were refit on the entire training set. For SIAMOR, we set everything as in [25], especially the size of the reference grid being 20 points per dimension.

Table 7.1 shows that both AutoML versions produce models that are not shape-compliant. In fact, we observe that all models produced during the cross-validation violate all three imposed shape constraints. Also the trained SIAMOR models violate one out of four shape constraints on average. As expected, the SIASCOR model satisfies all shape constraints. Figure 7.3 juxtaposes one-dimensional graphs of the SIASCOR and the auto-sklearn model at the same point. We can see graphically that the SIASCOR model is monotonically decreasing while the auto-sklearn model is not. However, training the SIASCOR model takes slightly longer than the other models, which can be traced back to the computationally expensive global optimization steps that guarantee shape compliance. Furthermore, we infer from Table 7.1 that the averaged error on the test sets during a cross-validation is more or less the same, considering the scale of the hardness (roughly between 300 and 500). The AutoML model has slightly lower test errors but considering the amount of data, this does not imply good generalization outside the data.

**Table 7.1** Press hardening case: Comparison of SIASCOR, Auto-Sklearn with default settings, Auto-Sklearn with custom settings, and SIAMOR. The table lists the average number of shape constraints that the models violate after being trained on each fold of a ten-fold cross-validation. We considered three shape constraints in total. Moreover, both training time and test error were measured in each fold and then averaged over the conducted cross-validation. In addition to the mean of the cross-validated (CV) test errors we also added the standard deviation. The training time is given in h:m:s and the error in root mean squared error (RMSE)

| Model | CV test error | Training time | Shape violations |
|-------|---------------|---------------|------------------|
| SIASCOR | $17.92 \pm 5.92$ | 00:03:09 | 0 out of 3 |
| AutoML1 | $16.06 \pm 6.26$ | 00:09:56 | 3 out of 3 |
| AutoML2 | $15.73 \pm 5.84$ | 00:09:56 | 3 out of 3 |
| SIAMOR | $17.92 \pm 5.92$ | 00:02:14 | 1 out of 3 |

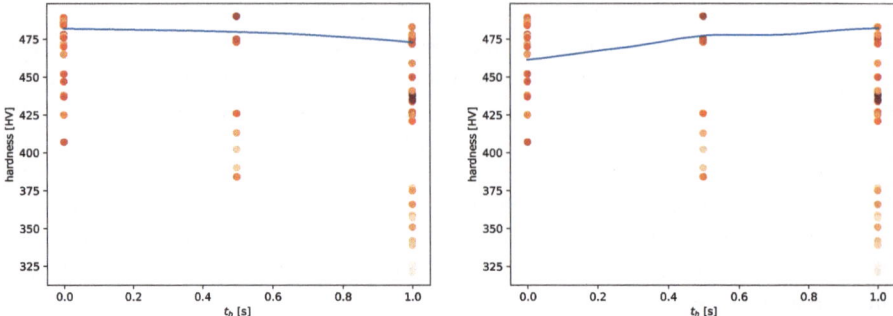

**Fig. 7.3** The left graph depicts the shape of the SIASCOR model and the right graph the shape of the auto-sklearn model both along the $t_c$ variable fixed at the point $x = (0.998, 0.506, 0.154, 0.334)$. The black points represent data points located along the axis and the orange points are the remaining data points projected onto the $\Delta t_h$ dimension. The darker the orange points are, the shorter the Euclidean distance is between the data point and its projection. Both models were trained on 90% of the data that are visible in the graph

### 7.4.2 Brushing

The brushing process is a metal-cutting process used for machining of surface structures with the help of brushes. In quality prediction, the goal is to predict the surface roughness given adjustable process parameters. In our example, we had a data set consisting of 125 points which were generated in an experimental setup where the average arithmetic roughness $R_a$ was measured for various settings of five process parameters: the diameter of the abrasive grits $Dia$, the cutting time $t_c$, number of revolutions of the brush $n_b$, number of revolutions of the work piece $n_w$, and the cutting depth $a_e$. Aside from the data, experts have prior knowledge about the behavior of these machining processes. In [31], we used the ISI method to formalize shape constraints from the expert knowledge. The shape constraints that the experts suggested are visualized in Fig. 7.2. In short, we had upper and lower bounds, monotonicity, convexity, and rebound constraints. For a mathematical description of the shape constraints, see [31].

Similar to the press hardening case, we scale the input data to the unit cube so that we can consider $X = [0, 1]^5$ and set the regularization parameter $\lambda = 0.01$. We choose $\phi$ to be an anisotropic polynomial feature map with maximal degrees (1, 5, 2, 2, 2) for the input dimensions $(Dia, t_c, n_b, n_w, a_e)$ and the parameter space to be $W = [-10^5, 10^5]^{136}$. Additionally, we integrated the same shape constraints as in [31]. With these settings, the SIP in (7.1) is specified for both SIAMOR and SIASCOR. Afterwards, we apply the algorithm from [47] to solve the SIP and set the optimality precision to $\delta = 0.0001$, whereas we use the same settings for SIAMOR as in the press hardening case.

Table 7.2 shows the results for the brushing case. Again, both AutoML approaches violate on average 7.8 out of 10 shape constraints. Even the final model of SIAMOR violates on average 1 out of 10 shape constraints. But, as expected, all

**Table 7.2** Brushing case: Comparison of SIASCOR, auto-sklearn with default settings, auto-sklearn with custom settings, and SIAMOR. The table lists the average number of shape constraints that the models violate after being trained on each fold of a ten-fold cross-validation. We considered ten shape constraints in total. Moreover, both training time and test error were measured in each fold and then averaged over the conducted cross-validation. In addition to the mean of the cross-validated (CV) test errors we also added the standard deviation. The training time is given in h:m:s and the error in root mean squared error (RMSE)

| Model | CV test error | Training time | Shape violations |
|-------|---------------|---------------|------------------|
| SIASCOR | $0.0272 \pm 0.008$ | 01:10:51 | 0 out of 10 |
| AutoML1 | $0.0216 \pm 0.006$ | 00:09:56 | 7.8 out of 10 |
| AutoML2 | $0.0217 \pm 0.006$ | 00:09:56 | 7.8 out of 10 |
| SIAMOR | $0.0267 \pm 0.008$ | 00:44:15 | 1 out of 10 |

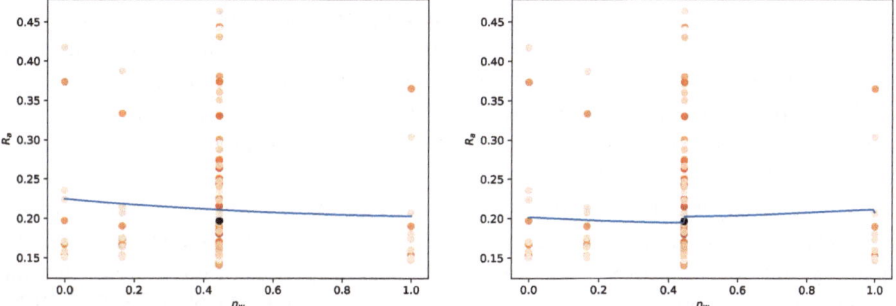

**Fig. 7.4** The left graph depicts the shape of the SIASCOR model and the right graph the shape of the auto-sklearn model both along the $n_w$ variable fixed at the point $x = (1, 0.5, 0.667, 0.444, 0.334)$. The black points represent the data points located along this axis and the orange points are the projected data points onto the $n_w$ dimension. The darker the orange points are, the shorter the Euclidean distance is between the data point and its projection. Both models were trained on 90% of the data that are visible in the graph

SIASCOR models are in accordance with all shape constraints. In Fig. 7.4, we see one exemplary shape violation of an AutoML model. On the right-hand side, the AutoML model violates the rebound and the convexity constraint and is, hence, not in accordance with physical laws. Although there are shape violations, the AutoML model used for Fig. 7.4 does not perform too badly. Similar to the convexity constraint in the graph, all other shape constraints were violated only in a very small region. This suggests an alternative measure for shape compliance that also takes into account the size of the region where violations occur. However, we do not go into that here. Anyway, this is also a representative example that AutoML may sometimes violate shape constraints but, nevertheless, not be entirely catastrophic. Moreover, the average training time of SIASCOR was higher compared to the other methods. The averaged test error is notably low for all models considering that the data ranges from 0.14 to 0.46333.

### 7.4.3  Milling

Milling processes are characterized by high flexibility and productivity, which is why they are used for precision applications with high performance [49]. Milling is a cutting process in which the tool rotates based on a geometrically determined definition and is subjected to chip removal from a workpiece due to the superposition of two effective directions of cutting and feed direction [9]. An assessment of the process is performed mainly on the basis of the quality with the help of roughness parameters (mainly $R_a$) and the mechanical loads on the basis of the parameter of the cutting force $F_c$. This addresses the decisive economic parameters for quality and energy consumption, so that their predictive capability is of high relevance. Technically, this is addressed by means of coating systems, which represent a central key to improving cutting properties due to the coating tribology [42]. Aluminum represents a central research field for optimization due to its wide range of applications and the alloy-dependent variability of technical properties during milling [50]. Using the example of milling various aluminum alloys with coated solid carbide tools and varying the technological parameters of cutting speed $v_c$ and tooth feed $f_z$, the influence on the arithmetic center-line depth $R_a$ and cutting force $F_c$ is analyzed. In addition, the parameter of the friction coefficient $\mu$ of the coating system was taken into account, although this parameter is a one-dimensional quantity and does not fully reflect the tribological properties of the system. In total, we consider two outputs ($R_a$ and $F_c$) and three materials ("EN-AW5754", "EN-AW6082", and "EN-AW7075"), resulting in six different informed learning tasks. For each material and each output, we had 80 data points. This milling use case has not been considered for shape-constrained regression before. Thus, we had no shape constraints available in the beginning. So, we applied the ISI methodology in order to gather the shape constraints for every task. In the first iteration of the methodology, we trained a purely data-driven polynomial model on the data. After the inspecting step, the expert provided us with an initial set of shape constraints. Then, we incorporated all these shape constraints into the prediction model using the SIASCOR method. During the inspection step in the second ISI iteration, however, we noticed that the model had a different shape than expected along the $\mu$-axis. After taking a closer look at the data, we realized that the imposed shape constraints and the data were in conflict. As said before, the one-dimensional variable $\mu$ does not fully capture the tribological properties and its influence on the variables $R_a$ and $F_c$ is hard to interpret. We therefore decided to drop the shape constraints along $\mu$ and left the formation of shape up to the data. In the inspection step of the following—third—ISI iteration, the expert was satisfied with the overall shape of the prediction function. And so, the iterative ISI procedure was terminated at that point.

Now we specify the parameter settings for SIASCOR used in the second and final iteration of ISI. First, we scale the input variables to the unit cube, according to the ranges $\mu \in [0.07, 0.5]$, $v_c \in [100, 1000]$, and $f_z \in [0.025, 0.25]$. The feature map is again an anisotropic polynomial with degrees (3, 2, 3) for ($\mu$, $v_c$, $f_z$) that induces the parameter space $W = [-10^5, 10^5]^{20}$. In addition, we set the regularization parameter to $\lambda = 0.00001$ for all models. Now let us consider the shape constraints for the models with output $F_c$. We impose a lower bound constraint with value 0 and an upper bound constraint with value 180, a decreasingness constraint along $v_c$, and convexity constraints along every dimension. Moreover, the models for material EN-AW6082 and EN-AW7075 had an additional increase constraint along $f_z$. For the output $R_a$, we impose a lower bound constraint with value 0, an upper bound constraint with value 6, and convexity constraints for all dimensions. Moreover, the models for material EN-AW5754 and EN-AW7075 had an increase constraint along $f_z$ and the model for material EN-AW7075 a decrease constraint along $v_c$. These settings specify the SIP in (7.1). Afterwards, we apply the simultaneous algorithm from [47] to solve the SIP with optimality precision $\delta = 0.0001$ for the models predicting $R_a$ and $\delta = 1$ for the ones predicting $F_c$. Furthermore, we used the same shape constraints as above. Apart from that, the SIAMOR and the AutoML models were trained with the same settings as in the other two application examples.

Tables 7.3 and 7.4 show the results for all trained models in the milling case. We see again that the two purely data-driven methods violate most of the expected shape behavior. Figure 7.5 shows one example of such a shape violation. Here we can see how severe the shape violations can be when shape constraints are not explicitly imposed. More specifically, the AutoML1 for material "EN-AW7075" and output $F_c$ violates both the monotonicity and the convexity constraint along the $f_z$ direction. This time, not only the SIASCOR model but also the SIAMOR model had no violations. Moreover, the training times for both expert-based methods were

**Table 7.3** Milling case for output $F_c$: Comparison of SIASCOR, auto-sklearn with default settings, auto-sklearn with custom settings, and SIAMOR for all three materials. The table lists the average number of shape constraints that the models violate after being trained on each fold of a ten-fold cross-validation. We considered six to seven shape constraints for the various materials. Moreover, both training time and test error were measured in each fold and then averaged over the conducted cross-validation. The training time is given in h:m:s and the error in root mean squared error (RMSE)

| Material | Model | CV test error | Training time | Shape violations |
|----------|-------|---------------|---------------|------------------|
| AW5754 | SIASCOR | $6.98 \pm 3.52$ | 00:00:19 | 0 out of 6 |
| AW5754 | AutoML1 | $6.99 \pm 3.73$ | 00:09:56 | 4 out of 6 |
| AW5754 | AutoML2 | $7.13 \pm 3.87$ | 00:09:56 | 4 out of 6 |
| AW5754 | SIAMOR | $7.01 \pm 3.50$ | 00:00:09 | 0.2 out of 6 |
| AW6082 | SIASCOR | $7.86 \pm 1.71$ | 00:00:09 | 0 out of 7 |
| AW6082 | AutoML1 | $7.60 \pm 4.16$ | 00:09:56 | 5 out of 7 |
| AW6082 | AutoML2 | $7.57 \pm 2.22$ | 00:09:56 | 5 out of 7 |
| AW6082 | SIAMOR | $7.90 \pm 1.69$ | 00:01:45 | 0.2 out of 7 |
| AW7075 | SIASCOR | $19.73 \pm 11.08$ | 00:00:12 | 0 out of 7 |
| AW7075 | AutoML1 | $20.27 \pm 11.22$ | 00:09:56 | 5 out of 7 |
| AW7075 | AutoML2 | $19.94 \pm 11.40$ | 00:09:56 | 5 out of 7 |
| AW7075 | SIAMOR | $19.74 \pm 11.05$ | 00:00:03 | 0 out of 7 |

**Table 7.4** Milling case for output $R_a$: Comparison of SIASCOR, auto-sklearn with default settings, auto-sklearn with custom settings, and SIAMOR for all three materials. The table lists the average number of shape constraints that the models violate after being trained on each fold of a ten-fold cross-validation. We considered five to seven shape constraints for the various materials. Moreover, both training time and test error were measured in each fold and then averaged over the conducted cross-validation. In addition to the mean of the cross-validated (CV) test errors we also added the standard deviation. The training time is given in h:m:s and the error in root mean squared error (RMSE)

| Material | Models | Test error | Training time | Shape violations |
|----------|--------|------------|---------------|------------------|
| AW5754 | SIASCOR | $0.3608 \pm 0.1934$ | 00:00:07 | 0 out of 6 |
| AW5754 | AutoML1 | $0.3163 \pm 0.1498$ | 00:09:58 | 4 out of 6 |
| AW5754 | AutoML2 | $0.3164 \pm 0.1474$ | 00:09:58 | 4 out of 6 |
| AW5754 | SIAMOR | $0.3611 \pm 0.1936$ | 00:00:02 | 0 out of 6 |
| AW6082 | SIASCOR | $0.2480 \pm 0.0607$ | 00:00:06 | 0 out of 5 |
| AW6082 | AutoML1 | $0.2292 \pm 0.0859$ | 00:09:58 | 2.9 out of 5 |
| AW6082 | AutoML2 | $0.2247 \pm 0.0871$ | 00:09:58 | 2.6 out of 5 |
| AW6082 | SIAMOR | $0.2497 \pm 0.0606$ | 00:04:00 | 0 out of 5 |
| AW7075 | SIASCOR | $0.7965 \pm 0.1710$ | 00:00:23 | 0 out of 7 |
| AW7075 | AutoML1 | $0.6071 \pm 0.2291$ | 00:09:58 | 4.8 out of 7 |
| AW7075 | AutoML2 | $0.6071 \pm 0.2265$ | 00:09:58 | 4.9 out of 7 |
| AW7075 | SIAMOR | $0.7967 \pm 0.1711$ | 00:00:03 | 0 out of 7 |

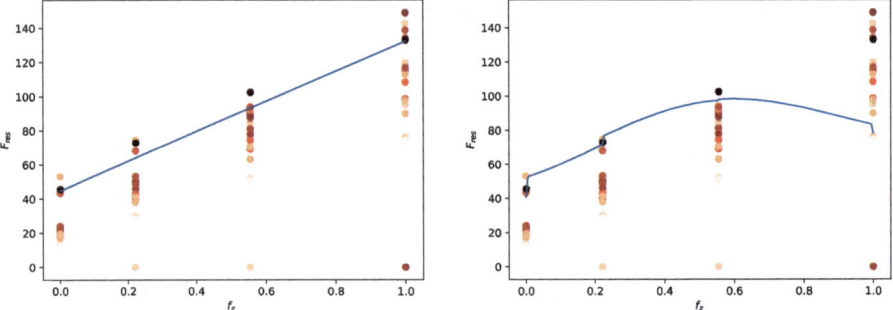

**Fig. 7.5** The left graph depicts the shape of the SIASCOR model and the right graph the shape of the auto-sklearn model both along the $f_z$ variable fixed at the point $x = (0, 0.11, 1)$. The black points represent the data points located along this axis and the orange points are the projected data points onto the $f_z$ dimension. The darker the points are, the shorter the Euclidean distance is between the data point and its projection. We can see that the auto-sklearn model strongly violates the monotonicity and convexity. Both models were trained on 90% of the data that are visible in the graph

too low to see a pattern. Both algorithms were fast due to the low dimensionality of $X$ and $W$. Again, the test errors were close to each other considering the ranges of the output.

## 7.5   Synthetic Example

In the previous section, we have considered real-world use cases of small data regression. We have seen that purely data-driven models fail to provide reliable models in terms of shape-compliance. Another indicator for the quality of models is the generalization error, i.e. the error on data points not contained in the training set. In this section, we examine the generalization error for SIASCOR and compare it to AutoML, SIAMOR, and Ridge regression.

One of the challenges with real small data problems is to compute good estimates of the generalization error with the limited data available. The standard approach is to conduct a cross-validation on the data set, as we have done in the previous real-world use cases. However, this approach limits the focus on the data set and we cannot infer the performance in regions with few or no data. Additionally, the authors in [35] show that cross-validation can lead to over-optimistic estimates, which may not be indicative of the actual generalization error.

To better understand the generalization error, we construct an synthetic application example where we can sample as many data points as we desire. Specifically, we construct a separable function given by $f^{\text{toy}} : [0, 1]^5 \to \mathbb{R}$ with

$$f^{\text{toy}}(x) = \sum_{i=1}^{5} f_i^{\text{toy}}(x_i) + 0.1 \, , \qquad (7.2)$$

where

$$f_1^{\text{toy}}(x_1) = 0.12(x_1 - 0.5)^3, \qquad\qquad f_2^{\text{toy}}(x_2) = 0.002/(x_2 + 0.1)^2,$$

$$f_3^{\text{toy}}(x_3) = 0.1(x_3 - 0.6)^2(x_3 - 2.4)^2, \quad f_4^{\text{toy}}(x_4) = 0.02(x_4 - 0.6)^2(x_3 - 2.4)^2,$$

$$f_5^{\text{toy}}(x_5) = 0.02(x_4 - 1.1)^2(x_3 - 3)^2.$$

$$(7.3)$$

The artificial function is polynomial in all but the second dimension and satisfies all the shape constraints of the brushing example, which can be verified easily.

As in the section before, we compare SIASCOR to SIAMOR and AutoML. The ansatz functions of SIASCOR and SIAMOR are polynomial, while the AutoML conducts a model selection over various ansatz function classes, including polynomial functions. In view of the nearly polynomial structure of the artificial function, we compare the shape-constrained models to unconstrained ridge polynomial regression. This way, we can demonstrate that it is the shape constraints that lead to better generalization power and not just the knowledge of the polynomial structure.

To generate the synthetic data set, we sample 30 points uniformly from the domain $X = [0, 1]^5$, evaluate $f^{\text{toy}}$ at these points, and add Gaussian noise with standard deviation $\sigma = 0.03408$ to the outputs. This corresponds to the expected error in the brushing case. Indeed, experts expect 5–10% error measuring the roughness. Choosing $\sigma$ as above implies that 95% of the data have an error below 10%.

Given the data set and the shape constraints, we formulate the shape-constrained regression problem based on Example 6.1 in [47]. First, we set the maximal degrees of our anisotropic polynomial as (1, 5, 2, 2, 2). Accordingly, we have $W = [10^{-5}, 10^5]^{136}$. Then, we set the regularization parameter as $\lambda = 0.01$ for SIASCOR and ridge and $\lambda = 0.05$ for SIAMOR after having conducted a hyperparameter optimization via a ten-fold cross-validation over the dataset. With these settings, the SIP is formulated and we can run SIASCOR and SIAMOR to solve it using the same, remaining settings as in the brushing section. In the case of SIASCOR, we solve the SIP with optimality precision $\delta = 10^{-5}$.

With the hyperparamter settings above, we first conduct another ten-fold cross-validation on the 30 data points. The test errors of the cross-validation are the predictive power estimates that we would have if we had only the data set available. For the cross-validation, we measure the RMSE in every fold and then compute the mean and standard deviation over all folds. Afterwards, we estimate the generalization error. For this, we retrain the models on the entire data set and measure the RMSE on 5000 points sampled from the ground truth function without noise. The results are listed in Table 7.5.

First and above all, we observe in Table 7.5 that the shape-constrained models achieve a lower generalization error than unconstrained models such as AutoML or pure ridge regression. Interestingly, the cross-validated test error on the data set is misleading in the case of SIAMOR because it appears to generalize better than SIASCOR. The generalization error is, however, lower for SIASCOR. This is in line with the observations in [35] where the authors have shown that cross-validation usually slightly underestimates the true generalization error. Nonetheless, for the unconstrained models the cross-validation error is indicative of the inferior generalization error.

In conclusion, the use of shape constraints is shown to lead to better generalization power in the synthetic data regression problem that we considered. This can be inferred from the lower generalization errors of SIASCOR and SIAMOR over AutoML. By considering Ridge regression that has the same ansatz functions, we

**Table 7.5** Synthetic Example: Comparison of SIASCOR, auto-sklearn with custom settings, SIAMOR, and Ridge regression. The models were trained on each fold of a 10-fold cross-validation. Then, we measured the average cross-validated (CV) test error. Moreover, we retrained the models on the entire data set and counted the shape violations. Ten shape constraints were considered. Lastly, we measured the generalization error by sampling 5000 points from the ground truth and computing the error to the prediction of the models

| Model | CV test error | Shape violations | Generalization error |
|-------|---------------|------------------|---------------------|
| SIASCOR | $0.04432 \pm 0.0153$ | 0 | 0.032 |
| AutoML2 | $0.05080 \pm 0.0194$ | 8 | 0.047 |
| SIAMOR | $0.04230 \pm 0.0153$ | 0 | 0.034 |
| Ridge | $0.05750 \pm 0.0156$ | 8 | 0.065 |

conclude that it is indeed the shape constraints that lead to a better performance and not the choice of ansatz functions. All in all, the evaluation of the generalization error on this artificial data set emphasizes another benefit of enforcing shape constraints.

## 7.6 Conclusion

Reliable models are important in situations with insufficient data. Take for instance the quality prediction in manufacturing. Data generation is expensive because it involves costly experiments. Therefore, data is scarce and, on top of that, noisy due to measurement errors. During adaptive process control, quality prediction is crucial to avoid wrong decision-making that comes along with high costs. Therefore, practitioners rely more on their prior knowledge than on models trained solely on the available data. Informed learning aims to get the best from both sides: inferring quantitative information from data while being in accordance with prior knowledge.

In this chapter, we incorporated shape knowledge into the training of machine learning models. First, we summarized the SIASCOR method from [31] by formulating the shape-constrained regression problem and discussing SIP algorithms for solving it. We also recalled the ISI methodology as a general method to inspect models for shape compliance or non-compliance, to elaborate and specify shape constraints, and to incorporate these shape constraints into the chosen regression model. We point out that SIASCOR is only one possible way of incorporating shape constraints. In principle, other methods of integrating shape knowledge can be used within the ISI methodology. Then, we considered three application examples from manufacturing: brushing, press hardening, and milling. The latter has not been applied to shape-constrained regression yet, so we used ISI to formalize shape constraints. Then, we applied the SIASCOR method with the simultaneous algorithms from [47] to obtain shape-compliant models that fit the data. The SIASCOR method was then compared to two purely data-driven AutoML approaches and another shape-constrained regression method SIAMOR that solves the corresponding SIP differently. During a ten-fold cross-validation, we created ten different model functions for each approach and compared their shape compliance (or more precisely the average number of shape constraints they violated), their average training time, and their average test error.

In the comparative study with real-world application examples, we have seen that, in general, purely data-driven models trained on a few data do not infer shape knowledge just from the data. Even methods like auto-sklearn that choose the best model out of many do not seem to perform satisfactorily outside the data set. In fact, the resulting models behave contrary to physical laws repeatedly. Consequently, we do not recommend using purely data-driven methods to create reliable models when the available data is limited. Furthermore, just by looking at the test errors computed with cross-validation, one is tempted to think that AutoML models generalize slightly better. But as we have seen, low test errors are misleading because the final models violate the shape constraints. In contrast, the final SIASCOR model was, as expected, shape-compliant.

In the comparative study with the synthetic example, we could also show that the use of shape constraints lead to more powerful models regarding generalization error. When trained on a small data set, purely data-driven methods, as AutoML or Ridge regression, do not perform well outside the data set.

In conclusion, we recommend leveraging shape knowledge when dealing with insufficient data. However, not all shape-knowledge-based approaches guarantee definite shape compliance of the final model. For instance, SIAMOR fails to rigorously impose shape constraints. From a theoretical point of view, this is to be expected but we have also seen it in practice. Besides, there are no mathematical guarantees for the termination of the algorithm and inferior properties for optimality. In contrast to SIAMOR, SIASCOR is a theoretically sound and practically reliable method to impose hard shape constraints as we have seen in the experiments. Aside from the shape compliance, we observed in our applications that for a higher number of inputs and parameters the SIAMOR is slightly faster than SIASCOR. However, the discrepancy between the training times was not large enough for proper assessment. Even if there was a significant gain in computation time, it would only justify using SIAMOR for exploring shape-constrained regression but not for imposing shape constraints on the final model.

A follow-up study should reconsider the adaptive feasible-point algorithm that solves the SIP to improve computation time. One theoretical aspect is to analyze the time complexity of the algorithms. This is crucial for better scalability with respect to an increasing number of model parameters and input variables. Another, more applied aspect is, of course, an efficient implementation. It is also interesting to see how the generalization error of SIASCOR changes with respect to different noise levels and to different numbers of data. This analysis would help to better estimate when to use the SIASCOR. An additional line of research can be to develop an elaborated way to find all regions where shape constraints are violated. This way, we cannot only visualize all shape violations but, on top of that, accelerate the process of finding the right shape constraints with ISI. One could also, in a comparative study, analyze if and to which extent SIASCOR improves extrapolation. Going in a similar direction, a tool that helps to assess whether some enforced shape constraints are in too much conflict with the data and should therefore be excluded would be very handy, as we have seen in the milling example. One further idea is to find a way to compute the confidence intervals of the SIASCOR model to spot regions of high and low variance.

**Acknowledgments** This contribution was supported by the Fraunhofer Cluster of Excellence "Cognitive Internet Technologies". Additionally, we gratefully acknowledge financial support from the Deutsche Forschungsgemeinschaft (DFG, German Research Foundation) within the research unit "FOR 5359: Deep Learning on Sparse Chemical Process Data".

# References

1. Altendorf, E.E., Restificar, A.C., Dieterich, T.G.: Learning from sparse data by exploiting monotonicity constraints (2012). https://doi.org/10.48550/ARXIV.1207.1364. URL https://arxiv.org/abs/1207.1364
2. Aubin-Frankowski, P.C., Szabo, Z.: Handling hard affine SDP shape constraints in RKHSs. URL https://arxiv.org/pdf/2101.01519
3. Bauckhage, C., Ojeda, C., Schücker, J., Sifa, R., Wrobel, S.: Informed machine learning through functional composition (2018)
4. Blankenship, J.W., Falk, J.E.: Infinitely constrained optimization problems. Journal of Optimization Theory and Applications **19**(2), 261–281 (1976). https://doi.org/10.1007/BF00934096
5. Chang, C.J., Dai, W.L., Chen, C.C.: A novel procedure for multimodel development using the grey silhouette coefficient for small-data-set forecasting. Journal of the Operational Research Society **66**(11), 1887–1894 (2015). https://doi.org/10.1057/jors.2015.17
6. Chuang, H.C., Chen, C.C., Li, S.T.: Incorporating monotonic domain knowledge in support vector learning for data mining regression problems. Neural Computing and Applications **32** (2020). https://doi.org/10.1007/s00521-019-04661-4
7. Cozad, A., Sahinidis, N.V., Miller, D.C.: A combined first-principles and data-driven approach to model building. Computers & Chemical Engineering **73**, 116–127 (2015). https://doi.org/10.1016/j.compchemeng.2014.11.010
8. Daw, A., Karpatne, A., Watkins, W., Read, J., Kumar, V.: Physics-guided neural networks (PGNN): An application in lake temperature modeling. URL https://arxiv.org/pdf/1710.11431
9. Deutsches Institut für Normung: Fertigungsverfahren spanen - teil 3: Fräsen; einordnung, unterteilung, begriffe. https://doi.org/10.31030/9500667
10. Diligenti, M., Roychowdhury, S., Gori, M.: Integrating prior knowledge into deep learning. In: 2017 16th IEEE International Conference on Machine Learning and Applications (ICMLA 2017), pp. 920–923. IEEE, Piscataway, NJ (2017). https://doi.org/10.1109/ICMLA.2017.00-37
11. Djelassi, H., Mitsos, A., Stein, O.: Recent advances in nonconvex semi-infinite programming: Applications and algorithms. EURO Journal on Computational Optimization **9**, 100006 (2021). https://doi.org/10.1016/j.ejco.2021.100006. URL https://www.sciencedirect.com/science/article/pii/S2192440621000034
12. Feurer, M., Eggensperger, K., Falkner, S., Lindauer, M., Hutter, F.: Auto-sklearn 2.0: Hands-free AutoML via meta-learning. URL https://arxiv.org/pdf/2007.04074
13. Feurer, M., Klein, A., Eggensperger, K., Springenberg, J., Blum, M., Hutter, F.: Efficient and robust automated machine learning. In: C. Cortes, N. Lawrence, D. Lee, M. Sugiyama, R. Garnett (eds.) Advances in Neural Information Processing Systems, vol. 28. Curran Associates, Inc. (2015). URL https://proceedings.neurips.cc/paper/2015/file/11d0e6287202fced83f79975ec59a3a6-Paper.pdf
14. Gupta, M., Cotter, A., Pfeifer, J., Voevodski, K., Canini, K., Mangylov, A., Moczydlowski, W., van Esbroeck, A.: Monotonic calibrated interpolated look-up tables. Journal of Machine Learning Research **17**(109), 1–47 (2016). URL http://jmlr.org/papers/v17/15-243.html
15. Gupta, M.R., Ilan, E.L., Mangylov, O., Morioka, N., Narayan, T., Zhao, S.: Multidimensional shape constraints. In: ICML (2020)
16. Haider, C., de Franca, F., Burlacu, B., Kronberger, G.: Shape-constrained multi-objective genetic programming for symbolic regression. Applied Soft Computing **132**, 109855 (2023). https://doi.org/10.1016/j.asoc.2022.109855. URL https://www.sciencedirect.com/science/article/pii/S1568494622009048
17. Hall, G.: Optimization over nonnegative and convex polynomials with and without semidefinite programming (2018). https://doi.org/10.48550/ARXIV.1806.06996. URL https://arxiv.org/abs/1806.06996

18. Hao, J., Zhou, M., Wang, G., Jia, L., Yan, Y.: Design optimization by integrating limited simulation data and shape engineering knowledge with bayesian optimization (BO-DK4do). Journal of Intelligent Manufacturing **31**(8), 2049–2067 (2020). https://doi.org/10.1007/s10845-020-01551-8

19. He, X., Zhao, K., Chu, X.: AutoML: A survey of the state-of-the-art. Knowledge-Based Systems **212**, 106622 (2021). https://doi.org/10.1016/j.knosys.2020.106622

20. He, Z., He, Y., Chen, Z., Zhao, Y., Lian, R.: Functional failure diagnosis approach based on bayesian network for manufacturing systems. In: 2019 Prognostics and System Health Management Conference (PHM-Qingdao), pp. 1–6. IEEE (2019). https://doi.org/10.1109/PHM-Qingdao46334.2019.8942813

21. Heese, R., Walczak, M., Morand, L., Helm, D., Bortz, M.: The good, the bad and the ugly: Augmenting a black-box model with expert knowledge. In: I.V. Tetko, V. Kůrková, P. Karpov, F. Theis (eds.) Artificial Neural Networks and Machine Learning – ICANN 2019: Workshop and Special Sessions, pp. 391–395. Springer International Publishing, Cham (2019)

22. Hettich, R., Kortanek, K.O.: Semi-infinite programming: Theory, methods, and applications. SIAM Review **35**(3), 380–429 (1993). https://doi.org/10.1137/1035089

23. Karpatne, A., Atluri, G., Faghmous, J.H., Steinbach, M., Banerjee, A., Ganguly, A., Shekhar, S., Samatova, N., Kumar, V.: Theory-guided data science: A new paradigm for scientific discovery from data. IEEE Transactions on Knowledge and Data Engineering **29**(10), 2318–2331 (2017). https://doi.org/10.1109/TKDE.2017.2720168

24. King, R., Hennigh, O., Mohan, A., Chertkov, M.: From deep to physics-informed learning of turbulence: Diagnostics (2018). https://doi.org/10.48550/ARXIV.1810.07785. URL https://arxiv.org/abs/1810.07785

25. Kurnatowski, M.v., Schmid, J., Link, P., Zache, R., Morand, L., Kraft, T., Schmidt, I., Schwientek, J., Stoll, A.: Compensating data shortages in manufacturing with monotonicity knowledge. Algorithms **14**(12) (2021). https://doi.org/10.3390/a14120345. URL https://www.mdpi.com/1999-4893/14/12/345

26. Ladický, L., Jeong, S., Solenthaler, B., Pollefeys, M., Gross, M.: Data-driven fluid simulations using regression forests. ACM Transactions on Graphics **34**(6), 1–9 (2015). https://doi.org/10.1145/2816795.2818129

27. Lauer, F., Bloch, G.: Incorporating prior knowledge in support vector regression. Machine Learning **70** (2008). https://doi.org/10.1007/s10994-007-5035-5

28. Li, D.C., Huang, W.T., Chen, C.C., Chang, C.J.: Employing virtual samples to build early high-dimensional manufacturing models. International Journal of Production Research **51**(11), 3206–3224 (2013). https://doi.org/10.1080/00207543.2012.746795

29. Li, D.C., Liu, C.W., Chen, W.C.: A multi-model approach to determine early manufacturing parameters for small-data-set prediction. International Journal of Production Research **50**(23), 6679–6690 (2012). https://doi.org/10.1080/00207543.2011.613867

30. Li, D.C., Wu, C.S., Tsai, T.I., Lina, Y.S.: Using mega-trend-diffusion and artificial samples in small data set learning for early flexible manufacturing system scheduling knowledge. Computers & Operations Research **34**(4), 966–982 (2007). https://doi.org/10.1016/j.cor.2005.05.019

31. Link, P., Poursanidis, M., Schmid, J., Zache, R., von Kurnatowski, M., Teicher, U., Ihlenfeldt, S.: Capturing and incorporating expert knowledge into machine learning models for quality prediction in manufacturing. Journal of Intelligent Manufacturing **33**(7), 2129–2142 (2022). https://doi.org/10.1007/s10845-022-01975-4

32. Lokrantz, A., Gustavsson, E., Jirstrand, M.: Root cause analysis of failures and quality deviations in manufacturing using machine learning. Procedia CIRP **72**(4–6), 1057–1062 (2018). https://doi.org/10.1016/j.procir.2018.03.229

33. López, M., Still, G.: Semi-infinite programming. European Journal of Operational Research **180**(2), 491–518 (2007). https://doi.org/10.1016/j.ejor.2006.08.045

34. Lu, Y., Rajora, M., Zou, P., Liang, S.: Physics-embedded machine learning: Case study with electrochemical micro-machining. Machines **5**(1), 4 (2017). https://doi.org/10.3390/machines5010004

35. Martens, H.A., Dardenne, P.: Validation and verification of regression in small data sets. Chemometrics and Intelligent Laboratory Systems **44**(1), 99–121 (1998). https://doi.org/10.1016/S0169-7439(98)00167-1. URL https://www.sciencedirect.com/science/article/pii/S0169743998001671

36. Mitsos, A.: Global optimization of semi-infinite programs via restriction of the right-hand side. Optimization **60**(10–11), 1291–1308 (2011). https://doi.org/10.1080/02331934.2010.527970

37. Muralidhar, N., Islam, M.R., Marwah, M., Karpatne, A., Ramakrishnan, N.: Incorporating prior domain knowledge into deep neural networks. In: 2018 IEEE International Conference on Big Data (Big Data), pp. 36–45. IEEE (122018). https://doi.org/10.1109/BigData.2018.8621955

38. Nagarajan, H.P.N., Mokhtarian, H., Jafarian, H., Dimassi, S., Bakrani-Balani, S., Hamedi, A., Coatanéa, E., Gary Wang, G., Haapala, K.R.: Knowledge-based design of artificial neural network topology for additive manufacturing process modeling: A new approach and case study for fused deposition modeling. Journal of Mechanical Design **141**(2), 442 (2019). https://doi.org/10.1115/1.4042084

39. Napoli, G., Xibilia, M.G.: Soft sensor design for a topping process in the case of small datasets. Computers & Chemical Engineering **35**(11), 2447–2456 (2011). https://doi.org/10.1016/j.compchemeng.2010.12.009

40. Neugebauer, R., Schieck, F., Polster, S., Mosel, A., Rautenstrauch, A., Schönherr, J., Pierschel, N.: Press hardening — an innovative and challenging technology. Archives of Civil and Mechanical Engineering **12**(2), 113–118 (2012). https://doi.org/10.1016/j.acme.2012.04.013

41. Pedregosa, F., Varoquaux, G., Gramfort, A., Michel, V., Thirion, B., Grisel, O., Blondel, M., Prettenhofer, P., Weiss, R., Dubourg, V., Vanderplas, J., Passos, A., Cournapeau, D., Brucher, M., Perrot, M., Duchesnay, E.: Scikit-learn: Machine learning in Python. Journal of Machine Learning Research **12**, 2825–2830 (2011)

42. Prengel, H., Pfouts, W., Santhanam, A.: State of the art in hard coatings for carbide cutting tools. Surface and Coatings Technology **102**(3), 183–190 (1998). https://doi.org/10.1016/S0257-8972(96)03061-7. URL https://www.sciencedirect.com/science/article/pii/S0257897296030617

43. Rembert Reemtsen, Jan-J. Rückmann: Semi-infinite programming (1998)

44. Riihimäki, J., Vehtari, A.: Gaussian processes with monotonicity information. Journal of Machine Learning Research - Proceedings Track **9**, 645–652 (2010)

45. Rueden, L.v., Mayer, S., Beckh, K., Georgiev, B., Giesselbach, S., Heese, R., Kirsch, B., Walczak, M., Pfrommer, J., Pick, A., Ramamurthy, R., Garcke, J., Bauckhage, C., Schuecker, J.: Informed machine learning - a taxonomy and survey of integrating prior knowledge into learning systems. IEEE Transactions on Knowledge and Data Engineering p. 1 (2021). https://doi.org/10.1109/TKDE.2021.3079836

46. Schmid, J.: Approximation, characterization, and continuity of multivariate monotonic regression functions. Analysis and Applications **20**(4) (2021). https://doi.org/10.1142/S0219530521500299

47. Schmid, J., Poursanidis, M.: Approximate solutions of convex semi-infinite optimization problems in finitely many iterations (2021). URL https://arxiv.org/abs/2105.08417

48. Stewart, R., Ermon, S.: Label-free supervision of neural networks with physics and domain knowledge. Proceedings of the AAAI Conference on Artificial Intelligence **31**(1) (2017). https://doi.org/10.1609/aaai.v31i1.10934. URL https://ojs.aaai.org/index.php/AAAI/article/view/10934

49. Teicher, U., Pirl, S., Nestler, A., Hellmich, A., Ihlenfeldt, S.: A novel hybrid clamping system for sheet metals and thin walled structures. Procedia Manufacturing **40**, 51–55 (2019). https://doi.org/10.1016/j.promfg.2020.02.010. URL https://www.sciencedirect.com/science/article/pii/S2351978920305412. 19th Machining Innovations Conference for Aerospace Industry 2019 (MIC 2019), 27–28 November 2019, Garbsen, Germany

50. Teicher, U., Pirl, S., Nestler, A., Hellmich, A., Ihlenfeldt, S.: Surface roughness and its prediction in high speed milling of aluminum alloys with pcd and cemented carbide tools. MM Science Journal **2019**, 3136–3141 (2019). https://doi.org/10.17973/MMSJ.2019_11_2019062

51. Torre, E., Marelli, S., Embrechts, P., Sudret, B.: Data-driven polynomial chaos expansion for machine learning regression. Journal of Computational Physics **388**(4), 601–623 (2019). https://doi.org/10.1016/j.jcp.2019.03.039
52. Tsai, T.I., Li, D.C.: Utilize bootstrap in small data set learning for pilot run modeling of manufacturing systems. Expert Systems with Applications **35**(3), 1293–1300 (2008). https://doi.org/10.1016/j.eswa.2007.08.043
53. Tsoukalas, A., Rustem, B.: A feasible point adaptation of the blankenship and falk algorithm for semi-infinite programming. Optimization Letters **5**(4), 705–716 (2011). https://doi.org/10.1007/s11590-010-0236-4
54. Weichert, D., Link, P., Stoll, A., Rüping, S., Ihlenfeldt, S., Wrobel, S.: A review of machine learning for the optimization of production processes. The International Journal of Advanced Manufacturing Technology **104**(5–8), 1889–1902 (2019). https://doi.org/10.1007/s00170-019-03988-5
55. Zhang, H., Roy, U., Tina Lee, Y.T.: Enriching analytics models with domain knowledge for smart manufacturing data analysis. International Journal of Production Research **58** (2020). https://doi.org/10.1080/00207543.2019.1680895

# Part III
# Neural Networks

# Chapter 8
# Predicting Properties of Oxide Glasses Using Informed Neural Networks

Gregor Maier, Jan Hamaekers, Dominik-Sergio Martilotti, and Benedikt Ziebarth

**Abstract** Many modern-day applications require the development of new materials with specific properties. In particular, the design of new glass compositions is of great industrial interest. Current machine learning methods for learning the composition-property relationship of glasses promise to save on expensive trial-and-error approaches. Even though quite large datasets on the composition of glasses and their properties already exist (i.e., with more than 350,000 samples), they cover only a very small fraction of the space of all possible glass compositions. This limits the applicability of purely data-driven models for property prediction purposes and necessitates the development of models with high extrapolation power.

In this chapter, we propose a neural network model which incorporates prior scientific and expert knowledge in its learning pipeline. This informed learning approach leads to an improved extrapolation power compared to blind (uninformed) neural network models. To demonstrate this, we train our models to predict three different material properties (glass transition temperature, Young's modulus (at room temperature) and shear modulus) of binary oxide glasses which do not contain sodium. As representatives for conventional blind neural network approaches we use five different feed-forward neural networks of varying widths and depths.

For each property, we set up model ensembles of multiple trained models and show that, on average, our proposed informed model performs better in extrapolating the three properties of previously unseen sodium borate glass samples than all five conventional blind models.

G. Maier (✉)
Fraunhofer SCAI, Sankt Augustin, Germany

University of Bonn, Bonn, Germany
e-mail: gregor.maier@scai.fraunhofer.de

J. Hamaekers · D.-S. Martilotti
Fraunhofer SCAI, Sankt Augustin, Germany
e-mail: jan.hamaekers@scai.fraunhofer.de; dominik-sergio.martilotti@scai.fraunhofer.de

B. Ziebarth
Schott AG, Mainz, Germany
e-mail: benedikt.ziebarth@schott.com

© The Author(s) 2025
D. Schulz, C. Bauckhage (eds.), *Informed Machine Learning*,
Cognitive Technologies, https://doi.org/10.1007/978-3-031-83097-6_8

## 8.1 Introduction

The development of new materials is essential for the modern-day progress in engineering applications and future-oriented technologies. Aside from ever new demands on physical and chemical materials properties, ecological issues, such as sustainability, long service life, environmental compatibility, and recyclability, are of great importance for product development in a variety of different fields. However, the common materials design process is still majorly based on the application of suitable empirical models, on past experiences and educated guesses, and on an extensive subsequent testing phase. The development of new glassy materials, in particular, would benefit to a large extent from a more resource-efficient, systematic, data-driven approach in contrast to the Edisonian trial-and-error approach which is still often used in traditional research and development [23].

The space of all possible glass compositions is very large as a glass can be made from the combination of 80 chemical elements, which leads to $10^{52}$ possible glass compositions [41]. Moreover, since the influencing parameters are usually known only qualitatively or not at all, the optimization of glass material properties is inherently challenging. A trial-and-error approach to find a glass composition with specific properties for a certain application is time-consuming and often not feasible in practice. An expert-guided approach with integrating experiences from the past is usually not sufficient as well since there are interesting glass properties that are extremely difficult to predict. Especially when properties show nonlinearities, caused, for example, by the so-called borate anomaly in alkali borate glasses [14], conventional exploration and exploitation strategies quickly reach their limits. Therefore, going beyond the area of known materials requires new approaches based on new and innovative methods. The field of Machine Learning (ML) provides such methods which allow to generate accurate models based on existing data in order to predict the properties of yet unseen materials.

### 8.1.1 Related Work

In recent years, ML techniques have been widely used for accelerating materials design [28, 36, 39]. In glass science, there have been several successful attempts to use ML to predict, i.a., optical, physical, and mechanical properties of glasses [1, 3, 8–10, 31]. Most ML models perform exceptionally well in interpolating the training data. However, given the high-dimensional search space of all possible glass compositions and its sparse coverage by experimental data, the search for new glass materials is majorly a question of designing models which possess a high extrapolation power. We refer to [21, 30] and references therein for reviews of the current status of ML in glass science and future challenges.

To address the lack of extrapolation power, ordinary ML methods can be extended by integrating prior knowledge which exists independently of the learning task. This idea is termed *Informed Machine Learning* and we refer to the recent survey [37] for a taxonomy and thorough overview of its application in current ML state-of-the-art use cases. For glass design, this idea is utilized, e.g., in [34] where the empirical MYEGA formula is integrated into a neural network architecture to predict the viscosity of a glass based on its compound fractions and temperature. In [7], this approach is developed further by additionally integrating prior chemical and physical knowledge of the glasses' elements into the training data. Similarly, in [2, 33], the authors use external chemical and physical knowledge to carefully design enriched descriptors of glass compositions which are used as inputs for ML models to predict properties of oxide glasses. In [20] and [22], the authors predict the dissolution kinetics of silicate glasses in an informed manner by suitably splitting the training data and using a descriptor which encodes the glasses' network structure, respectively. They demonstrate the superior performance of the informed approach compared to the uninformed approach. This superiority is also shown in [4], where the authors design a neural network model which is informed by statistical mechanics in order to predict structural properties of oxide glasses.

## 8.1.2  Contributions

In this chapter, we propose a new ML model based on neural networks for the property prediction of oxide glasses which integrates prior knowledge in order to achieve a high degree of extrapolation of the training data. We modify the ideas from [7] in order to predict three material properties, that is, the *glass transition temperature* $T_g$, the *Young's modulus E (at room temperature)*, and the *shear modulus G*. We focus our analysis on binary oxide glasses, that is, oxide glasses which consist of exactly two compounds. Our model is informed in the sense that we explicitly integrate prior knowledge into the design of our training data, the hypothesis set, and the final hypothesis at four major points in our learning pipeline. We place emphasis on explaining how this is done in detail in terms of the taxonomy in [37]. Especially the design of the network architecture to realize permutation invariance with respect to the input features seems, to the best of our knowledge, to be new in the field of glass materials modeling.

To examine the extrapolation power of our models, we train and validate them on glass samples which do not contain sodium in their compositions. The trained models are then used to predict the properties of sodium borate glass compositions with varying compound fractions. For each property, we train multiple models and study the average performance of the model ensemble. To demonstrate the superiority of the informed model ensemble compared to blind (uninformed) approaches, we perform the same experiments with five standard fully connected feed-forward neural networks of varying widths and depths without integration of any prior knowledge. We compare the results quantitatively in terms of error metrics and qualitatively in terms of a meaningful approximation of the respective composition-property curves.

**Outline** The remainder of this chapter is organized as follows: In Sect. 8.2, we explain our methodology. That is, in Sect. 8.2.1, we present our automated pipeline for collecting and preparing data for model training, validation, and testing. In Sect. 8.2.2, we describe the different model setups in the blind and the informed setting. In Sect. 8.2.3, we explain how we train and evaluate our models. We discuss the results of our experiments in Sect. 8.3 and conclude our findings in Sect. 8.4.

**Notation** For notational convenience, we use the letter $P$ whenever we refer to one of the three properties $T_g$, $E$, or $G$. Moreover, for all entities which exist for every property $P$, we use the prefix "$P$-" to specify the respective entity. For example, given $P$, we refer to the dataset that is used to train a model for predicting $P$ by "$P$-training set". Moreover, we use the symbols $\mathbb{N}$ and $\mathbb{R}$ to denote the set of positive integers and the set of real numbers, respectively.

## 8.2 Methodology

The prediction quality of any data-driven machine learning algorithm in the context of supervised learning is strongly dependent on the quantity and quality of the training data. Before presenting our neural network approach to the problem of glass property prediction in detail in Sects. 8.2.2 and 8.2.3, we therefore describe in the following Sect. 8.2.1 how we collect and prepare our data.

### 8.2.1 Data Collection and Preparation

We use data from the INTERGLAD Ver. 8 database [26] and the SciGlass database [12] and merge them together into a common glassmodel database. For the identification of oxide glasses we follow the same definition as in [1] and only consider glasses whose mole atomic fraction of oxygen is at least 0.3 and whose compounds do not contain the chemical elements S, H, C, Pt, Au, F, Cl, N, Br, and I, which could affect the balance of oxygen. The resulting glassmodel database of oxide glasses consists of 420,973 glass samples in total. It lists the mole atomic fractions of 118 chemical elements and the mole fractions of 439 compounds, i.e., the oxides that a glass composition consists of, together with the values of 87 material properties. However, among the 118 elements, only 66 elements appear with non-vanishing fraction in at least one glass sample. Among the 439 compounds, only 183 compounds appear with non-vanishing fraction in at least one glass sample.

To obtain clean data for training, validating, and testing our models, we apply a sequence of preprocessing steps which follows in parts the procedure described in [1, 10]. For each glass property $P$, we extract clean data from the "dirty" glassmodel database in an automated fashion in form of a preprocessing pipeline

**Fig. 8.1** Steps in the preprocessing pipeline as described in Sect. 8.2.1

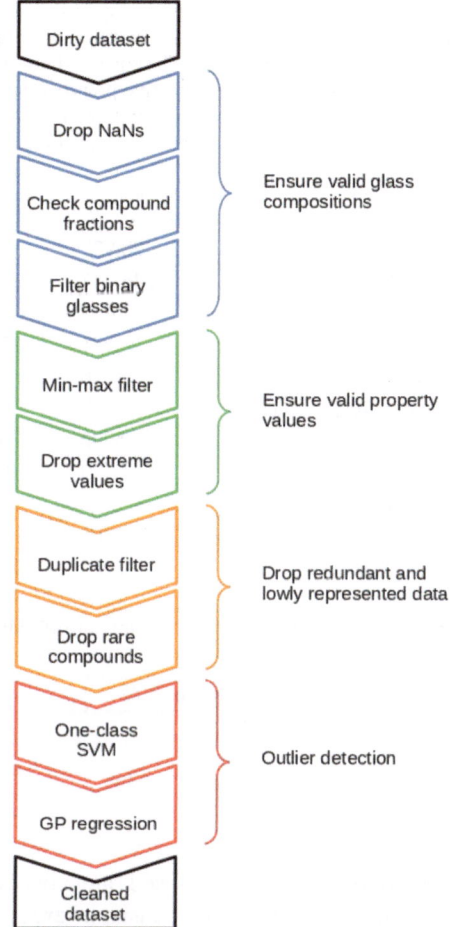

whose steps are schematically depicted in Fig. 8.1. The number of samples which are dropped in each step is shown in Table 8.1.

We begin with all samples from the entire glassmodel database. As a first step, we make sure that all glass samples have numerically valid entries. That is, we first remove glass samples which have a Not-a-Number (NaN) entry for at least one compound fraction. Moreover, we drop all glasses which have NaN entries for **P**.[1]

Next, we make sure that all glass samples are physically valid binary glass compositions. For this, we first discard glasses whose compound fractions do not

---

[1] At the end of this and all the following preprocessing steps, we always drop all compounds which do not appear in any of the glass samples that are present in the dataset at the respective preprocessing stage.

**Table 8.1** Data reduction in each step of the preprocessing pipeline. The first and last row show the number of samples (#S) and number of compounds (#C) which are present in the dirty and cleaned dataset, respectively. The rows in between show the number of samples and compounds which are dropped in the respective preprocessing steps. We remark that in the cleaned datasets, the number of elements appearing with non-vanishing fraction in at least one glass sample is 32 for $T_g$, 23 for $E$, and 24 for $G$ (and therefore coincides with #C)

| P | $T_g$ | | $E$ | | $G$ | |
|---|---|---|---|---|---|---|
| | #S | #C | #S | #C | #S | #C |
| Dirty dataset | 420,973 | 439 | 420,973 | 439 | 420,973 | 439 |
| Drop NaNs | 344,247 | 283 | 396,460 | 329 | 410,589 | 356 |
| Check compound fractions | 17,483 | 1 | 5482 | 0 | 1867 | 0 |
| Filter binary glasses | 50,681 | 83 | 16,651 | 57 | 6386 | 38 |
| Min-max filter | 16 | 0 | 21 | 0 | 15 | 0 |
| Drop extreme values | 10 | 0 | 4 | 0 | 4 | 0 |
| Duplicate filter | 5902 | 0 | 1577 | 0 | 1253 | 2 |
| Drop rare compounds | 229 | 40 | 79 | 30 | 67 | 19 |
| One-class SVM | 22 | 0 | 5 | 0 | 10 | 0 |
| GP regression | 205 | 0 | 62 | 0 | 63 | 0 |
| Cleaned dataset | 2178 | 32 | 632 | 23 | 719 | 24 |

**Table 8.2** Minimum and maximum cut-off values and duplicate thresholds used in the preprocessing pipeline

| P | Min. cut-off | Max. cut-off | Duplicate threshold |
|---|---|---|---|
| $T_g$ (°C) | 50 | $1.8 \times 10^3$ | 5.0 |
| $E$ (GPa) | 5.0 | $2.0 \times 10^2$ | 1.5 |
| $G$ (GPa) | 0.10 | $2.0 \times 10^2$ | 0.75 |

add up to a value in the closed range between 0.9999 and 1.0001. Then, we exclude all samples which do not consist of exactly two compounds.

To ensure physically valid property values, we fix a closed range of values between a minimum and maximum cut-off value for each property **P** (see Table 8.2). We determine these values by investigating the distribution of the glass samples with respect to their **P**-values in the datasets that result from the preprocessing pipeline up to this point. Property values outside of this range are considered non-physical but may be present in the database due to typos or other mistakes. Hence, we drop each glass sample with a **P**-value outside of the respective range.

As the minimum and maximum values are rather crude bounds, in a further step, we remove glasses with extreme **P**-values, which have a high chance of still appearing in the datasets again because of typos or other mistakes. To do so, we compute the 0.05th percentile and the 99.95th percentile of **P**-values among all remaining glass samples and subsequently discard all glasses with **P**-values below the lower percentile or above the upper percentile.

A lot of glass samples appear in both the INTERGLAD and the SciGlass database. Consequently, there may be many duplicates among the remaining data points at this stage of the preprocessing pipeline. We therefore apply a duplicate filter which consists of the following steps:

1. We group all glasses with the same (up to the fifth decimal place) compound fractions.
2. For each such group we do the following:

   2.1 We drop all but the first sample which agree *exactly* in their values of **P**.
   2.2 We compute the midpoint of the range of values of **P** among all remaining samples.
   2.3 If the **P**-value of every sample has a distance to the midpoint smaller than a certain **P**-dependent threshold (see Table 8.2), then, as a representative of the group of duplicates, we select the first sample in the group, assign to it the median of the **P**-values of the samples in the group, and drop all other samples. Otherwise, we discard the whole group of glass samples.

   The values for the duplicate thresholds are determined by using domain knowledge and investigating the average spread of **P**-values in a group of duplicates.

In the next step, we deal with compounds of low representability and iteratively drop compounds which appear in less than one percent of all remaining glass samples. This allows us to reduce the dimension of the compound space and leaves us only with glasses whose compounds are present in sufficiently many samples in order to use them for robust model training.[2]

As a final step, we apply an outlier detection based on a one-class support vector machine (SVM) followed by an outlier detection based on Gaussian process (GP) regression.[3]

The resulting cleaned datasets encompass all problem-specific information which is available for each glass property **P**. Given **P**, we denote the elements and compounds which are present (with non-vanishing value in at least one glass sample) in the corresponding cleaned dataset as **P**-elements and **P**-compounds, respectively. The cleaned datasets are subsequently split into training, validation, and test sets as described in Sect. 8.2.3.

## 8.2.2 Model Setups

We use neural networks for the approximation of the composition-property relationship of binary oxide glasses. The target quantity is given by one of the respective

---

[2] At the end of each iteration, we again drop those samples whose compound fractions do not add up to a value in the closed range between 0.9999 and 1.0001.

[3] We fit a Gaussian process to the data and drop samples with a too large deviation in their **P**-value from the respective mean curve.

properties **P**. The composition of a glass can be represented in various ways. Designing a representation in form of a feature vector that encodes a given glass composition in a way that is suitable as input for a neural network is an essential part of the modeling process and is one of the key differences between the blind (uninformed) and our informed learning approach. A second difference lies in the design of the model architecture where the black-box modeling approach of standard blind feed-forward neural networks can be leveraged in the informed setting by integrating prior scientific knowledge.

In the following Sects. 8.2.2.1 and 8.2.2.2, we describe in detail the choice of the feature vectors and the network architectures for the blind and informed models, respectively, and highlight their differences.

### 8.2.2.1   Blind Models

In the *blind* approach, only the available problem-specific data is used to design a suitable ML model for the composition-property relationship of oxide glasses. This approach is blind or uninformed in the sense that no prior knowledge that exists independently of the learning task is integrated into the model setup.

Feature Vectors

Each glass composition is, by definition, uniquely determined by its compound fractions. It is therefore natural to use the compound fractions, grouped together in a feature vector for a given glass composition, as input for a neural network model to predict one of the glass's properties.

Network Architectures

If no further information is available, the standard architectural design of a neural network is given by a (fully connected) feed-forward neural network (FFNN) [15]. This class of models satisfies the universal approximation theorem, that is, for any continuous function on a compact domain there exists a FFNN which approximates the function within a given arbitrary tolerance [11, 17, 29]. This result justifies the usage of the set of FFNNs as hypothesis space. In the context of glass materials research, this approach is followed for example in [8, 31] to model several different properties of oxide glasses.

A fully connected FFNN is characterized by (i) its input and output dimensions, i.e., the number of units in its input and output layer, respectively, (ii) its depth, i.e., the number of layers (without counting the input layer), and (iii) its width, i.e., the number of units, of each hidden layer. An example architecture of a fully connected FFNN is shown in Fig. 8.2.

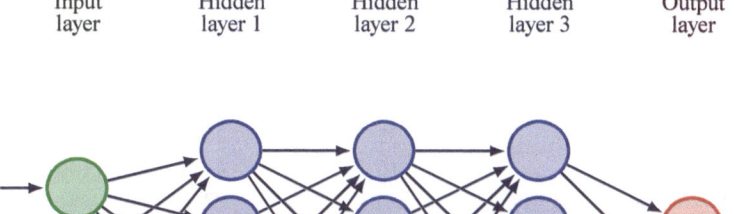

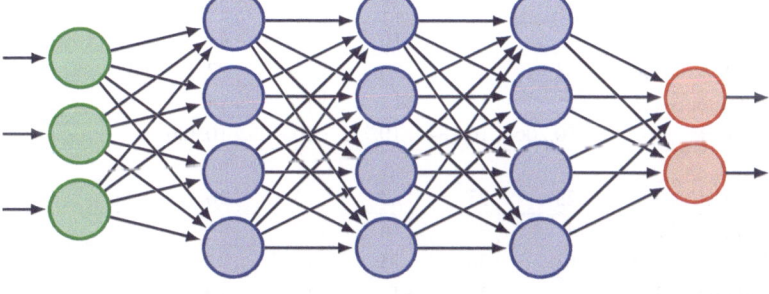

**Fig. 8.2** Schematic architecture of a fully connected FFNN with input dimension 3, output dimension 2, depth 4, and constant width 4. Each layer represents an affine linear function. The additive bias nodes are not shown. In case of the blind models, the input nodes store the compound fractions of a given glass sample and the output node provides the predicted value for **P**. Image adapted from [25]

We use a variety of different FFNNs as benchmark models which we compare our informed model to. To capture the main architectural trends of designing a FFNN and their effects on the prediction quality, we consider five different FFNNs with depths $L = 2, 4, 8, 16, 32$ for each property **P**. The input dimensions are determined by the number of respective **P**-compounds and the output dimension is always one as we predict scalar-valued properties. For each **P**-model, we choose the width to be constant for all hidden layers such that the total number of trainable parameters is roughly the same among all **P**-models including the informed model which we describe in Sect. 8.2.2.2. Each hidden layer is a linear layer with an additive bias term. In accordance to the universal approximation theorem, the output layer is a linear layer with no additive bias term. The exact dimensions of all models are summarized in Table 8.3. In all models, we use the rectified linear unit (ReLU) as activation function.

### 8.2.2.2 Informed Model

Rather than just using the compound fractions as input features for a neural network, we can increase the informational capacity of a glass sample's representation by utilizing characteristic chemical and physical quantities of each element which is present in the given glass sample and provide them as additional inputs to a neural network. Features which are carefully engineered in such an informed manner can lead to an improved prediction quality of the model, given that the model's expressive power is large enough. The latter issue is a question of the model's architecture. If there are too few parameters, the model will underfit the training

**Table 8.3** Model hyperparameters

|            | Blind |       |       |       |       | Informed |            |      |
|------------|-------|-------|-------|-------|-------|----------|------------|------|
|            |       |       |       |       |       | Former   | Non-former | Down |
| $T_g$      |       |       |       |       |       |          |            |      |
| Input dim. | 32    | 32    | 32    | 32    | 32    | 28       | 28         | 32   |
| Length $L$ | 2     | 4     | 8     | 16    | 32    | 4        | 4          | 4    |
| Width $W$  | 304   | 63    | 38    | 26    | 18    | 32       | 32         | 32   |
| Output dim.| 1     | 1     | 1     | 1     | 1     | 14       | 18         | 1    |
| #Parameters| 10,336| 10,206| 10,184| 10,712| 10,872| 10,336   |            |      |
| $E$        |       |       |       |       |       |          |            |      |
| Input dim. | 23    | 23    | 23    | 23    | 23    | 26       | 26         | 23   |
| Length $L$ | 2     | 4     | 8     | 16    | 32    | 4        | 4          | 4    |
| Width $W$  | 385   | 63    | 38    | 25    | 17    | 32       | 32         | 32   |
| Output dim.| 1     | 1     | 1     | 1     | 1     | 10       | 13         | 1    |
| #Parameters| 9625  | 9639  | 9842  | 9725  | 9605  | 9623     |            |      |
| $G$        |       |       |       |       |       |          |            |      |
| Input dim. | 24    | 24    | 24    | 24    | 24    | 26       | 26         | 24   |
| Length $L$ | 2     | 4     | 8     | 16    | 32    | 4        | 4          | 4    |
| Width $W$  | 372   | 63    | 38    | 25    | 17    | 32       | 32         | 32   |
| Output dim.| 1     | 1     | 1     | 1     | 1     | 10       | 14         | 1    |
| #Parameters| 9672  | 9702  | 9880  | 9750  | 9622  | 9688     |            |      |

data no matter how carefully we designed the input features. If there are too many parameters, however, the model might overfit the training data and pick up on spurious patterns and noise in the input features. In general, by building as much prior information as possible into the model's architecture we expect to obtain a more robust inference behavior, especially in the extrapolation regime.

## Feature Vectors

Compared to the uninformed approach, we change our viewpoint and identify a given glass composition not by the fractions of its compounds but by the mole atomic fractions of its elements. For each element, there is additional extensive scientific knowledge about its chemical and physical properties, which exists independently of our learning problem. According to the taxonomy in [37] this knowledge is represented as a weighted graph. Its nodes are given by elements and properties and each element node is connected with a property node via an edge which is weighted by the element's respective property value. We integrate this knowledge into our training data by designing feature vectors in a hybrid fashion. We partially follow the approach used in [7], where, for each element, the authors extract physical and chemical properties, i.a., from the Python library mendeleev [24]. They design and select feature vectors for a neural network model in a way that allows them to complement information from the space of chemical compositions with information from the space of chemical and physical properties.

In our case, we first extract for each glass property **P** and each **P**-element a list of characteristic chemical and physical properties from mendeleev. We drop properties with non-numeric values and only keep those which are available for all **P**-elements. We also drop properties which we consider to be unrelated to the elements' influence on the glass material properties, namely, the elements' *abundances in the earth crust* and the elements' *dipole polarizability uncertainties*. We refer to [24] and references therein for a detailed explanation of all the available properties in the mendeleev library.

Among the remaining properties, we drop those which are highly correlated. More specifically, we compute the standard pairwise Pearson correlation coefficients and iteratively, for each pair of properties whose coefficient is larger than 0.95, we only keep one of the two properties. The resulting list of properties is given in Table 8.4.

For each glass property **P** and each **P**-element, we group the resulting element properties into vectors, which we call **P**-element-vectors. These element-vectors represent all prior knowledge about the elements which exists independently of the learning problem.

In the next step, we complement this information with our given problem-specific data. For each glass property **P** and a given glass sample, we consider the collection of **P**-element-vectors corresponding to the sample's elements and extend each of them by one more component which lists the mole atomic fraction of the respective element in the glass sample. Eventually, we obtain for each glass sample a collection of feature vectors $(v_1, \ldots, v_M) \in (\mathbb{R}^d)^M$, where we arrange the tuple in lexicographical order of the elements' symbols. Here, $M = M(\mathbf{P})$ is the number of **P**-elements and $d = d(\mathbf{P})$ is the number of properties resulting from the property extraction process described above (including the entry with the mole atomic fraction). Each glass sample can thus be represented as a point in a subspace $\Omega = \Omega(\mathbf{P}) \subset (\mathbb{R}^d)^M$.

Network Architecture

For each glass property **P**, we want to design a neural network which approximates the functional relationship $f : \Omega \to \mathbb{R}, \mathbf{V} \mapsto \mathbf{P}(\mathbf{V})$, where $\mathbf{P}(\mathbf{V})$ is the value of property **P** for the glass sample with representation $\mathbf{V} = (v_1, \ldots, v_M)$. We design the architecture of the network by two leading principles in the spirit of informed learning.

First, we observe that the order in which the feature vectors are passed to the function $f$ actually does not matter. That is, the function $f$ is *permutation invariant* with respect to the order of the input vectors. More specifically, $f$ is a function on sets of the form $\{v_1, \ldots, v_M\}$. In terms of the taxonomy in [37] we use this scientific knowledge, which is represented as a spatial invariance, and directly integrate it into the architecture of the network which we use to approximate $f$. It is shown in [40]

**Table 8.4** Chemical and physical properties extracted from the `mendeleev` library which are used for the correlation study described in Sect. 8.2.2.2. Properties that are marked with ✗ are dropped before the correlation study as they are not available for all respective **P**-elements. Properties that are marked with ✓ are not highly correlated among each other and are used as final features. All unmarked properties are dropped due to too high correlation with other properties. We refer to [24] and the references therein for detailed explanations of the properties

| Properties from `mendeleev` | $T_g$ | $E$ | $G$ |
|---|---|---|---|
| Atomic number | ✓ | ✓ | ✓ |
| Atomic radius | ✓ | ✓ | ✓ |
| Atomic radius by Rahm et al. | ✓ | ✓ | ✓ |
| Atomic volume | ✓ | ✓ | ✓ |
| Atomic weight | | | |
| Boiling temperature | ✓ | ✓ | ✓ |
| $C_6$ dispersion coefficient by Gould and Bučko | ✓ | ✓ | ✓ |
| Covalent radius by Cordero et al. | ✓ | | |
| Single bond covalent radius by Pyykko et al. | | | |
| Double bond covalent radius by Pyykko et al. | | | |
| Density | ✓ | ✓ | ✓ |
| Dipole polarizability | | | |
| Electron affinity | ✓ | ✓ | ✓ |
| Electron affinity in the Allen scale | ✗ | ✓ | ✓ |
| Electron affinity in the Ghosh scale | ✓ | ✓ | ✓ |
| Electron affinity in the Pauling scale | ✓ | | |
| Glawe's number | ✓ | ✓ | ✓ |
| Group in periodic table | ✓ | ✓ | ✓ |
| Heat of formation | ✓ | ✓ | ✓ |
| First ionization energy | ✓ | ✓ | ✓ |
| Lattice constant | ✓ | ✓ | ✓ |
| Maximum coordination number | ✓ | ✓ | ✓ |
| Maximum oxidation state | ✓ | ✓ | ✓ |
| Melting temperature | ✓ | | |
| Mendeleev's number | | | |
| Minimum coordination number | ✓ | ✓ | ✓ |
| Minimum oxidation state | ✓ | ✓ | ✓ |
| Period in periodic table | ✓ | ✓ | ✓ |
| Pettifor scale | | | |
| Index to chemical series | ✓ | ✓ | ✓ |
| Number of valence electrons | ✓ | ✓ | ✓ |
| Van der Waals radius | ✓ | ✓ | ✓ |
| Van der Waals radius according to Alvarez | ✓ | ✓ | ✓ |
| Van der Waals radius according to Batsanov | | | |
| Van der Waals radius from the MM3 FF | | | |
| Van der Waals radius from the UFF | ✓ | ✓ | ✓ |
| Number $d$ of all features (including mole atomic fractions) | 28 | 26 | 26 |

that such a function $f$ on sets can be written as

$$f(\{v_1, \ldots, v_M\}) = \psi \left( \sum_{i=1}^{M} \phi(v_i) \right), \qquad (8.1)$$

where $\phi : \mathbb{R}^d \to \mathbb{R}^N$ denotes an inner embedding function with $N \in \mathbb{N}$ being an appropriately chosen embedding dimension, and $\psi : \mathbb{R}^N \to \mathbb{R}$ denotes an outer (downstream) function. Here, $\phi$ and $\psi$ can be approximated by neural networks. Using the universal approximation theorem of neural networks, the right-hand-side of (8.1) yields architectures of neural networks which, in principal, can approximate $f$ arbitrarily well.

In our specific use case, we can refine the network's architecture even further by integrating prior chemical knowledge. Glass oxides can be categorized in three groups [5]. *Glass formers* are oxides that can readily form a glassy material and build the backbone of a glass's network structure. *Glass modifiers* are oxides that cannot form a glassy material by themselves but influence its material properties when mixed with a glass former. *Glass intermediates* are oxides which can act both as a glass former as well as a glass modifier depending on the respective cation's oxidation number. For our purposes, we only differentiate between oxides which are glass formers and oxides which are not glass formers. We refer to the latter group as *glass non-formers*. We use the classification proposed in [5] to determine for every element whether its oxide is a glass former or a glass non-former. The classification is shown in Table 8.5.

The scientific knowledge whether an element's oxide has glass-forming or glass-non-forming ability is naturally represented as a simple knowledge graph, where each element is represented by a node. There is also a glass former node and a glass

**Table 8.5** Classification of elements based on the glass-forming and glass-non-forming properties of their oxides. The last row shows all elements whose oxides are glass formers according to the classification in [5]. The second column lists, for given **P**, the respective **P**-elements whose oxides are glass formers and which are a subset of the elements in the last row. The fourth column lists all respective **P**-elements whose oxides are glass non-formers. We classify oxygen as a glass non-former

| **P** | Formers | #Formers | Non-formers | #Non-formers |
|---|---|---|---|---|
| $T_g$ | As, B, Bi, Ge, Mo, P, Pb, Sb, Si, Sn, Te, Tl, V, W | 14 | Ag, Al, Ba, Ca, Cs, Cu, Fe, Ga, K, La, Li, Mg, Na, O, Rb, Sr, Ti, Zn | 18 |
| $E$ | B, Bi, Ge, Mo, Nb, P, Pb, Si, Te, V | 10 | Al, Ba, Ca, Co, Cs, K, Li, Mg, Na, O, Sr, Ti, Zn | 13 |
| $G$ | B, Bi, Ge, Mo, Nb, P, Pb, Si, Te, V | 10 | Al, Ba, Ca, Co, Cs, K, Li, Mg, Na, O, Rb, Sr, Ti, Zn | 14 |
| All | As, B, Bi, Ge, Mo, Nb, P, Pb, Sb, Se, Si, Sn, Ta, Te, Tl, V, W | 17 | | |

non-former node. Each element node is connected via an edge with the glass former node or the glass non-former node depending on whether the element's oxide is a glass former or a glass non-former. Due to the largely different influence on a glass's properties, we integrate this prior knowledge additionally into our hypothesis set by using two functions to treat glass formers and non-formers separately. The *glass former network* receives as input only feature vectors of elements whose oxides are glass formers. The *glass non-former network* receives as input all other elements whose oxides, by definition, are glass non-formers. The outputs of the glass former network are added together, as are the outputs of the glass non-former network. The results are concatenated and then used as input for the *downstream network* which yields the final prediction for the respective property **P**.

More specifically, let $\Omega = \Omega_f \cup \Omega_{nf}$ be the decomposition of $\Omega$ into the space $\Omega_f$ of representations of glass formers and the space $\Omega_{nf}$ of representations of glass non-formers. We then replace the inner function $\phi$ in (8.1) by two separate functions, $\phi_f : \Omega_f \to \mathbb{R}^{N_f}$, $N_f \in \mathbb{N}$, for the glass former network and $\phi_{nf} : \Omega_{nf} \to \mathbb{R}^{N_{nf}}$, $N_{nf} \in \mathbb{N}$, for the glass non-former network. Permutation invariance then holds only within the feature vectors $v_1, \ldots, v_{M_f}$ corresponding to the $M_f$ glass formers and within the $M_{nf} := M - M_f$ feature vectors $v_{M_f+1}, \ldots, v_M$ corresponding to the glass non-formers. The resulting representation of $f$ then has the following form,

$$f(\{v_1, \ldots, v_M\}) = \psi \left( \sum_{i=1}^{M_f} \phi_f(v_i), \sum_{i=M_f+1}^{M} \phi_{nf}(v_i) \right), \qquad (8.2)$$

where, under slight abuse of notation, we used the same notation $\psi$ for the downstream function as in (8.1). As glass former network, glass non-former network, and downstream network we use three separate ReLU-FFNNs to approximate the functions $\phi_f$, $\phi_{nf}$, and $\psi$ in (8.2), respectively. Their widths and depths are listed in Table 8.3. The overall network architecture of our informed model is illustrated in Fig. 8.3.

The embedding dimensions $N_f$ and $N_{nf}$ are hyperparameters of the glass former and non-former network, respectively. It is shown in [38] that in the scalar case, $d = 1$, the choice $N = M$ in (8.1) is a sufficient and necessary condition in order to approximate the function $f$ arbitrarily well by a neural network whose architecture is given by the right-hand side in (8.1). In the vector-valued case, $d > 1$, to the best of our knowledge, no non-trivial necessary condition on the embedding dimension is known so far. In [16], the authors prove a sufficient condition in form of an upper bound on the embedding dimension, which, however, is very pessimistic. Based on the results in the one-dimensional case, we choose $N_f = M_f$ and $N_{nf} = M_{nf}$ in (8.2).

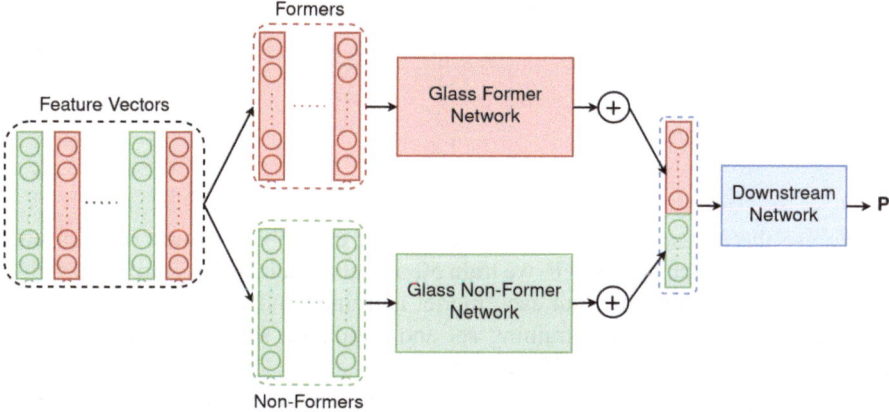

**Fig. 8.3** Architecture of the informed model. The feature vectors of the elements are split according to whether the elements' oxides are glass formers or glass non-formers and input to separate neural networks. The results of the latter are first summed individually and then concatenated to a vector which is used as input for the final downstream neural network that predicts a value for property **P**

### 8.2.3  Model Training and Evaluation

Recall that we consider three different glass material properties: glass transition temperature $T_g$, Young's modulus $E$ at room temperature, and shear modulus $G$. For each of these properties **P**, we split the cleaned datasets from Sect. 8.2.1 further into datasets for training, validation, and testing. We then apply the blind models and the informed model discussed in Sects. 8.2.2.1 and 8.2.2.2.

For data management and visualization, we use the Python libraries `pandas` [35], `Scikit-learn` [27], and `Matplotlib` [18], respectively. The neural network models are built using the `PyTorch-Lightning` [13] module.

We describe the data splitting in a bit more detail. For each property **P**, we apply the following steps. First, we apply the preprocessing pipeline described in Sect. 8.2.1. Then, we apply the feature design and selection processes for the blind and the informed models as described in Sects. 8.2.2.1 and 8.2.2.2, respectively. Next, we split up the resulting cleaned dataset into those glass samples which contain sodium and those which do not contain sodium. From the samples which contain sodium we extract only those binary oxides which consist of $B_2O_3$ as glass former and $Na_2O$ as glass non-former. The resulting dataset is our **P**-test set. The other glass samples, which do not contain sodium, are randomly split for each model into a **P**-training set and a **P**-validation set using a 80%/20% ratio. The dimensions of the resulting datasets are shown in Table 8.6. We emphasize that sodium as an element is totally absent in the training and validation sets and only present in the test sets. Examining the performance of the trained models on the test sets therefore allows us to properly evaluate their extrapolation power.

**Table 8.6** Dimensions of the training, validation, and test sets

| P | #Training samples | #Validation samples | #Test samples |
|---|---|---|---|
| $T_g$ | 1385 | 347 | 125 |
| E | 415 | 104 | 42 |
| G | 477 | 120 | 73 |

We use the *bagging* method from the field of ensemble learning [6]. For each model setup and each property **P**, we train 50 models. Their architectures and weight initializations are the same, but each model is trained and validated on a different random 80%/20%-split into training set and validation set. We therefore end up with a model ensemble of 50 different models.

For training, we use the ADAM optimizer with default settings [19] and weight decay of $10^{-5}$ and train for a maximum of 1000 epochs with a batch size of 8. We start with a learning rate of 0.001 and multiply it by a factor of 0.5 if the model's performance on the validation set in terms of the mean squared error (MSE) does not improve over the course of 50 epochs. Moreover, to avoid overfitting, we use the early stopping criterion and stop training if the model's MSE on the validation set does not improve over the course of 100 epochs.

After training, we apply a post-processing step. For each property **P**, we discard those models whose predictions for **P** on the whole test set can be considered to be constant. More specifically, we first compute for each sample in the test set the mean value of the models' predictions for **P**. Then, we drop those models where the deviation of the predicted property values for all samples in the test set from the respective mean value is less than or equal to the **P**-duplicate threshold from Table 8.2. This is in alignment with informed learning since we know a priori that for each property **P**, not all $Na_2O$-$B_2O_3$ glass samples have the same **P**-value. In terms of the taxonomy in [37], we thus use this expert knowledge, which is represented as algebraic equations, i.e., being of non-constant value, and integrate it into our final hypothesis.

Among the remaining models we compute the mean and the 95%-confidence interval of the predictions. This yields the final prediction of the model ensemble and quantifies its uncertainty. We compare the ensembles' performances quantitatively in terms of their root mean squared errors (RMSE), mean absolute errors (MAE), and maximum errors (MAX) on the respective **P**-test sets, which are summarized in Table 8.7. All blind 32-layer networks yield constant predictions for all three properties and are therefore discarded as non-physical in the post-processing step. Nevertheless, we still record their respective ensemble errors in Table 8.7 to get a more conclusive picture. However, when talking about the best and worst error values, we *only consider the ensembles of blind models with depths $L = 2, 4, 8, 16$* and neglect the values of the models with 32 layers.

To also get a qualitative picture of the ensembles' extrapolation performances, we plot the composition-property curves of the ensembles' averaged predictions on the **P**-test sets in Figs. 8.4, 8.5, and 8.6. Recall that the test sets consist of the

**Table 8.7** Results and errors of the model ensembles' averaged predictions on the test sets. Among the blind ensembles with depths $L = 2, 4, 8, 16$, bold blue numbers denote the lowest error values, bold red numbers the highest ones. Bold black numbers denote the lowest error values among all model ensembles. The last column shows the relative improvement in the error when comparing the error value of the informed ensemble to the lowest error value (blue) among the blind ensembles. The blind models with 32 layers are not considered in the error analysis as these models only yield constant predictions and are therefore discarded as non-physical

| | Blind | | | | | | Informed | Rel. improv. |
|---|---|---|---|---|---|---|---|---|
| | Depth | 2 | 4 | 8 | 16 | 32 | | |
| $T_g$ (°C) | | | | | | | | |
| RMSE | | **86.2** | 111 | **112** | 109 | 86.5 | **62.5** | 27% |
| MAE | | **63.0** | 77.2 | 81.7 | **84.4** | 73.6 | **44.4** | 30% |
| MAX | | **265** | **341** | 338 | 311 | 205 | **186** | 30% |
| #Non-const. predictions | | 50 | 50 | 50 | 48 | 0 | 24 | |
| $E$ (GPa) | | | | | | | | |
| RMSE | | **9.88** | 10.7 | **12.0** | 11.6 | 15.6 | **4.67** | 53% |
| MAE | | **8.58** | 9.44 | **10.7** | 10.3 | 12.5 | **3.50** | 59% |
| MAX | | **16.8** | 18.2 | 19.5 | **19.8** | 33.7 | **12.5** | 26% |
| #Non-const. predictions | | 50 | 50 | 50 | 43 | 0 | 48 | |
| $G$ (GPa) | | | | | | | | |
| RMSE | | **2.92** | 3.65 | **4.02** | 3.81 | 5.94 | **1.39** | 52% |
| MAE | | **2.43** | 3.09 | **3.39** | 3.13 | 5.08 | **1.12** | 54% |
| MAX | | **6.10** | 7.24 | 7.89 | **8.37** | 12.2 | **2.95** | 52% |
| #Non-const. predictions | | 50 | 50 | 50 | 39 | 0 | 48 | |

$Na_2O$-$B_2O_3$ glass samples together with their respective **P**-values. We also plot the predictions of the best and worst performing model of each **P**-ensemble in terms of the RMSE on the test set. Moreover, we plot the property values of all available alkali borate glasses, that is, binary glasses which consist of $B_2O_3$ as glass former and $Na_2O$, $Li_2O$, and $Rb_2O$ as glass non-former, respectively. It is known that these glass compositions have similar material properties [14].

Concerning Figs. 8.4, 8.5, and 8.6, a few remarks are in order. First, there are no $Rb_2O$-$B_2O_3$ glass samples available for $E$. Next, since we only consider binary oxide glasses, knowing the compound fraction of $B_2O_3$ completely determines the compound fraction of the respective alkali oxide as glass non-former as well. Finally, since the blind 32-layer networks are discarded, their predictions are not shown.

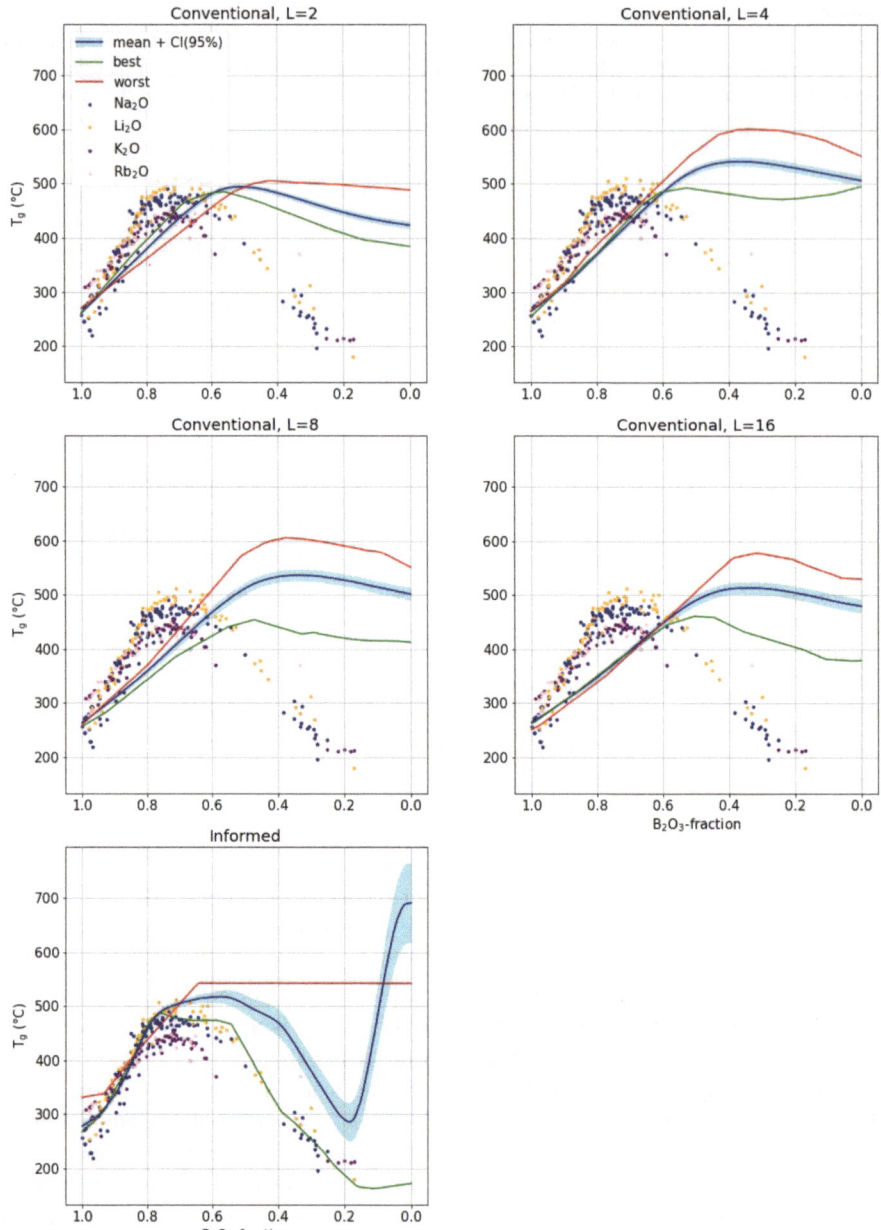

**Fig. 8.4** $T_g$-values of binary alkali borate glasses. Scattered points represent $T_g$-values given in the cleaned $T_g$-dataset. Solid lines show the predictions for the $T_g$-value of Na$_2$O-B$_2$O$_3$ glass samples of the blind and informed models, respectively. The model ensembles' mean curves are shown as blue solid lines with the shaded blue area depicting the 95% confidence band. The predictions of the best and worst performing models in the ensembles are shown as green and red solid lines, respectively

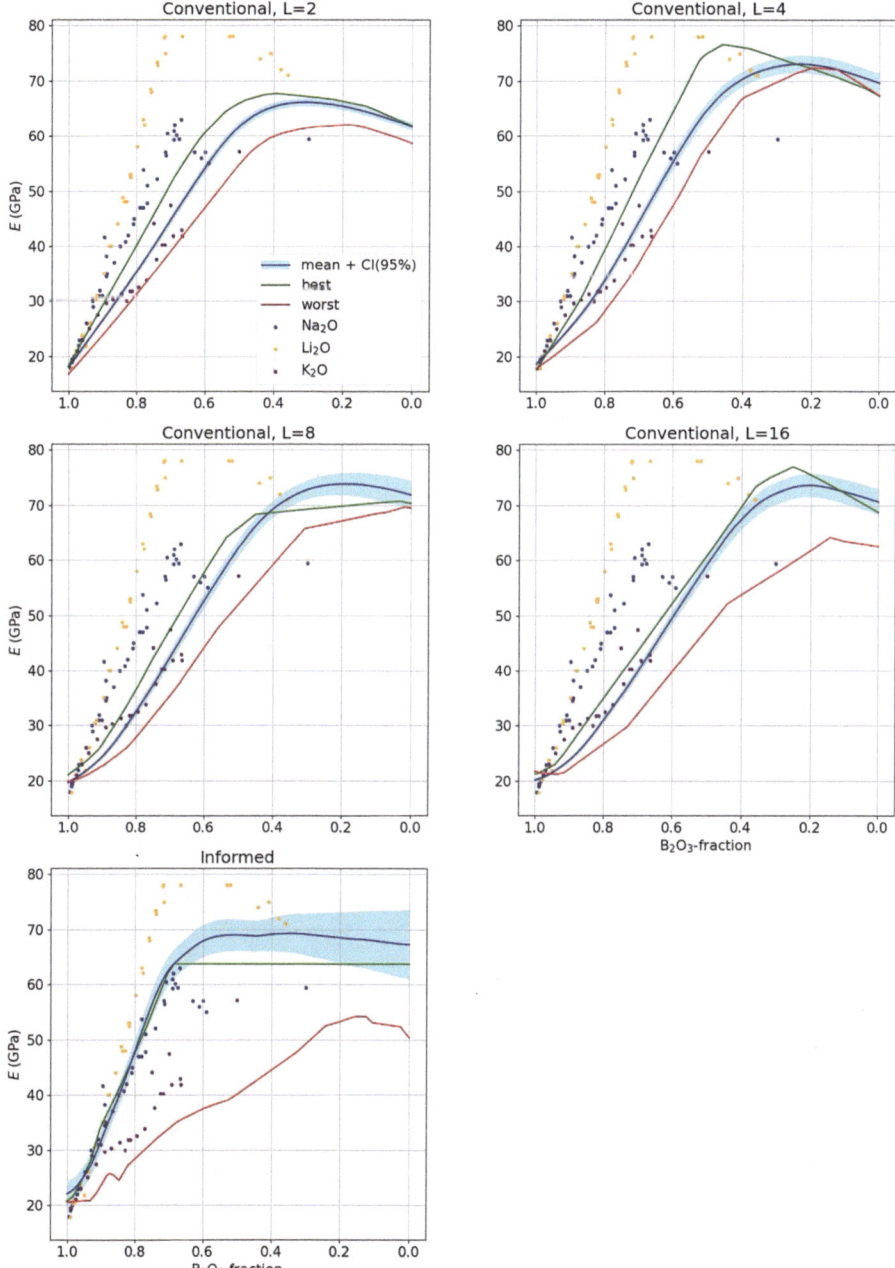

**Fig. 8.5** $E$-values of binary alkali borate glasses. Scattered points represent $E$-values given in the cleaned $E$-dataset. Solid lines show the predictions for the $E$-value of $Na_2O$-$B_2O_3$ glass samples of the blind and informed models, respectively. The model ensembles' mean curves are shown as blue solid lines with the shaded blue area depicting the 95% confidence band. The predictions of the best and worst performing models in the ensembles are shown as green and red solid lines, respectively

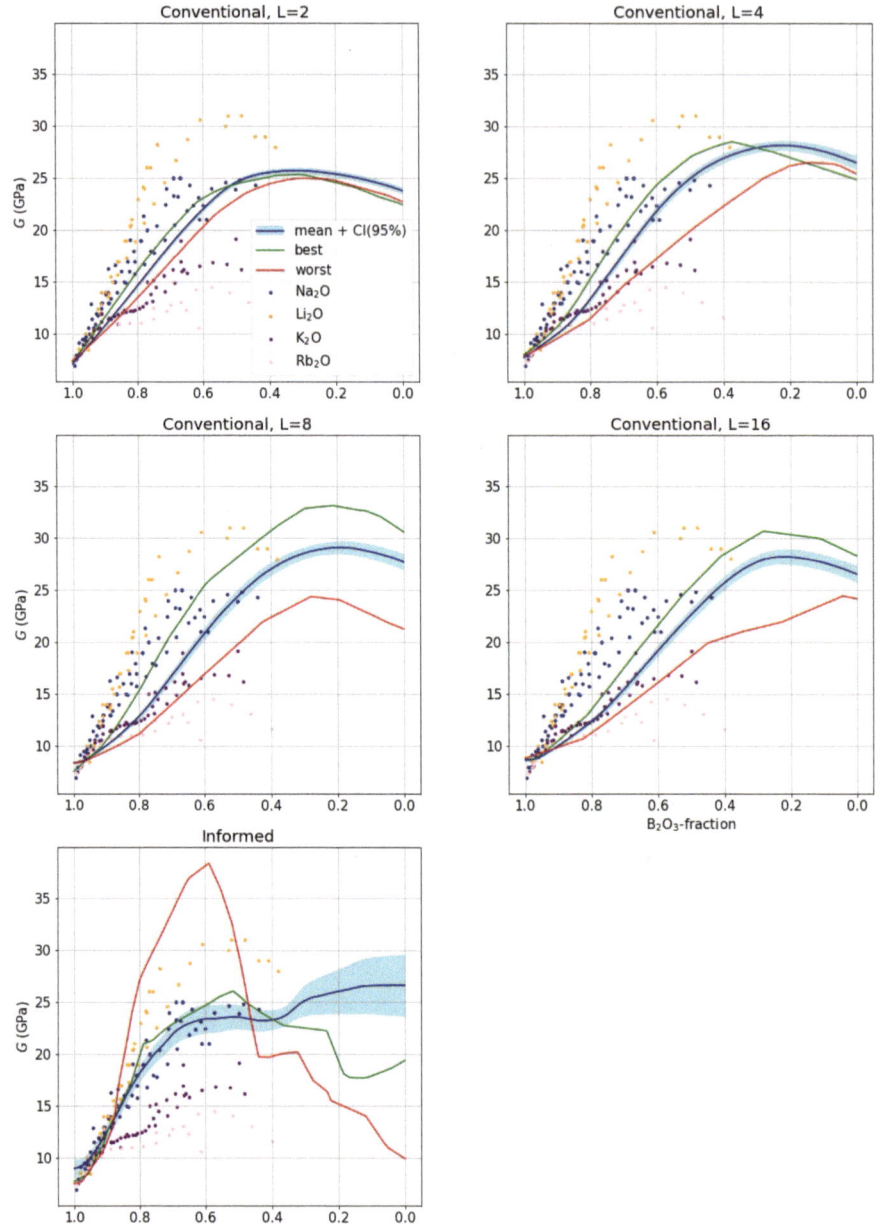

**Fig. 8.6** $G$-values of binary alkali borate glasses. Scattered points represent $G$-values given in the cleaned $G$-dataset. Solid lines show the predictions for the $G$-value of $Na_2O$-$B_2O_3$ glass samples of the blind and informed models, respectively. The model ensembles' mean curves are shown as blue solid lines with the shaded blue area depicting the 95% confidence band. The predictions of the best and worst performing models in the ensembles are shown as green and red solid lines, respectively

## 8.3 Results and Discussion

We first compare the models' performances quantitatively in terms of their average errors in Table 8.7. Among the blind models, we note that the ensembles of shallow two-layer networks perform best for all three properties in terms of RMSE, MAE, and MAX. Considering the worst performing ensembles, we note that the deeper networks with 8 and 16 layers perform worst, on average, in terms of almost all three error metrics for all three properties. Only for $T_g$ in the case of MAX, the ensemble of models with only four layers performs worst. We note that in this case, the ensemble of models with 32 layers actually performs best in terms of MAX. In all other cases, however, the 32-layer network ensembles perform worst for all three properties when compared to the other blind models. We conclude that, in general, increasing network complexity in terms of increasing depth tends to lead to worse performing models.

The number of models with non-constant predictions clearly decays with increasing network depth for all three properties. Whereas the networks with 2, 4, and 8 layers lead to no constant predictions, the 16-layer networks lead to some constant predictions. There is a steep decay when increasing the number of layers from 16 to 32, where all models for all properties lead to only constant predictions. A possible explanation for this phenomenon could be that the models' loss landscapes become more and more rugged with increasing network depth yielding constant predictions to be local minima which are hard to escape during the optimization routine. This matches the observation from above that increasing network depth generally tends to lead to worse performing models.

Invoking now the errors of the informed models, we see that they perform best, on average, in terms of all three error metrics for all three properties. They lead to a relative improvement in the errors between 26% up to 59%. For $E$ and $G$, only two models yield constant predictions, whereas for $T_g$ more than half of all models do. Again, this could indicate that the loss landscape of the informed networks is much more rugged for $T_g$ than for the other two properties.

To get a more conclusive qualitative picture of the extrapolation behavior of all models, we take a closer look at Figs. 8.4, 8.5, and 8.6. We observe that the blind networks in terms of the ensembles' means as well as the best and worst performing models are not able to qualitatively capture the trend of the $Na_2O$-$B_2O_3$ curves correctly and instead generally deviate from the test points to a large extent. However, the ensembles' predictions for all blind networks seem to be quite close to each other for all three properties. This is reflected by the small width of the confidence band around the mean curves as well as the similar shape of the mean curves and the curves of the best and worst performing models. This indicates that the blind models are robust with respect to training.

The mean curves of the informed model ensembles qualitatively capture the trend of the $Na_2O$-$B_2O_3$ curves to a more acceptable degree. This is most noticeable in the cases of $T_g$ and $G$ where — in contrast to the blind networks — the mean curves capture the nonlinearity of the respective $Na_2O$-$B_2O_3$ curve. For $E$, the informed model ensemble yields more accurate trajectories than the blind ensembles in the linear regime with $B_2O_3$-fractions between 0.7 and 1.0, but the kink in the $Na_2O$-$B_2O_3$ curve at a $B_2O_3$-fraction of around 0.7 is not captured. This could, in parts, be due to the small training and validation sets which are available for $E$ (see Table 8.6) and, in particular, to the lack of $Rb_2O$-$B_2O_3$ glass samples which the models could base their predictions on. We explain the latter point in more detail below. Nevertheless, whereas the mean curve for $E$ shows at least a physically reasonable trajectory in the region of low $B_2O_3$-fractions, where there are no data points of alkali borate glasses available, the mean curves for $T_g$ and $G$ show a non-physical incline for glasses of $B_2O_3$-fractions of less than 0.2 and 0.4, respectively. As a further observation, we note that, in general, for all three properties, the uncertainty of the model ensembles' predictions in terms of the width of the confidence band around the mean curve is much higher than in the blind settings, especially in the regions of low $B_2O_3$-fractions where there are only few or no alkali borate glass samples available. This indicates less robustness with respect to training the models and is also most noticeably reflected by the large deviation of the worst performing model's curve from the mean curve for all three properties.

As most probable explanation, we suspect these observations to be caused by the choice of our training and test sets. As already indicated in Sect. 8.2.3, we note that the curves of all alkali borate glasses show a similar trajectory for all three properties since these glasses have similar material properties. We also note that only $Na_2O$-$B_2O_3$ glasses are not present in the training and validation sets. In regions where the other alkali borate glasses are available in the training and validation sets, the models are thus, in principal, able to learn the properties of the $Na_2O$-$B_2O_3$ glasses based on the other alkali borate glass samples. In regions where there are many of these samples available and where their property curves are very close to the $Na_2O$-$B_2O_3$ curve, the informed models' predictions thus tend to be quite accurate. In regions where only few or no alkali borate glasses are available in the training and validation sets, the models are prone to base their predictions on other spurious or noisy features. This leads to non-physical predictions with a high uncertainty. This phenomenon is amplified by a large feature set and is therefore pronounced to a much higher degree in the informed setting than in the blind one.

In summary, the informed model shows, on average in the ensemble setting, clear superior performance to all considered blind (uninformed) models in extrapolating the property curves of $Na_2O$-$B_2O_3$ binary glasses for all three properties $T_g$, $E$, and $G$. This is in terms of quantitative error measurements on the test sets as well as in the qualitative approximation of the property curves.

Finally, we emphasize the importance of the ensemble setting. Whereas single models might yield bad predictions, averaging multiple trained models, as we observe in our specific use case, often yields good approximations of the target quantities [32].

## 8.4    Conclusion and Outlook

In this chapter, we presented an informed neural network approach for the prediction of three material properties of binary oxide glasses, that is, glass transition temperature $T_g$, Young's modulus $E$ (at room temperature), and shear modulus $G$. We compared this approach to five different blind (uninformed) models for all three properties and demonstrated its superior average extrapolation power when applied in an ensemble setting to alkali borate glass samples which contain sodium as previously unseen element.

In terms of the taxonomy of Informed Machine Learning introduced in [37], we integrated prior knowledge into our learning pipeline at four major points. We integrated scientific knowledge, represented as a weighted graph, knowledge graph, and spatial invariance in the training data and in the hypothesis set, respectively. Moreover, we integrated expert knowledge, represented as algebraic equations, into the final hypothesis.

Our informed neural network model could be improved in various ways. First, the list of chemical and physical element features could be extended. Second, instead of classifying glass oxides into formers and non-formers, we could follow the refined classification into formers, modifiers, and intermediates and treat these three classes by three separate neural networks. Third, in this chapter, we did not tune any of the models' hyperparameters. A thorough hyperparameter study probably leads to improved model performance. Finally, by relying on further expert knowledge, we could potentially filter out even more predicted property curves in the post-processing step than just constant predictions. This might improve the final predictions even further.

Our results show that our informed neural network model is capable of meaningfully extrapolating various properties of binary glass samples with previously unseen compounds. As a next step, we plan to scale up our approach in order to make it applicable to oxide glass samples with three or more compounds. We also plan to make it more universal such that it can accurately predict more material properties (Table 8.4).

**Acknowledgments** This work was supported in part by the BMBF-project 05M2AAA MaGriDo (Mathematics for Machine Learning Methods for Graph-Based Data with Integrated Domain Knowledge), by the Fraunhofer Cluster of Excellence "Cognitive Internet Technologies", and by the Deutsche Forschungsgemeinschaft (DFG, German Research Foundation) via project 390685813 - GZ 2047/1 - Hausdorff Center for Mathematics (HCM).

## References

1. Alcobaca, E., Mastelini, S.M., Botari, T., Pimentel, B.A., Cassar, D.R., de Leon Ferreira, A.C.P., Zanotto, E.D., et al.: Explainable machine learning algorithms for predicting glass transition temperatures. Acta Materialia **188**, 92–100 (2020)
2. Bishnoi, S., Badge, S., Krishnan, N.A., et al.: Predicting oxide glass properties with low complexity neural network and physical and chemical descriptors. Journal of Non-Crystalline Solids **616**, 122488 (2023)

3. Bishnoi, S., Singh, S., Ravinder, R., Bauchy, M., Gosvami, N.N., Kodamana, H., Krishnan, N.A.: Predicting Young's modulus of oxide glasses with sparse datasets using machine learning. Journal of Non-Crystalline Solids **524**, 119643 (2019)

4. Bødker, M.L., Bauchy, M., Du, T., Mauro, J.C., Smedskjaer, M.M.: Predicting glass structure by physics-informed machine learning. npj Computational Materials **8**(1), 192 (2022)

5. Boubata, N., Roula, A., Moussaoui, I.: Thermodynamic and relative approach to compute glass-forming ability of oxides. Bulletin of Materials Science **36**, 457–460 (2013)

6. Breiman, L.: Bagging predictors. Machine Learning **24**, 123–140 (1996)

7. Cassar, D.R.: ViscNet: Neural network for predicting the fragility index and the temperature-dependency of viscosity. Acta Materialia **206**, 116602 (2021)

8. Cassar, D.R., de Carvalho, A.C., Zanotto, E.D.: Predicting glass transition temperatures using neural networks. Acta Materialia **159**, 249–256 (2018)

9. Cassar, D.R., Mastelini, S.M., Botari, T., Alcobaca, E., de Carvalho, A.C., Zanotto, E.D.: Predicting and interpreting oxide glass properties by machine learning using large datasets. Ceramics International **47**(17), 23958–23972 (2021)

10. Cassar, D.R., Santos, G.G., Zanotto, E.D.: Designing optical glasses by machine learning coupled with a genetic algorithm. Ceramics International **47**(8), 10555–10564 (2021)

11. Cybenko, G.: Approximation by superpositions of a sigmoidal function. Mathematics of Control, Signals and Systems **2**(4), 303–314 (1989)

12. EPAM Systems: epam/SciGlass. https://github.com/epam/SciGlass (2019). License: MIT License

13. Falcon, W., The PyTorch Lightning team: PyTorch Lightning (Version 1.4). https://github.com/Lightning-AI/lightning (2019). License: Apache-2.0

14. Feller, S.: Borate glasses. In: Springer Handbook of Glass, pp. 505–524. Springer (2019)

15. Goodfellow, I., Bengio, Y., Courville, A.: Deep learning. MIT Press (2016). http://www.deeplearningbook.org

16. Han, J., Li, Y., Lin, L., Lu, J., Zhang, J., Zhang, L.: Universal approximation of symmetric and anti-symmetric functions. Communications in Mathematical Sciences **20**(5), 1397–1408 (2022)

17. Hornik, K.: Approximation capabilities of multilayer feedforward networks. Neural Networks **4**(2), 251–257 (1991)

18. Hunter, J.D.: Matplotlib: A 2D graphics environment. Computing in Science & Engineering **9**(3), 90–95 (2007)

19. Kingma, D.P., Ba, J.: Adam: A method for stochastic optimization. In: 3rd International Conference on Learning Representations, ICLR 2015, San Diego, CA, USA, May 7–9, 2015, Conference Track Proceedings (2015)

20. Krishnan, N.A., Mangalathu, S., Smedskjaer, M.M., Tandia, A., Burton, H., Bauchy, M.: Predicting the dissolution kinetics of silicate glasses using machine learning. Journal of Non-Crystalline Solids **487**, 37–45 (2018)

21. Liu, H., Fu, Z., Yang, K., Xu, X., Bauchy, M.: Machine learning for glass science and engineering: A review. Journal of Non-Crystalline Solids **557**, 119419 (2021)

22. Liu, H., Zhang, T., Krishnan, N.A., Smedskjaer, M.M., Ryan, J.V., Gin, S., Bauchy, M.: Predicting the dissolution kinetics of silicate glasses by topology-informed machine learning. npj Materials Degradation **3**(1), 32 (2019)

23. Mauro, J.C.: Decoding the glass genome. Current Opinion in Solid State and Materials Science **22**(2), 58–64 (2018)

24. Mentel, Ł.: mendeleev – A Python resource for properties of chemical elements, ions and isotopes (Version 0.12.1). https://github.com/lmmentel/mendeleev (2014–). License: MIT License

25. Neutelings, I.: Neural networks. https://tikz.net/neural_networks. License: Attribution-ShareAlike 4.0 International (CC BY-SA 4.0), Accessed: 2023-04-05

26. New Glass Forum: International Glass Database System INTERGLAD Ver. 8. https://www.newglass.jp/interglad_n/gaiyo/info_e.html

27. Pedregosa, F., Varoquaux, G., Gramfort, A., Michel, V., Thirion, B., Grisel, O., Blondel, M., Prettenhofer, P., Weiss, R., Dubourg, V., Vanderplas, J., Passos, A., Cournapeau, D., Brucher, M., Perrot, M., Duchesnay, E.: Scikit-learn: Machine learning in Python. Journal of Machine Learning Research **12**, 2825–2830 (2011)

28. Pilania, G.: Machine learning in materials science: From explainable predictions to autonomous design. Computational Materials Science **193**, 110360 (2021)

29. Pinkus, A.: Approximation theory of the MLP model in neural networks. Acta Numerica **8**, 143–195 (1999)

30. Ravinder, Venugopal, V., Bishnoi, S., Singh, S., Zaki, M., Grover, H.S., Bauchy, M., Agarwal, M., Krishnan, N.A.: Artificial intelligence and machine learning in glass science and technology: 21 challenges for the 21st century. International Journal of Applied Glass Science **12**(3), 277–292 (2021)

31. Ravinder, R., Sridhara, K.H., Bishnoi, S., Grover, H.S., Bauchy, M., Jayadeva, Kodamana, H., Krishnan, N.A.: Deep learning aided rational design of oxide glasses. Materials Horizons **7**(7), 1819–1827 (2020)

32. Rokach, L.: Ensemble-based classifiers. Artificial Intelligence Review **33**, 1–39 (2010)

33. Shih, Y.T., Shi, Y., Huang, L.: Predicting glass properties by using physics- and chemistry-informed machine learning models. Journal of Non-Crystalline Solids **584**, 121511 (2022)

34. Tandia, A., Onbasli, M.C., Mauro, J.C.: Machine learning for glass modeling. In: Springer Handbook of Glass, pp. 1157–1192. Springer (2019)

35. The pandas development team: pandas-dev/pandas: Pandas. https://github.com/pandas-dev/pandas. License: BSD-3-Clause

36. Vasudevan, R.K., Choudhary, K., Mehta, A., Smith, R., Kusne, G., Tavazza, F., Vlcek, L., Ziatdinov, M., Kalinin, S.V., Hattrick-Simpers, J.: Materials science in the artificial intelligence age: high-throughput library generation, machine learning, and a pathway from correlations to the underpinning physics. MRS communications **9**(3), 821–838 (2019)

37. Von Rueden, L., Mayer, S., Beckh, K., Georgiev, B., Giesselbach, S., Heese, R., Kirsch, B., Pfrommer, J., Pick, A., Ramamurthy, R., et al.: Informed machine learning–A taxonomy and survey of integrating prior knowledge into learning systems. IEEE Transactions on Knowledge and Data Engineering **35**(1), 614–633 (2021)

38. Wagstaff, E., Fuchs, F.B., Engelcke, M., Osborne, M.A., Posner, I.: Universal approximation of functions on sets. Journal of Machine Learning Research **23**(151), 1–56 (2022)

39. Wang, A.Y.T., Murdock, R.J., Kauwe, S.K., Oliynyk, A.O., Gurlo, A., Brgoch, J., Persson, K.A., Sparks, T.D.: Machine learning for materials scientists: An introductory guide toward best practices. Chemistry of Materials **32**(12), 4954–4965 (2020)

40. Zaheer, M., Kottur, S., Ravanbakhsh, S., Poczos, B., Salakhutdinov, R.R., Smola, A.J.: Deep sets. Advances in Neural Information Processing Systems **30** (2017)

41. Zanotto, E., Coutinho, F.: How many non-crystalline solids can be made from all the elements of the periodic table? Journal of Non-Crystalline Solids **347**(1–3), 285–288 (2004)

# Chapter 9
# Graph Neural Networks for Predicting Side Effects and New Indications of Drugs Using Electronic Health Records

**Jayant Sharma, Manuel Lentzen, Sophia Krix, Thomas Linden, Sumit Madan, Van Dinh Tran, and Holger Fröhlich**

**Abstract** Drug development is a costly and time-intensive process. However, promising strategies such as drug repositioning and side effect prediction can help to overcome these challenges. Repurposing approved drugs can significantly reduce the time and resources required for preclinical and clinical trials. Furthermore, early detection of potential safety issues is crucial for both drug development programs and the wider healthcare system. For both goals, drug repositioning and side effect prediction, existing machine learning (ML) approaches mainly rely on data collected in preclinical phases, which is not necessarily representative of the real-world situation faced by patients. In this chapter, we construct a knowledge graph based on diagnoses, prescriptions and diagnostic procedures found in large-scale electronic health records, as well as secondary information from different databases, such as drug side effects and chemical compound structure. We show that modern Graph Neural Networks (GNNs) allow for an accurate and interpretable prediction of novel drug-indication and drug-side effect associations in the knowledge graph.

J. Sharma
Fraunhofer SCAI, Sankt Augustin, Germany
e-mail: jayant.sharma@scai.fraunhofer.de

M. Lentzen · S. Krix · T. Linden · H. Fröhlich (✉)
Fraunhofer SCAI, Sankt Augustin, Germany

b-it University of Bonn, Bonn, Germany
e-mail: manuel.lentzen@scai.fraunhofer.de; sophia.krix@scai.fraunhofer.de;
thomas.linden@scai.fraunhofer.de; holger.froehlich@scai.fraunhofer.de

S. Madan
Fraunhofer SCAI, Sankt Augustin, Germany

Computer Science, University of Bonn, Bonn, Germany
e-mail: sumit.madan@scai.fraunhofer.de

V. D. Tran
Computer Science, University of Freiburg, Freiburg, Germany
e-mail: dinh@informatik.uni-freiburg.de

© The Author(s) 2025
D. Schulz, C. Bauckhage (eds.), *Informed Machine Learning*,
Cognitive Technologies, https://doi.org/10.1007/978-3-031-83097-6_9

Altogether, our work demonstrates the potential of GNNs for knowledge-informed ML in healthcare.

## 9.1 Introduction

In recent years, there has been a growing trend in utilizing large-scale, structured electronic health records (EHRs), including administrative claims data, to gain insights into real-world patient trajectories and to develop predictive machine learning models for various health-related outcomes. These outcomes include drug resistance [3], heart failure [16], COVID-19 risk factors [20], risk factors of epilepsy-related comorbidities [23], dementia risk [32], Parkinson's Disease [56] and many more. In addition to demographic information, administrative claims data consists of time-stamped codes for diagnoses, prescriptions and diagnostic procedures.

The main advantage of this type of data is that it is collected routinely in large quantities in many healthcare systems worldwide. However, some significant challenges are associated with this type of data, including bias in diagnosis coding due to economic reasons. Since there is no unique mapping of a diagnosis to a coding scheme such as *International Statistical Classification of Diseases and Related Health Problems* (ICD),[1] physicians may select codes that provide the highest economic benefit rather than the most accurate diagnosis code. Moreover, medications are often coded on a product and not on a chemical substance level, which can lead to inconsistencies as several drugs might have the same chemical substance. Further challenges arise due to the irregular nature of the time series data, as the length of medical history and the time intervals between doctor visits differ between patients. Despite these challenges, EHRs remain a valuable source of information to gain insights into real-world patient trajectories and develop predictive machine learning models for various health-related outcomes.

One promising use of EHR data is to construct knowledge graphs that can subsequently be used for a variety of predictive machine learning models. These graphs can be based on an individual patient's medical history or constructed from aggregated data from a large number of patients. A couple of studies have explored the latter approach. For instance, a study by Rotmensch et al. utilized machine learning techniques to automatically generate a knowledge graph linking diseases to symptoms from aggregated EHR data [35]. The resulting graph was then compared to the Google Health knowledge graph. Another example is the study from Cho et al. in which a knowledge graph was derived from EHRs and used for knowledge graph embedding with the graph convolutional HinSAGE network [14]. The learned embeddings were then used to predict the onset of cardiovascular disease.

---

[1] https://www.who.int/standards/classifications/classification-of-diseases.

Generally, knowledge graph embedding has been increasingly applied throughout recent years and multiple approaches exist. One of the earliest models for this purpose was introduced by Bordes et al. and is called TransE [9]. TransE learns low-dimensional embeddings for entities and relations in a knowledge graph, such that the embedding of a head entity plus the embedding of a relation is close to the embedding of a tail entity. Another model proposed by Zhang et al. is CrossE, which incorporates an interaction matrix to model the interactions between entities and relations [57]. The Graph Convolutional Network [21], introduced by Kipf and Welling, is another method for generating knowledge graph embeddings. GCNs use a convolution operation, applied to a node and its neighbors, to compute a new node representation. The authors have shown that GCNs can be used for semi-supervised node classification on several benchmark datasets. To handle multi-relational graphs, Schlichtkrull et al. proposed an extension of GCNs named Relational Graph Convolutional Network (R-GCN) [37]. In R-GCNs, each relation type in the graph is treated as a distinct edge type, and the model learns separate weight matrices for each relation type. The performance of R-GCNs was demonstrated on link prediction and entity classification tasks.

The Graph Attention Network (GAT) [42], introduced by Velickovic et al., is another approach that uses attention mechanisms to learn node representations. GATs learn separate attention weights for each neighbor of a node, allowing the model to focus on the most relevant neighbors for each node. The authors have demonstrated that GATs outperform GCNs on several benchmark datasets. Most recently, Liu et al. extended this approach by proposing a Relation-Aware Graph ATtention network (RAGAT) [25]. Compared to the standard GAT, the attention mechanism is extended to consider relations between nodes in addition to their features and edge connections. To accomplish this, a relation matrix is incorporated into the attention mechanism, which encodes the relationships between pairs of nodes in the graph. This matrix is learned during training, and the attention weights are then computed based on the features, edge connections, and relations between the nodes.

In this chapter, we aim to perform Informed Machine Learning [44] by integrating existing scientific knowledge and administrative claims data to develop a machine learning (ML) framework based on the RAGAT model for predicting (a) novel indication areas of existing drugs and (b) potential safety issues of drugs. To our knowledge, this idea has not been well explored so far. Paik et al. manually engineered similarity measures between drugs and disease to repurpose the drug terbutaline sulfate for amyotrophic lateral sclerosis [31]. Other authors have used traditional statistical/pharmaco-epidemiological methods for the same purpose [11, 53], see [54] for an overview. The contribution of this chapter is to show that (a) it is possible to construct a knowledge graph based on EHR data plus additional information obtained from databases and (b) to integrate this knowledge graph via modern Graph Neural Networks (GNN) into a learning algorithm, which allows for predicting novel links with high accuracy, hence opening new opportunities to leverage administrative claims data for drug repurposing and drug safety monitoring.

## 9.2   Methods

In this section, we first provide an overview of the data utilized in our study and the process of normalizing and mapping diagnosis and medication codes. Following this, we describe the construction of our knowledge graphs and the integration of existing scientific knowledge. Finally, we will provide information on the models and evaluation methods employed in this study.

### 9.2.1   Overview About Data

For this study, we utilized data from the IBM Explorys® Therapeutic Dataset. The dataset available to us comprises over 700 million records of 4.5 million patients from all over the USA from 2010 until mid of 2021. It includes records of time-stamped codes of prescribed drugs, noted diagnoses and performed procedures. To avoid potential biases, we only focused on patients with one year of medical history in this chapter. In the case of patients tested later on as COVID positive, we only considered encounters one year prior to their first COVID diagnosis to avoid unwanted bias induced by the pandemic outbreak in the healthcare setting. Altogether, we ended up with more than 44 million records of 3.8 million patients.

### 9.2.2   Code Normalization and Mapping

The raw data includes different versions of ICD codes to represent diagnoses that are not compatible with each other. Furthermore, individual ICD codes typically appear rather rarely across patients, partially because the mapping of observed diagnoses to ICD codes is non-unique. To address these aspects, we decided to normalize diagnosis codes by mapping them to the PheWAS ontology [43], which has a far lower number of terms compared to ICD.

Medications were mapped from RXNorm [30] to the widely employed Anatomical-Chemical-Classification (ATC)[2] system (5th level) scheme the World Health Organisation provided, which pools chemically related substances together.

---

[2] https://www.whocc.no/atc_ddd_index_and_guidelines/atc_ddd_index/.

### 9.2.3  Initial Knowledge Graph Construction

Next to the code normalized data we described before, we integrated prior scientific knowledge from the MEDI [48] and SIDER [22] databases to construct an initial knowledge graph $G = (V, R, E)$ comprising entities $V$ connected via relations $R$ and edges $E$ as follows:

- Two diagnoses $d$, $f$ were connected via an undirected edge of relation type *co-occurrence*, if they were observed together within the same visit/encounter more often than expected by chance ($p$-value $\leq 0.05$, as explained below).
- A drug $a$ was connected via a directed edge of relation type *treated with* to diagnosis $d$, if (i) the prescription of $a$ was observed together within the same visit as the diagnosis $d$ more often than expected by chance, and (ii) $d$ has been reported as indication area for $a$ in the MEDI database.
- A drug $a$ was connected via a directed edge of relation type *causes* to side effect $s$, if $s$ has been reported as side effect for $a$ in the SIDER database.
- A medical procedure $p$ was connected with a directed edge of relation type *finds out* to diagnosis $d$, if the procedure and diagnosis were observed together within the same visit more often than expected by chance ($p$-value $\leq 0.05$).

Figure 9.1 exemplifies this knowledge graph construction. Importantly, several relation types were constructed based on two entities being observed within the same visit more often than expected by chance to avoid spurious or false positive edges. This was done by performing a hyper-geometric or one-sided Fisher's exact test: Let $A, B$ be two entities in the graph. Then a contingency table can be constructed based on relative frequencies observed in the data:

| A co-occurs with B | A occurs, but not B |
|---|---|
| B occurs, but not A | neither A nor B occurs |

Based on this table, we can reject the null hypothesis that A and B occur statistically independently of each other. We corrected resulting $p$-values for multiple testing using Holm's method, which controls the family-wise error rate [18]. Only edges with a multiple testing corrected $p$-value below 0.05 were considered for further analysis. In total, the resulting knowledge graph had 30,736 triples, connecting 3087 nodes. Among these, 952 were diagnosis nodes, 858 drug nodes, 631 side effect nodes, and 646 procedure nodes. There were 8573 drug-indication and 5239 drug-side effect relationships.

### 9.2.4  Extended Knowledge Graph Construction

In addition to the initial knowledge graph construction, we explored whether adding more prior knowledge would enhance the link prediction performance. In

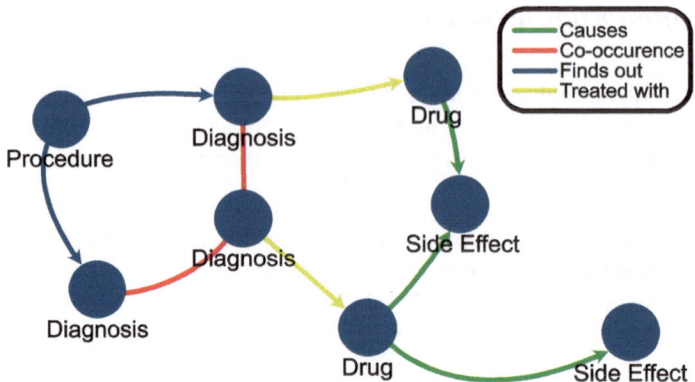

**Fig. 9.1** Example about initial knowledge graph construction based on EHR data. Apart from diagnosis-diagnosis, all other relations are unidirectional. The non-directional diagnosis co-occurrence is established by training the model with flipped diagnosis relations and observing the similarity in predictions performance metrics

particular, we used information on chemical compound similarity between drugs and information on disease categories to enrich the existing knowledge graph. The details are described below.

### 9.2.4.1 Chemical Compound Similarities

Based on the mapping of drugs to ATC codes, we retrieved SMILES string representations of chemical compounds from DrugBank [50]. Subsequently, we calculated Extended and Functional Connectivity Fingerprints (ECFP4, FCFP) describing the pharmacophore and topological graph structure of each compound [34], respectively. Both of these fingerprints result in a 1024 dimensional binary vector representation of each compound. The similarity between each pair of fingerprints $C_1, C_2$ was then assessed via the Tanimoto-Jaccard coefficient

$$T(C_1, C_2) = \frac{|C_1 \cap C_2|}{|C_1 \cup C_2|},$$  (9.1)

where $C_1 \cap C_2$ is the intersection between $C_1$ and $C_2$ and $C_1 \cup C_2$ their union.

Following common convention in the chemoinformatics literature, we considered two compounds as highly similar if $T(C_1, C_2) > 0.85$ for either class of fingerprint.[3] In such a case we connected $C_1$ and $C_2$ with an edge of relation

---

[3] https://chem.libretexts.org/Courses/Intercollegiate_Courses/Cheminformatics_OLCC_(2019)/6%3A_Molecular_Similarity/6.2%3A_Similarity_Coefficients.

type *chemically similar* in the knowledge graph. Altogether, 1095 drug-drug relationships were added to the knowledge graph in this manner.

### 9.2.4.2 Use of Diagnosis-Diagnosis Relationships

To encode relationships between diagnoses, we introduced a further relation type *same disease class*. That means we set an edge between two diagnoses $d_1$, $d_2$, if they shared the same parent in the PheWAS ontology. For example, "bacterial enteritis" and "viral enteritis" share the parent "intestinal infection". 1272 diagnosis-diagnosis relationships were added to the knowledge graph in this manner.

## 9.2.5 Relation Aware Graph Attention Networks

In this study, we aimed to reposition drugs and predict adverse drug events using a graph-based approach. To achieve this, we utilized a heterogeneous graph representation with different relations and node types, on which we performed link prediction.

We utilized a neural network architecture called Relation Aware Graph Attention Networks (RAGATs) [25], which were recently introduced as an extension of Graph Attention Networks (GATs) [42]. GATs are neural networks designed to analyze and interpret graph-structured data by focusing on the graph's most relevant nodes and edges. This approach enables the network to generate more accurate and expressive representations of the underlying graph structure. RAGATs take the GAT approach one step further by considering the different relationships between nodes in a graph, which the original GAT architecture did not account for. RAGATs do this by using relation-aware message functions that calculate relation-specific attention coefficients for each node in the graph. These coefficients are then combined with the node's feature to generate a new and more informative node representation. It is worth noting that to the best of our knowledge, RAGATs have not been used before for analyzing healthcare data. In the following paragraphs, we provide more details on how we employed RAGATs to analyze the graph-structured data in our study.

For each relation type $r$, a learnable weight matrix $W_r$ is initiated to capture common features associated with specific relations. Additionally, we augment the knowledge graph by adding the inverse of each relationship and self-loops as separate directions. To encode a triple $(u, r, v)$ into a vector $c^r_{(u,r,v)}$, we use

$$c^r_{(u,r,v)} = W_r e_u + W_r (e_u \circ e_v), \qquad (9.2)$$

where $e_u, e_v \in R^d$ are the embeddings of entities $u$ and $v$, and $\circ$ denotes the Hadamard product.

Next, we calculate the message $m_{(u,r,v)}$ using the relation-type specific weight $W_{\mathrm{dir(r)}}$ for either the original, inverse, or self-loop directions as

$$m_{(u,r,v)} = W_{\mathrm{dir(r)}}c^r_{(u,r,v)}. \tag{9.3}$$

To compute the absolute attention coefficients $b_{u,r}$ for each message $e_{u \to v}$, we use a learnable weight matrix $W_{\mathrm{att}}$ and the LeakyReLU activation function, as shown in (9.4). Then, we calculate the relative attention values using (9.5).

$$b_{(u,r)} = \mathrm{LeakyReLU}(W_{\mathrm{att}}m_{(u,r,v)}) \tag{9.4}$$

$$\alpha_{(u,r)} = \frac{\exp(b_{(u,r)})}{\sum_{(i \in N_v)} \sum_{r \in R_{i,u}} \exp(b_{(i,r)})} \tag{9.5}$$

Here, $N_v$ is the set of neighbor nodes of $v$, and $R_{i,u}$ is the set of relation types from node $i$ to node $u$.

We use a multi-head attention mechanism to generate node representations. For a given node $v$, we compute the node embedding $e'_v$ by computing a weighted sum of the messages passed from its neighbors using the attention coefficients $\alpha^h_{u,r}$ and message embeddings $m^h_{(u,r,v)}$ across $H$ attention heads, as shown in (9.6) where tanh is the hyperbolic tangent

$$e'_v = \tanh\left(\frac{1}{H}\sum_{h=1}^{H}\sum_{(u,r) \in N_v} \alpha^h_{u,r} m^h_{(u,r,v)}\right). \tag{9.6}$$

We also linearly project each relation embedding $e_r$ to have the same dimension as the updated node embedding $e'_v$ using a trainable weight matrix $W_{\mathrm{rel}}$ and

$$e'_r = W_{rel}e_r. \tag{9.7}$$

For decoding, we use InteractE [41], which employs random feature permutations, reshaping of permuted features, and circular convolution. The decoder produces a probability $p_{(u,r,v)}$ for each triple. During the training of the RAGAT model, we employ the binary cross entropy loss function given by

$$\mathcal{L} = -\frac{1}{N}\sum_i \left(t_i \cdot \log(p_i) + (1 - t_i) \cdot \log(1 - p_i)\right), \tag{9.8}$$

where $i$ runs over all triples, $p_i$ is the corresponding probability predicted by the model, and $t_i$ corresponds to the true class label indicating the existence or non-existence of the triple in the training data.

## 9.2.6   Evaluation against Alternative Methods

We compared RAGAT's against TransE [10], DistMult [55] and ComplEx [40], which are well-known shallow, geometric knowledge graph embedding approaches. TransE regards relations as translations from subject to object entities. The intuition is that the embedding of an object $o$ should be close to that of the subject $s$ plus that of the relation type $r$, if $(s, r, o)$ holds

$$f(s, r, o) = ||e_s + e_r - e_o||. \tag{9.9}$$

DistMult relies on the following scoring function:

$$f(s, r, o) = ||e_s \circ e_r \circ e_o||_1. \tag{9.10}$$

ComplEx is an extension of DistMult, which employs an embedding of $s, r, o$ into the complex space and then uses

$$f(s, r, o) = Re(e_s \circ e_r \circ \bar{e}_o) \tag{9.11}$$

as scoring function.

We trained and tested RAGAT and competing methods all on the same data: We performed a stratified split of the overall set of triples into 64% for training, 16% for validation and 20% for testing. We performed Bayesian hyperparameter optimization using Optuna [1] (version 2.10.0) on the validation set. During the tuning process, the values of hyperparameters were sampled by Tree-structured Parzen Estimator (TPE) [7, 8] from a user-defined search space. We used the Mean Reciprocal Rank (MRR) as objective function for hyperparameter optimization. Table 9.1 provides the list of hyperparameters and their search space. A total of 65 trials were run.

## 9.2.7   Performance Measures

The evaluation of RAGAT relies on an "open world" assumption, meaning that non-existing relations may not necessarily be considered negatives. In this study,

**Table 9.1** RAGAT hyper-parameters and their ranges, in which they were optimized during Bayesian hyperparameter optimization

| Hyperparameter | Range |
|---|---|
| Learning rate | $[10^{-5}, 10^{-1}]$ |
| Mini batch size | $[56, 1024]$ |
| Label smoothing | $[0.04, 0.4]$ |
| Attention heads | 1, 2, 3 |
| Drop-out ratio | $[0, 0.5]$ |

we evaluate the performance of the compared models on an unseen test set using common rank-based measures used in positive-unlabeled learning: Hits@K, mean reciprocal rank, and the area under the precision-recall curve. These metrics are particularly useful for large and sparse knowledge graphs, where numerous negative triples are unobserved. Hits@K and MRR are well-established evaluation metrics for knowledge graph completion models, making them ideal for models trained using positive-unlabeled learning, where only positive triples are labeled, and the aim is to predict the missing links in the knowledge graph.

- **Hits@k** Hits@k measures the proportion of correct entities that are included in the top k predictions generated by the model. Specifically, for a given set of test triples (u, r, v), where the model predicts the probability of each triple being true, the entities are ranked based on their predicted probabilities. If the correct entity is among the top $k$ predicted entities, we count it as a hit. The Hits@k metric is defined as the proportion of test triples for which at least one of the correct entities appears in the top k predicted entities.
- **Mean Reciprocal Rank (MRR)** MRR is calculated by taking the average of the reciprocal ranks of the correct entities. For each test triple (u, r, v), we rank the entities based on their predicted probabilities and calculate the reciprocal rank of the correct entity, i.e., 1 if the correct entity is ranked first, 1/2 if it is ranked second, 1/3 if it is ranked third, and so on. The MRR metric is defined as the average of the reciprocal ranks over all test triples. Let $r_i$ indicate the rank for triple $i$. Then the $MRR \in (0, 1]$ is defined as

$$MRR = \frac{1}{n} \sum_i \frac{1}{r_i}. \tag{9.12}$$

- **Area under Precision Recall Curve (AUPRC)** To determine the AUPRC, we rank the test triples based on their predicted probabilities and gradually increase the classification threshold from 0 to 1. We compute the precision and recall values based on the top-ranked triples at each threshold. Precision is the ratio of correct predictions to the total number of positive predictions, while recall is the ratio of correct predictions to the total number of positive instances in the test set. We can create a precision-recall curve by plotting precision against recall at each threshold and calculating the AUPRC as the area under this curve. A perfect classifier would have an AUPRC of 1, while a random classifier would have an AUPRC of 0.5. The AUPRC score ranges from 0 to 1.

## 9.3  Results

This section presents the outcomes of our experiments. We first describe the performance of various models in predicting drug-side effects and drug-indication relations. Subsequently, we examine a newly predicted indication and adverse effects in light of existing literature.

### 9.3.1 Performance Comparison

We compare the performance of our RAGAT model with baseline models TransE, DistMult, and ComplEx. RAGAT was trained on both the initial and the extended knowledge graphs, and the following sections highlight these comparisons.

#### 9.3.1.1 Initial Knowledge Graph

Comparing our RAGAT model against TransE, DistMult, and ComplEx demonstrated an apparent increase in prediction performance for all performance measures (Fig. 9.2). This impression was confirmed when focusing on the prediction performances for drug–side effect and drug–indication relations separately (Table 9.2). Notably, the prediction performance for drug–side effect links was lower than for drug–indication relationships because of the comparably smaller number of side effect links in our dataset.

#### 9.3.1.2 Extended Knowledge Graph

We compared the prediction performance of our RAGAT model trained on the initial knowledge graph with the one trained on the extended knowledge graph. Only a

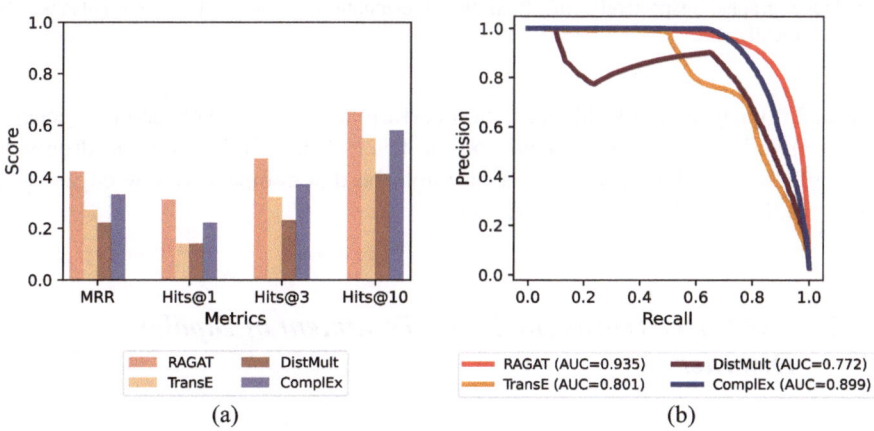

**Fig. 9.2** Overall performance evaluation of tested models. (**a**) Rank-based evaluation metrics: MRR and Hits@k. (**b**) Precision recall curve

**Table 9.2** Relation-type specific prediction performance for side effects and novel indications

| Relation type | Model | MRR | Hits@10 | Hits@3 | Hits@1 | AUPR |
|---|---|---|---|---|---|---|
| Drug-indication | RAGAT (initial KG) | 0.61 | 0.8 | 0.65 | 0.52 | 0.96 |
| | RAGAT (extended KG) | 0.61 | 0.8 | 0.67 | 0.51 | 0.96 |
| | TransE | 0.42 | 0.66 | 0.46 | 0.3 | 0.94 |
| | DistMult | 0.40 | 0.68 | 0.45 | 0.27 | 0.93 |
| | ComplEx | 0.53 | 0.77 | 0.58 | 0.41 | 0.95 |
| Drug-side effect | RAGAT (initial KG) | 0.32 | 0.56 | 0.36 | 0.21 | 0.91 |
| | RAGAT (extended KG) | 0.34 | 0.58 | 0.4 | 0.23 | 0.87 |
| | TransE | 0.25 | 0.46 | 0.26 | 0.15 | 0.84 |
| | DistMult | 0.28 | 0.46 | 0.29 | 0.20 | 0.88 |
| | ComplEx | 0.36 | 0.56 | 0.39 | 0.27 | 0.89 |

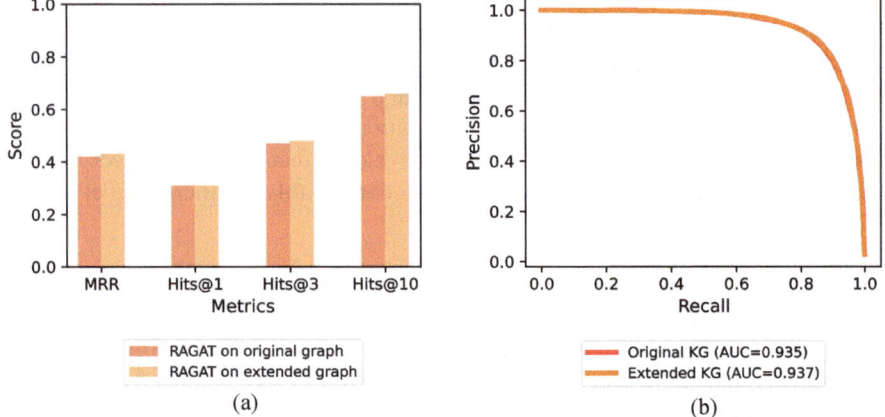

**Fig. 9.3** Overall performance evaluation of RAGAT models trained on the initial and the extended knowledge graphs, respectively. (**a**) Rank-based evaluation metrics: MRR and Hits@k. (**b**) Precision recall curve

marginal improvement of link prediction performance could be observed in general (Fig. 9.3) and also more specifically on the level of drug-indication and drug-side effect relations (Table 9.2). The reason might be that comparably few edges were added via the knowledge graph extension.

### 9.3.2 Use Case: Trazodone in the Treatment of Bipolar Disorder

Due to the limited enhancement of prediction performance by the extended knowledge graph for the following use case, we employed the RAGAT model trained on the initial knowledge graph. The model predicted a new link between bipolar disorder and the drug Trazodone, a triazolopyridine compound with antidepressant, anxiolytic, sedative, and hypnotic properties. Figure 9.4 shows the corresponding subgraph of our knowledge graph with the predicted link in orange. The inferred edge establishes an association between Trazodone to bipolar disorder due to the path

$$Bipolar\ Disorder \xrightarrow{co\text{-}occurence} Anxiety\ Disorder \xrightarrow{treated\ with} Trazodone.$$

That means this path is one possible explanation of the newly predicted triple (Bipolar Disorder, treated with, Trazodone). Further paths supporting this triple are

$$Bipolar\ Disorder \xrightarrow{co\text{-}occurence} Alcoholism \xrightarrow{treated\ with} Trazodone$$

and

$$Bipolar\ Disorder \xrightarrow{co\text{-}occurence} Essential\ Hypertension \xrightarrow{treated\ with} Trazodone.$$

Figure 9.5 visualizes the weights learned by the model in the last attention layer. The figure shows that the model puts strong attention (dark blue hues) on the direct neighbors of the node "Bipolar Disorder" that are directly connected to Trazodone.

In the literature, Trazodone has been found to reduce acute psychomotor activation in patients with bipolar disorder [5]. Furthermore, low doses of Trazodone

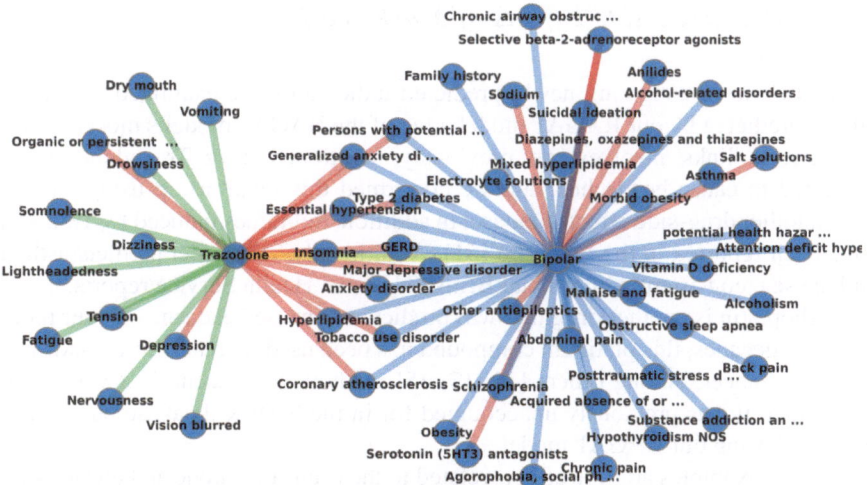

**Fig. 9.4** Repositioning of Trazodone for treatment of bipolar disorder: The figure shows a zoom into the knowledge graph. The predicted association is shown as a thick yellow edge. Known associations between drugs and side effects are colored red; known associations between drugs and indications are colored green. Associations between diagnoses are depicted in blue

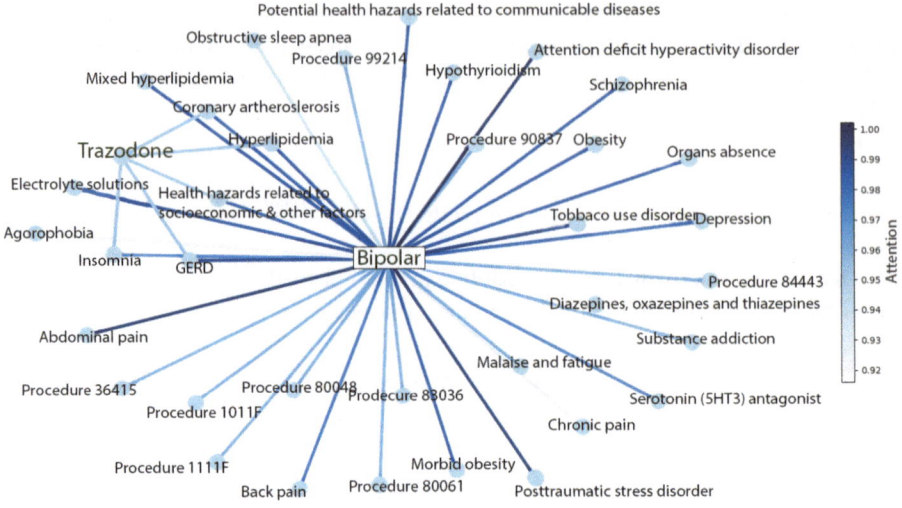

**Fig. 9.5** Attention weights learned by RAGAT model. A darker color indicates a higher attention weight

in combination with a mood stabilizer have been reported as a safe treatment of insomnia in patients with bipolar disorder [49].

### 9.3.3  Predicted Side Effects of Marketed Drugs

In addition to investigating newly predicted indications, we examined the adverse effects predicted by our RAGAT model. One of the RAGAT model's most confident side effects links involves the previously mentioned drug Trazodone. It was predicted to cause headaches, and we confirmed this relationship using NSIDES [39], another drug-side effect database. In addition, migraines induced by Trazodone have been reported in the literature [51]. Additionally, the model predicts asthenia and nausea, consistent with official British National Health Service reports.[4]

Carboplatin is another drug that was predicted to cause headaches by our model. In recent decades, this platinum compound has been used to treat ovarian and small-cell lung cancer, among others [13, 17, 45]. Although headache is a well-known side effect, it was previously unaccounted for in the SIDER database but could be predicted using our RAGAT model.

Further examples are side effects related to the drugs tolcapone and eltrombopag. Tolcapone, which was predicted to induce orthostatic hypotension, is a medication

---

[4] https://www.nhs.uk/medicines/trazodone/side-effects-of-trazodone/, https://www.medicines.org.uk/emc/product/4976.

used alongside levodopa to treat Parkinson's disease [4]. However, several adverse effects, including orthostatic hypotension, are now known in literature, and the drug is no longer used [27]. The last example of successful side effect prediction by our RAGAT model is the link between eltrombopag and muscle cramps. The drug is used, for instance, to treat Thrombocytopenia, but several side effects, including the predicted muscle cramps, have been reported [19].

## 9.4 Discussion

Our study demonstrates one possible strategy of Informed Machine Learning [44] by utilizing large-scale EHR data and existing knowledge from MEDI, SIDER, DrugBank, and PheWAS to construct a rich knowledge graph that incorporates diagnoses, drugs, their indications, known side effects, and chemical compound similarities. We showed that a recently published GNN variant (RAGAT) can be trained on such a graph to predict new links between drugs and indications as well as between drugs and side effects. Our RAGAT model outperformed classical geometric knowledge embedding techniques, such as TransE, DistMult and ComplEx, demonstrating the potential of such an approach for optimizing resource use by suggesting novel indications for existing drugs and detecting potential side effects in a timely manner.

The idea of repositioning existing drugs for new indications has recently gained a lot of attention in the context of the ongoing COVID-19 pandemic [24, 36]. In general, the approach is motivated by the fact that the development of a new drug typically lasts 10 to 15 years and can cost over a billion dollars [12, 52]. Moreover, serious side effects do not only impose a strong risk for failure of drug development programs (with respective financial consequences) but can later on also result in elevated public health costs [58]. Indeed, unwanted side effects of drugs have been estimated to be responsible for almost 5% of all hospital admissions worldwide [6].

While there is a large body of literature on various ML approaches for drug repositioning [2, 26, 28, 29, 33, 38, 46], these techniques typically rely on data that is very different from the one studied in this chapter because it is collected in a preclinical stage via biological assays (e.g., expression of genes in a cell line). However, the question lies in how far such research data is representative of the real-world situation in a patient. Routinely collected EHR data can provide valuable insights into real-world patient trajectories, but using such data for ML and in particular for drug repositioning and side effect prediction is not trivial. Our work showcases the potential of EHR data for GNN-based drug repositioning and safety assessment and clearly distinguishes itself from recently published studies, which use GNNs for different applications such as predicting disease outcomes of an individual patient [15, 47].

## 9.5 Conclusion

The present work demonstrates the potential of GNNs for integrating a knowledge graph into a learning algorithm, which is one possible strategy of Informed Machine Learning [44]. In particular, we showed that in this way it is possible to use modern GNNs for predicting side effects and new indications of existing drugs on the basis of EHRs.

In future work, a comparison of GNN/ML based predictions with findings obtained via conventional statistical/pharmaco-epidemiological techniques would be interesting. Finally, there is the important open question, of how far our models and the insights derived from these models would generalize to other healthcare systems outside the USA. Here, the clear limiting factor is currently the availability and accessibility of corresponding data.

**Acknowledgments** This contribution was partially supported by the Fraunhofer Cluster of Excellence "Cognitive Internet Technologies".

## References

1. Akiba, T., Sano, S., Yanase, T., Ohta, T., Koyama, M.: Optuna: A next-generation hyperparameter optimization framework. In: Proceedings of the 25th ACM SIGKDD International Conference on Knowledge Discovery and Data Mining (2019)
2. Aliper, A., Plis, S., Artemov, A., Ulloa, A., Mamoshina, P., Zhavoronkov, A.: Deep Learning Applications for Predicting Pharmacological Properties of Drugs and Drug Repurposing Using Transcriptomic Data. Molecular Pharmaceutics **13**(7), 2524–2530 (2016). https://doi.org/10.1021/acs.molpharmaceut.6b00248
3. An, S., Malhotra, K., Dilley, C., Han-Burgess, E., Valdez, J.N., Robertson, J., Clark, C., Westover, M.B., Sun, J.: Predicting drug-resistant epilepsy — A machine learning approach based on administrative claims data. Epilepsy & Behavior **89**, 118–125 (2018). https://doi.org/10.1016/j.yebeh.2018.10.013
4. Aronson, J.: Meyler's Side Effects of Drugs: The International Encyclopedia of Adverse Drug Reactions and Interactions. ISSN. Elsevier Science. URL https://books.google.de/books?id=NOKoBAAAQBAJ
5. Ballerio, M., Politi, P., Crapanzano, C., Emanuele, E., Cuomo, A., Goracci, A., Fagiolini, A.: Clinical effectiveness of parenteral trazodone for the management of psychomotor activation in patients with bipolar disorder. Neuro endocrinology letters **39**(3), 205–208 (2018)
6. Beijer, H., de Blaey, C.: Hospitalisations caused by adverse drug reactions (ADR): A meta-analysis of observational studies. Pharmacy World and Science **24**(2), 46–54 (2002). https://doi.org/10.1023/A:1015570104121
7. Bergstra, J., Bardenet, R., Bengio, Y., Kégl, B.: Algorithms for hyper-parameter optimization. In: Proceedings of the 24th International Conference on Neural Information Processing Systems, NIPS'11, p. 2546–2554. Curran Associates Inc., Red Hook, NY, USA (2011)
8. Bergstra, J., Yamins, D., Cox, D.: Making a Science of Model Search: Hyperparameter Optimization in Hundreds of Dimensions for Vision Architectures. In: Proceedings of the 30th International Conference on Machine Learning, pp. 115–123. PMLR (2013)
9. Bordes, A., Usunier, N., Garcia-Durán, A., Weston, J., Yakhnenko, O.: Translating embeddings for modeling multi-relational data. In: Proceedings of the 26th International Conference

on Neural Information Processing Systems - Volume 2, NIPS'13, p. 2787–2795. Curran Associates Inc., Red Hook, NY, USA (2013)

10. Bordes, A., Usunier, N., Garcia-Duran, A., Weston, J., Yakhnenko, O.: Translating Embeddings for Modeling Multi-relational Data. In: Advances in Neural Information Processing Systems, vol. 26. Curran Associates, Inc. (2013)

11. Brilliant, M.H., Vaziri, K., Connor, T.B., Schwartz, S.G., Carroll, J.J., McCarty, C.A., Schrodi, S.J., Hebbring, S.J., Kishor, K.S., Flynn, H.W., Moshfeghi, A.A., Moshfeghi, D.M., Fini, M.E., McKay, B.S.: Mining Retrospective Data for Virtual Prospective Drug Repurposing: L-DOPA and Age-related Macular Degeneration. The American Journal of Medicine 129(3), 292–298 (2016). https://doi.org/10.1016/j.amjmed.2015.10.015

12. Brown, D.G., Wobst, H.J., Kapoor, A., Kenna, L.A., Southall, N.: Clinical development times for innovative drugs. Nature Reviews Drug Discovery (2021). https://doi.org/10.1038/d41573-021-00190-9

13. Buckingham, R., Fitt, J., Sitziaz, J.: Patients' experiences of chemotherapy: side-effects of carboplatin in the treatment of carcinoma of the ovary 6(1), 59–71. https://doi.org/10.1111/j.1365-2354.1997.tb00270.x. URL https://onlinelibrary.wiley.com/doi/abs/10.1111/j.1365-2354.1997.tb00270.x. _eprint: https://onlinelibrary.wiley.com/doi/pdf/10.1111/j.1365-2354.1997.tb00270.x

14. Cho, H.N., Ahn, I., Gwon, H., Kang, H.J., Kim, Y., Seo, H., Choi, H., Kim, M., Han, J., Kee, G., Jun, T.J., Kim, Y.H.: Heterogeneous graph construction and hinsage learning from electronic medical records 12(1), 21152. https://doi.org/10.1038/s41598-022-25693-2

15. Choi, E., Xu, Z., Li, Y., Dusenberry, M.W., Flores, G., Xue, Y., Dai, A.M.: Learning the Graphical Structure of Electronic Health Records with Graph Convolutional Transformer. arXiv:1906.04716 [cs, stat] (2020)

16. Desai, R.J., Wang, S.V., Vaduganathan, M., Evers, T., Schneeweiss, S.: Comparison of Machine Learning Methods With Traditional Models for Use of Administrative Claims With Electronic Medical Records to Predict Heart Failure Outcomes. JAMA Network Open 3(1), e1918962 (2020). https://doi.org/10.1001/jamanetworkopen.2019.18962

17. Ford, J., Osborn, C., Barton, T., Bleehen, N.M.: A phase i study of intravenous RMP-7 with carboplatin in patients with progression of malignant glioma 34(11), 1807–1811. https://doi.org/10.1016/S0959-8049(98)00155-5. URL https://www.sciencedirect.com/science/article/pii/S0959804998001555

18. Holm, S.: A Simple Sequentially Rejective Multiple Test Procedure. Scandinavian Journal of Statistics 6(2), 65–70 (1979)

19. Hong, Y., Li, X., Wan, B., Li, N., Chen, Y.: Efficacy and safety of eltrombopag for aplastic anemia: A systematic review and meta-analysis 39(2), 141–156. https://doi.org/10.1007/s40261-018-0725-2

20. Jucknewitz, R., Weidinger, O., Schramm, A.: Covid-19 risk factors: Statistical learning from German healthcare claims data. Infectious Diseases (London, England) 54(2), 110–119 (2022). https://doi.org/10.1080/23744235.2021.1982141

21. Kipf, T.N., Welling, M.: Semi-Supervised Classification with Graph Convolutional Networks. arXiv:1609.02907 [cs, stat] (2017)

22. Kuhn, M., Letunic, I., Jensen, L.J., Bork, P.: The SIDER database of drugs and side effects. Nucleic Acids Research 44(Database issue), D1075–D1079 (2016). https://doi.org/10.1093/nar/gkv1075

23. Linden, T., De Jong, J., Lu, C., Kiri, V., Haeffs, K., Fröhlich, H.: An Explainable Multimodal Neural Network Architecture for Predicting Epilepsy Comorbidities Based on Administrative Claims Data. Frontiers in Artificial Intelligence 4, 58 (2021). https://doi.org/10.3389/frai.2021.610197

24. Linden, T., Hanses, F., Domingo-Fernández, D., DeLong, L.N., Kodamullil, A.T., Schneider, J., Vehreschild, M.J.G.T., Lanznaster, J., Ruethrich, M.M., Borgmann, S., Hower, M., Wille, K., Feldt, T., Rieg, S., Hertenstein, B., Wyen, C., Roemmele, C., Vehreschild, J.J., Jakob, C.E.M., Stecher, M., Kuzikov, M., Zaliani, A., Fröhlich, H.: Machine Learning Based Prediction of COVID-19 Mortality Suggests Repositioning of Anticancer Drug for Treating Severe Cases.

Artificial Intelligence in the Life Sciences p. 100020 (2021). https://doi.org/10.1016/j.ailsci.2021.100020

25. Liu, X., Tan, H., Chen, Q., Lin, G.: RAGAT: Relation Aware Graph Attention Network for Knowledge Graph Completion. IEEE Access **9**, 20840–20849 (2021). https://doi.org/10.1109/ACCESS.2021.3055529

26. Lotfi Shahreza, M., Ghadiri, N., Mousavi, S.R., Varshosaz, J., Green, J.R.: Heter-LP: A heterogeneous label propagation algorithm and its application in drug repositioning. Journal of Biomedical Informatics **68**, 167–183 (2017). https://doi.org/10.1016/j.jbi.2017.03.006

27. Micek, S.T., Ernst, M.E.: Tolcapone: a novel approach to parkinson's disease **56**(21), 2195–2205. https://doi.org/10.1093/ajhp/56.21.2195

28. Moridi, M., Ghadirinia, M., Sharifi-Zarchi, A., Zare-Mirakabad, F.: The assessment of efficient representation of drug features using deep learning for drug repositioning. BMC Bioinformatics **20**(1), 577 (2019). https://doi.org/10.1186/s12859-019-3165-y

29. Napolitano, F., Zhao, Y., Moreira, V.M., Tagliaferri, R., Kere, J., D'Amato, M., Greco, D.: Drug repositioning: A machine-learning approach through data integration. Journal of Cheminformatics **5**, 30 (2013). https://doi.org/10.1186/1758-2946-5-30

30. Nelson, S.J., Zeng, K., Kilbourne, J., Powell, T., Moore, R.: Normalized names for clinical drugs: RxNorm at 6 years. Journal of the American Medical Informatics Association **18**(4), 441–448 (2011). https://doi.org/10.1136/amiajnl-2011-000116

31. Paik, H., Chung, A.Y., Park, H.C., Park, R.W., Suk, K., Kim, J., Kim, H., Lee, K., Butte, A.J.: Repurpose terbutaline sulfate for amyotrophic lateral sclerosis using electronic medical records. Scientific Reports **5**(1), 8580 (2015). https://doi.org/10.1038/srep08580

32. Park, J.H., Cho, H.E., Kim, J.H., Wall, M.M., Stern, Y., Lim, H., Yoo, S., Kim, H.S., Cha, J.: Machine learning prediction of incidence of Alzheimer's disease using large-scale administrative health data. npj Digital Medicine **3**(1), 1–7 (2020). https://doi.org/10.1038/s41746-020-0256-0

33. Pham, T.H., Qiu, Y., Zeng, J., Xie, L., Zhang, P.: A deep learning framework for high-throughput mechanism-driven phenotype compound screening and its application to COVID-19 drug repurposing. Nature Machine Intelligence **3**(3), 247–257 (2021). https://doi.org/10.1038/s42256-020-00285-9

34. Rogers, D., Hahn, M.: Extended-Connectivity Fingerprints. Journal of Chemical Information and Modeling **50**(5), 742–754 (2010). https://doi.org/10.1021/ci100050t

35. Rotmensch, M., Halpern, Y., Tlimat, A., Horng, S., Sontag, D.: Learning a health knowledge graph from electronic medical records. **7**, 5994

36. Santos, S.d.S., Torres, M., Galeano, D., Sánchez, M.d.M., Cernuzzi, L., Paccanaro, A.: Machine learning and network medicine approaches for drug repositioning for COVID-19. Patterns **3**(1), 100396 (2022). https://doi.org/10.1016/j.patter.2021.100396

37. Schlichtkrull, M., Kipf, T.N., Bloem, P., van den Berg, R., Titov, I., Welling, M.: Modeling Relational Data with Graph Convolutional Networks. arXiv:1703.06103 [cs, stat] (2017)

38. Schultz, B., Zaliani, A., Ebeling, C., Reinshagen, J., Bojkova, D., Lage-Rupprecht, V., Karki, R., Lukassen, S., Gadiya, Y., Ravindra, N.G., Das, S., Baksi, S., Domingo-Fernández, D., Lentzen, M., Strivens, M., Raschka, T., Cinatl, J., DeLong, L.N., Gribbon, P., Geisslinger, G., Ciesek, S., van Dijk, D., Gardner, S., Kodamullil, A.T., Fröhlich, H., Peitsch, M., Jacobs, M., Hoeng, J., Eils, R., Claussen, C., Hofmann-Apitius, M.: A method for the rational selection of drug repurposing candidates from multimodal knowledge harmonization. Scientific Reports **11**(1), 11049 (2021). https://doi.org/10.1038/s41598-021-90296-2

39. Tatonetti, N.P., Ye, P.P., Daneshjou, R., Altman, R.B.: Data-Driven Prediction of Drug Effects and Interactions. Science Translational Medicine **4**(125), 125ra31–125ra31 (2012). https://doi.org/10.1126/scitranslmed.3003377

40. Trouillon, T., Welbl, J., Riedel, S., Gaussier, É., Bouchard, G.: Complex embeddings for simple link prediction. In: Proceedings of the 33rd International Conference on International Conference on Machine Learning - Volume 48, ICML'16, pp. 2071–2080. JMLR.org, New York, NY, USA (2016)

41. Vashishth, S., Sanyal, S., Nitin, V., Agrawal, N., Talukdar, P.: InteractE: Improving Convolution-based Knowledge Graph Embeddings by Increasing Feature Interactions (2020). https://doi.org/10.48550/arXiv.1911.00219

42. Veličković, P., Cucurull, G., Casanova, A., Romero, A., Liò, P., Bengio, Y.: Graph Attention Networks (2018). https://doi.org/10.48550/arXiv.1710.10903

43. Verma, A., Bang, L., Miller, J.E., Zhang, Y., Lee, M.T.M., Zhang, Y., Byrska-Bishop, M., Carey, D.J., Ritchie, M.D., Pendergrass, S.A., Kim, D.: Human-Disease Phenotype Map Derived from PheWAS across 38,682 Individuals. American Journal of Human Genetics **104**(1), 55–64 (2019). https://doi.org/10.1016/j.ajhg.2018.11.006

44. von Rueden, L., Mayer, S., Beckh, K., Georgiev, B., Giesselbach, S., Heese, R., Kirsch, B., Walczak, M., Pfrommer, J., Pick, A., Ramamurthy, R., Garcke, J., Bauckhage, C., Schuecker, J.: Informed Machine Learning - A Taxonomy and Survey of Integrating Prior Knowledge into Learning Systems. IEEE Transactions on Knowledge and Data Engineering pp. 1–1 (2021). https://doi.org/10.1109/TKDE.2021.3079836

45. Wagstaff, A.J., Ward, A., Benfield, P., Heel, R.C.: Carboplatin **37**(2), 162–190. https://doi.org/10.2165/00003495-198937020-00005

46. Wang, X., Ji, H., Shi, C., Wang, B., Cui, P., Yu, P., Ye, Y.: Heterogeneous Graph Attention Network. arXiv:1903.07293 [cs] (2021)

47. Wanyan, T., Honarvar, H., Azad, A., Ding, Y., Glicksberg, B.S.: Deep Learning with Heterogeneous Graph Embeddings for Mortality Prediction from Electronic Health Records. Data Intelligence **3**(3), 329–339 (2021). https://doi.org/10.1162/dint_a_00097

48. Wei, W.Q., Cronin, R.M., Xu, H., Lasko, T.A., Bastarache, L., Denny, J.C.: Development and evaluation of an ensemble resource linking medications to their indications. Journal of the American Medical Informatics Association: JAMIA **20**(5), 954–961 (2013 Sep-Oct). https://doi.org/10.1136/amiajnl-2012-001431

49. Wichniak, A., Jarkiewicz, M., Okruszek, Ł., Wierzbicka, A., Holka-Pokorska, J., Rybakowski, J.K.: Low Risk for Switch to Mania during Treatment with Sleep Promoting Antidepressants. Pharmacopsychiatry **48**(3), 83–88 (2015). https://doi.org/10.1055/s-0034-1396802

50. Wishart, D.S., Knox, C., Guo, A.C., Shrivastava, S., Hassanali, M., Stothard, P., Chang, Z., Woolsey, J.: DrugBank: A comprehensive resource for in silico drug discovery and exploration. Nucleic Acids Research **34**(Database issue), D668–672 (2006). https://doi.org/10.1093/nar/gkj067

51. Workman, E.A., Tellian, F., Short, D.: Trazodone induction of migraine headache through mCPP. The American Journal of Psychiatry **149**(5), 712 (1992). https://doi.org/10.1176/ajp.149.5.712b

52. Wouters, O.J., McKee, M., Luyten, J.: Estimated Research and Development Investment Needed to Bring a New Medicine to Market, 2009-2018. JAMA **323**(9), 844–853 (2020). https://doi.org/10.1001/jama.2020.1166

53. Wu, Y., Warner, J.L., Wang, L., Jiang, M., Xu, J., Chen, Q., Nian, H., Dai, Q., Du, X., Yang, P., Denny, J.C., Liu, H., Xu, H.: Discovery of Noncancer Drug Effects on Survival in Electronic Health Records of Patients With Cancer: A New Paradigm for Drug Repurposing. JCO clinical cancer informatics **3**, 1–9 (2019). https://doi.org/10.1200/CCI.19.00001

54. Xu, H., Li, J., Jiang, X., Chen, Q.: Electronic Health Records for Drug Repurposing: Current Status, Challenges, and Future Directions. Clinical Pharmacology & Therapeutics **107**(4), 712–714 (2020). https://doi.org/10.1002/cpt.1769

55. Yang, B., Yih, W.t., He, X., Gao, J., Deng, L.: Embedding Entities and Relations for Learning and Inference in Knowledge Bases (2015). https://doi.org/10.48550/arXiv.1412.6575

56. Yuan, W., Beaulieu-Jones, B., Krolewski, R., Palmer, N., Veyrat-Follet, C., Frau, F., Cohen, C., Bozzi, S., Cogswell, M., Kumar, D., Coulouvrat, C., Leroy, B., Fischer, T.Z., Sardi, S.P., Chandross, K.J., Rubin, L.L., Wills, A.M., Kohane, I., Lipnick, S.L.: Accelerating diagnosis of Parkinson's disease through risk prediction. BMC Neurology **21**(1), 201 (2021). https://doi.org/10.1186/s12883-021-02226-4

57. Zhang, W., Paudel, B., Zhang, W., Bernstein, A., Chen, H.: Interaction embeddings for prediction and explanation in knowledge graphs. In: Proceedings of the Twelfth ACM

International Conference on Web Search and Data Mining, WSDM '19, p. 96–104. Association for Computing Machinery, New York, NY, USA (2019). https://doi.org/10.1145/3289600.3291014

58. Zhao, J., Henriksson, A., Asker, L., Boström, H.: Predictive modeling of structured electronic health records for adverse drug event detection. BMC Medical Informatics and Decision Making **15**(Suppl 4), S1 (2015). https://doi.org/10.1186/1472-6947-15-S4-S1

# Chapter 10
# On the Interplay of Subset Selection and Informed Graph Neural Networks

**Niklas Breustedt, Paolo Climaco, Jochen Garcke, Jan Hamaekers,**
**Gitta Kutyniok, Dirk A. Lorenz, Rick Oerder, and Chirag Varun Shukla**

**Abstract** Machine learning techniques paired with the availability of massive datasets dramatically enhance our ability to explore the chemical compound space by providing fast and accurate predictions of molecular properties. However, learning on large datasets is strongly limited by the availability of computational resources and can be infeasible in some scenarios. Moreover, the instances in the datasets may not yet be labelled and generating the labels can be costly, as in the case of quantum chemistry computations. Thus, there is a need to select small training subsets from large pools of unlabeled data points and to develop reliable ML

N. Breustedt
IAA, Technical University Braunschweig, Braunschweig, Germany
e-mail: n.breustedt@tu-braunschweig.de

P. Climaco
INS, University of Bonn, Bonn, Germany
e-mail: climaco@ins.uni-bonn.de

J. Garcke (✉)
Fraunhofer SCAI, Sankt Augustin, Germany
INS, University of Bonn, Bonn, Germany
e-mail: jochen.garcke@scai.fraunhofer.de

J. Hamaekers · R. Oerder
Fraunhofer SCAI, Sankt Augustin, Germany
e-mail: jan.hamaekers@scai.fraunhofer.de; rick.oerder@scai.fraunhofer.de

G. Kutyniok
LMU Munich, Munich, Germany
University of Tromsø, Tromsø, Norway
DLR - German Aerospace Center, Wessling, Germany
e-mail: kutyniok@math.lmu.de

D. A. Lorenz
Center for Industrial Mathematics, Faculty 3, Universität Bremen, Bremen, Germany
e-mail: d.lorenz@uni-bremen.de

C. V. Shukla
LMU Munich, Munich, Germany
e-mail: shukla@math.lmu.de

© The Author(s) 2025
D. Schulz, C. Bauckhage (eds.), *Informed Machine Learning*,
Cognitive Technologies, https://doi.org/10.1007/978-3-031-83097-6_10

methods that can effectively learn from small training sets. This chapter focuses on predicting the molecules' atomization energy in the QM9 dataset. We investigate the advantages of employing domain knowledge-based data sampling methods for an efficient training set selection combined with informed ML techniques. In particular, we show how maximizing molecular diversity in the training set selection process increases the robustness of linear and nonlinear regression techniques such as kernel methods and graph neural networks. We also check the reliability of the predictions made by the graph neural network with a model-agnostic explainer based on the rate-distortion explanation framework.

## 10.1  Introduction

Modelling the relationship between molecules and their properties is of great interest in several research areas, such as computational drug design [37], material discovery [43] and battery development [3]. The field of computational chemistry offers powerful *ab initio* methods to compute physical and chemical properties of atomic systems.

Unfortunately, these approaches are often limited by their high computational complexity, which restricts their practical applicability to only small sets of molecules. Therefore, machine learning (ML) methods for molecular property prediction have recently gained increased attention in molecular and material science because of their computational efficiency and accuracy on par with established first principle methods [4, 5, 11]. However, to effectively employ ML in real-world problems, there is a need for labelled datasets that can effectively represent the chemical space of interest, i.e., sets of molecules for which the target properties have already been computed using ab initio methods. Thus, on the one hand, accurately choosing which data points to label in the analyzed chemical space is crucial to avoid creating a dataset with redundant information and limiting the required amount of ab initio calculations. On the other hand, it is critical to develop data-efficient ML methods that perform accurate predictions.

Integrating domain knowledge of physical and chemical principles into the dataset selection process and the development of ML techniques is a primary goal of the chemical and material science ML community [31]. Physical and chemical principles, such as spatial invariances, symmetries, algebraic equations and chemical properties, can increase the robustness, reliability and effectiveness of ML methods while reducing the required training data [6, 50].

This chapter focuses on predicting the atomization energies of molecules in the QM9 dataset [47, 49] and shows how to exploit domain knowledge to select training sets according to specific criteria and how different ML methods may benefit from training on sets selected through such criteria. Specifically, by using Mordred [42], a publicly available library, we generate knowledge-based vector representations of molecules based on their SMILES representation [59] without requiring any ab initio computations. Further, based on such a molecular vector-based representation, we define a training set selection process and can observe that a diversity in the selected subset can increase the reliability of ML methods, indicated by the reduction of the maximum absolute error of the prediction. The maximum absolute error can be interpreted as a measure of robustness, and it is a helpful metric to evaluate ML methods in chemical and material science [65] since the average error alone gives an incomplete impression [19, 56]. Furthermore, this chapter shows how diversity reduces the gap between the predictive robustness of linear regression-based approaches relying only on the molecular topological information, such as kernel ridge regression (KRR) [32], and non-linear approaches relying on molecular geometric representations obtained through ab initio computations, such as graph neural networks (GNN) [17, 24, 26]. We compare the effectiveness of a diversity-based selection with that of random sampling and of an alternative selection approach based on domain knowledge that focuses on representativeness, i.e., the distribution of chosen properties of the dataset should be present with the same amount in the selected training sets.

Finally, we note that our GNNs are inherently opaque (i.e. the logic flow to the decision-making process of the neural network is obscured). This inherently opaque nature of common deep neural network architectures has led to a rise in demand for trustworthy explanation techniques, which vary in their meaning and validity [48]. Unlike other modalities in computer vision and natural language processing, the non-Euclidean nature of graph-structured data poses a significant challenge to trustworthy and interpretable explanation generation. To this end, there exist a variety of explanation techniques and explanation types [12, 21, 36, 46, 52, 60, 62, 64], the most popular of which are subgraph explanation techniques.

We probe the domain knowledge learned/retained by our GNNs for different sampling strategies through the application of a novel *post-hoc* model-agnostic explanation technique, graph rate-distortion explainer (GRDE). GRDE builds on the existing rate-distortion explanation (RDE) framework [27, 38] to generate *instance-level* subgraph explanations on the input graphs, which highlight the substructures and features in the graph that are most relevant towards the GNNs' predictions.

After describing related work, we give in the following first an overview on three ML models that are designed for the prediction of molecular properties but are based on different underlying working principles. In this way, we hope that our results yield insights for a variety of methods that are used in practical applications. Following that, we discuss two ways of sampling subsets from a larger dataset, one aiming to maximize the diversity of the selected samples and the other seeking to choose a collection of points representative of the set from which we sample. Afterwards, we test the introduced methods, namely the SchNet, KRR and the spatial 3-hop convolution network which is proposed in this chapter, by performing numerical experiments on the QM9 dataset while putting special emphasis on the effects of the sampling strategies. After a discussion of the numerical results and a comparison between the different ML models, we seek explanations of the model predictions by applying GRDE to one of the employed graph neural networks.

## 10.2   Related Work

In recent years, there has been growing interest in incorporating domain-specific knowledge into the selection of training data and the development of learning algorithms, which is referred to as Informed Machine Learning [50]. Ideally, the training data selection process should be based purely on the data's features, as labels may be expensive to compute, and should be model-independent so that the selected training data is beneficial for multiple learning models rather than just one. This allows for greater flexibility in model selection and avoids the need for repeating the dataset selection process for each model. Considering these practical aspects, it is clear that a feature-based and model-independent selection process is desirable for efficient and effective Machine Learning. This section reviews some of the relevant work in this area. Coreset approaches [14] are among the most popular strategies for feature-based and model-independent selection of training datasets. Several of these approaches involve incorporating domain-knowledge into the selection of training data by selecting data points that are representative of the distribution of the target points for which we want to predict the new labels. The simplest and yet one of the most common coreset approaches is uniform sampling, which involves selecting a random subset of data points from the larger dataset. Uniform sampling is also considered a benchmark for every other selection approach. Unfortunately, uniform sampling does not exploit domain knowledge and can lead to biased results if the dataset is imbalanced or if certain data points are more important than others. To address this issue, importance sampling [7] is an approach that exploits domain knowledge to assign weights to each data point based on its importance or relevance to the problem at hand. The weights are then used for a nonuniform selection of the training set that privileges more important data points. Another class of methods are the grid-based approaches [2], which involve dividing the feature space into a grid and selecting one or more representative points from each grid's cell. This can be useful for problems with a high-dimensional feature space or when there is a need for a more structured selection of data points. Greedy constructions are coreset approaches that iteratively select the most informative data points based on a pre-defined criterion. For instance, well-known greedy selection methods are submodular function maximization algorithms [30]. Greedy approaches can be effective for selecting a small subset of highly informative data points, but they may be computationally expensive for large datasets. Overall, the choice of coreset approach depends on the specific problem and dataset characteristics, as well as computational constraints. See [14] for a more detailed review of coreset approaches. Finally, the field of experimental design [61] offers additional sampling strategies to perform a feature-based selection of the training set that can benefit specific regression model classes, e.g., linear models.

In this chapter, incorporating domain knowledge in the learning of algorithms refers to methods which are known as informed graph neural networks. While graph neural networks recently gained increasing attention by the works from Gori et al. [18] and Scarselli et al. [51], the question of how to use domain knowledge to improve the performance of learning methods dates back to the last century (e.g. see [23] or [29]). More recently, physics informed neural networks, which address supervised learning tasks complying with the known laws from physics, are a hot topic in several applications, e.g. to find surface breaking cracks in a metal plate [54] or to solve inverse heat transfer problems [8]. For graph neural networks, based on the message passing principle, i.e. the process of updating so called states or representations attached to each node of a graph using the node's neighbourhood, many different models were proposed (e.g. ChebNet [10], Gated Graph Neural Networks [34], Graph Attention Networks [58]), the most popular being the graph convolutional model by Kipf and Welling [26] which is motivated by an approximation of spectral graph convolutions. Combining incorporating domain knowledge with graph neural networks leads to the very recent informed graph neural networks. In [20] the authors combine theory from thermodynamics with graph neural networks to predict the behaviour of dynamical systems and in [25] combine physical properties of molecules are combined with graph neural networks to predict the cetane number of possible alternative fuels. For more detailed overviews on GNNs or informed neural networks we refer to the book [35] and a recent review [9].

We further build upon the interpretability of graph neural networks in this chapter by introducing a method akin to perturbation techniques on image data to graph-structured data. The main goal of interpretability is to invoke transparency in the otherwise opaque prediction process of neural networks, and is further applicable in the detection of bias as well as to explain incorrect classifications in the predictive model. Previous work in interpretability for other modalities such as audio and images [27, 38] has shown great success in identifying a neural network's sensitivity to specific subsets of the input data. More specifically, among the variety of local and global interpretability techniques, perturbation [27, 28] and gradient-based [55] techniques have been shown to accurately capture a predictive model's sensitivityto some concepts in the input data. These techniques generally seek to optimize a heatmap over the input data such that high-intensity zones are the most relevant to the model's prediction for the given data point. We further discuss this in detail with respect to graphs in Sect. 10.3.4.

Inspired by the exhaustive work on interpretability for other modalities, several methods [36, 46, 52, 60] have also been proposed for graph-structured data, with perturbation techniques such as GNNExplainer [60] being the baseline for comparison. For a detailed overview of GNN interpretability, we refer to [63]. These techniques, however, have been shown to suffer from unfaithfulness on large graphs since they optimise masks only for small graphs as well as manually threshold their relevance scores. See [1] for a detailed review on the current issues with graph interpretability.

## 10.3 Methods and Sampling Strategies

This section introduces the approaches we use for predicting the atomization energy, explaining the GNN output and sampling the training data. Section 10.3.1 introduces the benchmark regression model SchNet, a GNN that uses 3-dimensional positional information to predict chemical properties. Next, Sects. 10.3.2 and 10.3.3 describe KRR and the spatial 3-hop convolution network, respectively. Both these approaches only exploit topological information encoded in the SMILES to perform the energy prediction task. Section 10.3.4 presents the rate-distortion explanation framework for graph data that we use to showcase the domain knowledge learned by the 3-hop convolution network. Finally, Sect. 10.3.5 introduces the approaches we use for the selection of training sets.

### 10.3.1 SchNet

SchNet is a symmetry-informed neural network model, designed for the prediction of chemical properties by Schütt et al. [53]. In contrast to the methods presented in Sects. 10.3.2 and 10.3.3, it is trained and evaluated on 3-dimensional structural information describing the atomic systems of interest. Usually, the positional information is obtained from computational methods such as density functional theory (DFT).

More formally, for an atomic system with N atoms, SchNet can be used to predict scalar properties as a function $f$ of $3N$ atomic coordinates (nuclear positions) and on $N$ atomic numbers of the corresponding atoms

$$f : \mathbb{R}^{3N} \times \mathbb{N}^{N} \rightarrow \mathbb{R}. \tag{10.1}$$

Internally, SchNet operates on a distance-based neighborhood graph, defined by a cutoff radius $r_{\text{cut}}$, in which nodes correspond to the atoms in the atomic system. In this scenario, edges do not necessarily correspond to chemical bonds but merely indicate whether two atoms are closer than the chosen cutoff radius. Hence, the chosen cutoff radius has a direct influence on the graph shown to the model. Similar to other GNNs [17, 26], SchNet operates in a layer-wise fashion by iteratively updating feature representations. At the $l$-th layer each atom, indexed by $i \in \{1, 2, ..., N\}$, is represented by a feature vector $x_i^l \in \mathbb{R}^F$ where $F$ is a hyperparameter. The main layer introduced by Schütt et al. is the continuous-filter convolutional layer: Denoting the atomic positions by $r_i \in \mathbb{R}^3$, this layer updates the atomic features as follows

$$x_i^{l+1} = \sum_{j \in \mathcal{N}(i)} x_j^l \circ W^l \left( r_i - r_j \right), \tag{10.2}$$

where $W^l : \mathbb{R}^3 \to \mathbb{R}^F$ is a trainable filter-generating function and $\circ$ denotes element-wise multiplication. In detail, $W^l$ is given as the composition $W^l = \tilde{W}^l \circ \varphi$ of a distance-based radial basis expansion

$$\varphi : r_i - r_j \mapsto \bigoplus_{k=1}^{N_{\text{radial}}} \exp\left(-\gamma\left(\|r_i - r_j\|_2 - \mu_k\right)^2\right) \tag{10.3}$$

and a trainable neural network $\tilde{W}^l$, where $0\,\text{Å} \leq \mu_k \leq 30\,\text{Å}$ are equidistributed centers and $\gamma = 10\,\text{Å}$. Here, $\bigoplus$ denotes the direct sum that concatenates the scalar outputs of the radial basis functions to a feature vector in $\mathbb{R}^{N_{\text{radial}}}$ which is then passed into $\tilde{W}^l$. Note that $\varphi$ is invariant with respect to actions of the orthogonal group $O(3)$ which assures that the predictions of SchNet are invariant with respect to translations, rotations and reflections of the input structure as well. Depending on the atomic species, initial embeddings $x_i^0$ are sampled from an $F$-dimensional standard normal distribution and optimized during the training process. In addition, non-linear layers such as dense feed-forward neural networks can be applied to the node features in order to increase the expressiveness of the model.

By summing over the images of a trainable readout function $R : \mathbb{R}^F \to \mathbb{R}$, the final node features in the last layer $L$ are transformed into a prediction of the target property $\hat{y}$:

$$\hat{y} = \sum_{j=1}^{N} R\left(x_j^L\right) \tag{10.4}$$

Involving only permutation-invariant operations such as the summation over adjacent atoms, the output is invariant with respect to mutual permutations of the atomic positions and atomic species. For more details on the model architecture see [53].

## 10.3.2 Kernel Ridge Regression

In kernel ridge regression, a vector-based representation of the molecules is mapped into a high-dimensional space using a non-linear map that is implicitly determined by defining a kernel function, which provides a measure of similarity between the molecular representations. The structure-energy relationship is learned in the high-dimensional space. In this chapter, we use the so-called Gaussian kernel

$$k(x_i, x_j) := e^{-\frac{\|x_i - x_j\|_2^2}{2\nu^2}}, \tag{10.5}$$

where $\|\cdot\|_2$ is the $L_2$-norm and $\nu \in \mathbb{R}$ a kernel hyperparameter to be selected through an optimization process. The kernel ridge regression model is constructed using the selected training set $\{x_i, y(x_i)\}_{i=1}^{P}$, where $\{x_i\}_{i=1}^{P}$ are the Mordred [42] based vector representations of the molecules and $\{y(x_i)\}_{i=1}^{P}$ the associated atomization energies. Once the regression model has been constructed, the predicted energies are given by the scalar values $\tilde{y}(x)$ defined as follows

$$\tilde{y}(x) := \sum_{i=1}^{p} \alpha_i k(x, x_i) , \tag{10.6}$$

where the vector $\boldsymbol{\alpha} = [\alpha_1, \alpha_2, \ldots, \alpha_p]^T \in \mathbb{R}^p$ is the solution of the following minimization problem

$$\boldsymbol{\alpha} = \underset{\tilde{\alpha}}{\operatorname{argmin}} \sum_{i=1}^{p} (\tilde{y}(x_i) - y(x_i))^2 + \lambda \tilde{\boldsymbol{\alpha}}^T K \tilde{\boldsymbol{\alpha}}. \tag{10.7}$$

Here, $K$ is the kernel matrix, i.e., $K_{i,j} = k(x_i, x_j)$, and the parameter $\lambda \in \mathbb{R}$ is the so-called regularization parameter that penalizes larger weights. The analytic solution to the minimization problem in (10.7) is given by

$$\boldsymbol{\alpha} = (K + \lambda I)^{-1} \tilde{y} , \tag{10.8}$$

where $\tilde{y} = [\tilde{y}(x_1), \tilde{y}(x_2), \ldots, \tilde{y}(x_p)]^T$. Once the training process has been concluded and the regression parameters $\{\alpha_i\}_{i=1}^{P}$ have been learned, the energy predictions for molecules not included in the training set can be computed using (10.6).

### 10.3.3   Spatial 3-Hop Convolution Network

In addition to the two previous approaches, we propose a third approach which builds on a newly developed spatial graph convolution structure. We call this approach spatial 3-hop convolution network. This approach exploits the graph structure, the node features and optionally edge features for regression or classification but does not need 3-dimensional structural information as is the case for SchNet.

A commonly used graph convolutional network by Kipf and Welling [26] is motivated by an approximation of a spectral convolution. Thereby, they consider spectral convolutions as

$$w \star x = U w U^\top x, \tag{10.9}$$

where $\boldsymbol{w} = diag(\theta) \in \mathbb{R}^{n \times n}$ is a filter, $\boldsymbol{x} \in \mathbb{R}^n$ is a graph signal on a graph with $n$ nodes, $\star$ denotes the spectral graph convolution operator and $\boldsymbol{U}$ is the matrix of eigenvectors from the eigendecomposition of the normalized graph Laplacian $\boldsymbol{I}_n - \boldsymbol{D}^{-\frac{1}{2}} \boldsymbol{A} \boldsymbol{D}^{-\frac{1}{2}}$. Moreover, $\boldsymbol{A}$ is the adjacency matrix of the underlying graph, $\boldsymbol{D}$ is the corresponding degree matrix and $\boldsymbol{I}_n$ is the $n \times n$ identity matrix. This convolution is approximated and generalized to matrix-valued graph signals which leads to the update of the graph convolutional network

$$\boldsymbol{H}^{(l+1)} = \sigma(\boldsymbol{H}^{(l)} \boldsymbol{W}_0 + \boldsymbol{D}^{-\frac{1}{2}} \boldsymbol{A} \boldsymbol{D}^{-\frac{1}{2}} \boldsymbol{H}^{(l)} \boldsymbol{W}_1), \tag{10.10}$$

where $\boldsymbol{H}^{(l)}$ is the matrix of hidden representations of the $l$-th layer, $\boldsymbol{W}_0$ and $\boldsymbol{W}_1$ are learnable parameters and $\sigma$ denotes the elementwise ReLU function. For the spatial 3-hop convolution layer we do not consider spectral graph convolutions but an intuitive spatial convolution using powers of the graphs adjacency matrix to calculate so called path matrices. Within these, for each node the number of paths of a certain length to every other node is stored. By defining a spatial convolution with path matrices and building a layer of the graph neural network using the convolution, we consider the number of paths of a given length from node $v$ to node $u$ as a measure for the impact of node $v$ on node $u$. Thus, nodes with more paths to the considered node will be taken into account more during the update.

For a graph $G$ with $n$ nodes a path is defined as a sequence of nodes $(1, \ldots, k)$ with $k < n$ such that for any $i, j \in (1, \ldots, k)$ it is $i \neq j$, i.e. no node appears twice.

With that, we define a spatial $k$-hop graph convolution of a graph signal $\boldsymbol{x} \in \mathbb{R}^n$ with a filter $\boldsymbol{w} \in \mathbb{R}^k$ on an undirected graph $G$ with $n$ nodes as

$$\boldsymbol{w} \star_k \boldsymbol{x} := \sum_{i=0}^{k} \boldsymbol{w}_i \boldsymbol{T}^{(i)} \boldsymbol{x},$$

where $\boldsymbol{T}^{(i)}$ is a path matrix such that $\boldsymbol{T}^{(i)}_{vu}$ is the number of paths of length $i$ from node $v$ to node $u$.

An approach to computing the needed path matrices is a recursion that starts with the adjacency matrix. Since the adjacency matrix equals the path matrix for paths of length one it is $\boldsymbol{T}^{(1)} = \boldsymbol{A}$. For every node $i$ and $u$ a neighbor of it, the number of paths of length two from node $i$ to node $j$ equals the number of paths of length one from $u$ to $j$ in which $i$ is not a part of. More generally, the number of paths of length $k$ from a node $i$ to a different node $j$ equals the sum of all paths from node $u$ to $j$ of length $k - 1$ over all $u \in \mathcal{N}(i)$ in which $i$ does not appear. Using this, it can be shown that $\boldsymbol{T}^{(2)} = \boldsymbol{A}^2 - \boldsymbol{D}$ and $\boldsymbol{T}^{(3)} = \boldsymbol{A}^3 - \boldsymbol{\Sigma} \circ \boldsymbol{A}$, where $\boldsymbol{A}$ and $\boldsymbol{D}$ are as above and $\boldsymbol{\Sigma}$ is an $n \times n$ matrix with $\boldsymbol{\Sigma}_{ij} = \boldsymbol{D}_{ii} + \boldsymbol{D}_{jj}$. This shows that the 3-hop spatial graph convolution is given by

$$\boldsymbol{w} \star_3 \boldsymbol{x} = (\boldsymbol{w}_0 \boldsymbol{I}_n + \boldsymbol{w}_1 \boldsymbol{A} + \boldsymbol{w}_2 (\boldsymbol{A}^2 - \boldsymbol{D}) + \boldsymbol{w}_3 (\boldsymbol{A}^3 - \boldsymbol{\Sigma} \circ \boldsymbol{A})) \boldsymbol{x}.$$

Note that the $w_k$'s can be seen as weights for the $k$-hop neighborhoods. A generalization of the former discussion to a signal $X \in \mathbb{R}^{n \times d}$ with $c$ node features for each node (analogously to Kipf and Welling [26]) leads to

$$H = X W_0 + A X W_1 + (A^2 - D) X W_2 + (A^3 - \Sigma \circ A) X W_3$$

which results in the spatial 3-hop convolution layer, the message passing layer of the spatial 3-hop convolution network,

$$H^{(l+1)} = \sigma(H^{(l)} W_0 + A H^{(l)} W_1 + (A^2 - D) H^{(l)} W_2 + (A^3 - \Sigma \circ A) H^{(l)} W_3),$$

where $\sigma$ is, again, the element-wise ReLU function.

## 10.3.4 Graph Rate-Distortion Explanations

We now present a formulation for the rate-distortion explanation framework [27, 38] for graph data. Given a pre-trained GNN model, $\Phi : \mathbb{R}^{n \times c} \longrightarrow \mathbb{R}^m$ and a set of attributed graphs $G = \{G_1, G_2, ..., G_p\}$ such that $G_i = (V_i, E_i, X_i)$ for all $i \in [1, p]$, our task is to explain the model decision over the set $G$, or more locally, $\Phi(G_i)$. This leads us to the two general branches of explanation techniques: global and local explanations. Global explanation techniques focus on explaining the underlying function learned by the model, $\Phi$. This can be done in a multitude of ways, such as testing the model's sensitivity to a concept [40] or reconstructing graphs from the embedding space learned by the model to reveal important motifs [62]. In general, global explanation techniques, while useful, are hard to construct and are unable to detect finer details on local data points. On the other hand, local explanation techniques, which are the more popular alternative, focus on explaining $\Phi$ for local instances, i.e. $\Phi(G_i)$. Similar to global explanations, there exist a variety of approaches, such as perturbation-based methods [36, 52, 60], surrogate methods [21], gradient-based methods [46], and additive methods [12, 64], each with their benefits and limitations. These techniques aim to extract information from $G_i$ that is most relevant to the local prediction $\Phi(G_i)$. More concretely, given a graph $G_i = (A_i, X_i)$, local explanation techniques commonly attempt to extract a subgraph $\hat{G}_i = (\hat{A}_i, \hat{X}_i) \subseteq G_i$ that is most relevant to the model for its prediction $\Phi(G_i)$. The rate-distortion framework for explaining graphs is a local, post-hoc, model-agnostic explanation technique that comes under the umbrella of perturbation-based graph explainers. Given the pre-trained model $\Phi$ and graph $G_i$, GRDE optimizes a binary deletion mask $S = (S_A, S_X)$ over $G_i$ to obtain a subgraph $\hat{G}_i$ such that $\Phi(\hat{G}_i)$ approximates $\Phi(G_i)$. Mask $S$ thus retains only the edges and features that are most relevant to the model's prediction on $G_i$. Given $A_i \in \mathbb{R}^{n \times n}$ and $X_i \in \mathbb{R}^{n \times f}$, where $n$ is the number of nodes and $f$ is the number of node features, our goal is to optimize masks $S_A \in [0, 1]^{n \times n}$ and $S_X \in [0, 1]^{n \times f}$. Let $\mathcal{V}_S = (\mathcal{V}_{S_A}, \mathcal{V}_{S_X})$ be probability distributions that can either be chosen manually or learned from the graph dataset. Then the obfuscation on $G_i$, i.e. the subgraph $\hat{G}_i$, can be defined as

$$\hat{G}_i = (\hat{A}_i, \hat{X}_i) = (A_i \odot S_A + (1 - S_A) \odot v_{S_A}, X_i \odot S_X + (1 - S_X) \odot v_{S_X}), \quad (10.11)$$

where $v_{S_A} \in \mathcal{V}_{S_A}$, $v_{S_X} \in \mathcal{V}_{S_X}$, and $\odot$ denotes element-wise multiplication. Intuitively, this implies that the masks $S$ keep some of the elements in $G_i$ while the elements that are not selected by $S$ are replaced with values from the probability distribution $\mathcal{V}_S$ as 'noise'. In general, the choice of $\mathcal{V}_S$ should be such that the resulting subgraph $\hat{G}_i$ remains within the data manifold, provided that the data manifold is known. Depending on the information in $G_i$, we can use a variety of probability distributions for $(\mathcal{V}_{S_A}, \mathcal{V}_{S_X})$. For example, in the case of a binary adjacency matrix, $\mathcal{V}_{S_A}$ can be the Gumbel-Softmax distribution, whereas for real-valued adjacency matrices and node feature matrices, $\mathcal{V}_S$ can be Gaussian distributions. We can also learn the probability distributions $\mathcal{V}_S$ from the data manifold itself, as previous attempts have shown success with inpainting GANs [27] for this strategy on other data modalities.

Furthermore, we define the expected distortion on $G_i$ with respect to the masks $S$ and perturbation distributions $\mathcal{V}_S$ as

$$\mathcal{D}(G_i, S, \mathcal{V}_S, \Phi) = \mathop{\mathbb{E}}_{v_{S_A} \in \mathcal{V}_{S_A}, v_{S_X} \in \mathcal{V}_{S_X}} \left[ d(\Phi(G_i), \Phi(\hat{G}_i)) \right], \quad (10.12)$$

where $d : \mathbb{R}^m \times \mathbb{R}^m \longrightarrow \mathbb{R}_+$ is the measure of distortion between the two model outputs. Commonly, we can set $d$ as the $\mathcal{L}^2$ distance or the KL-divergence between the two model outputs. Thus, we can define the rate-distortion explanation on $G_i$ as the optimal subgraph $\hat{G}_i$ that solves the minimization problem

$$\min_{S=(S_A, S_X)} \mathcal{D}(G_i, S, \mathcal{V}_S, \Phi) \text{ s.t. } \|S_A\|_0 \le j, \|S_X\|_0 \le k, \quad (10.13)$$

where $j, k$ are the desired levels of sparsity for $S_A, S_X$ respectively.

Note that solving (10.13) is $\mathcal{NP}$-hard [38]. Thus, we use an $l_1$ relaxation on (10.13) to get the relaxed optimization problem given by

$$\min_{S=(S_A, S_X)} \mathcal{D}(G_i, S, \mathcal{V}_S, \Phi) + \lambda_A \|S_A\|_1 + \lambda_X \|S_X\|_1, \quad (10.14)$$

where $\lambda_A, \lambda_X > 0$ are hyperparameters to control the sparsity level of the masks. We can further relax the binary masks $S$ by sampling them from the concrete distribution [39] or Gumbel Softmax distribution [22]. This allows us to solve the optimization problem in (10.14) with differentiable techniques such as stochastic gradient descent.

## 10.3.5 Sampling Strategies

We now introduce two approaches for sampling a set of points from a large dataset. The first method focuses on maximising the diversity of the selected set, while the second aims to select a set that is representative of the whole dataset.

### 10.3.5.1 Diversity

In short, diverse subsets are iteratively selected from $\Omega \subset \mathbb{R}^d$ using the *farthest point sampling* (FPS) algorithm [13], where the resulting subset is a sub-optimal minimizer of the *fill distance*. We denote this approach by FPS.

To maximize diversity of the selection we consider the concept of fill distance. Given a dataset $\Omega \subset \mathbb{R}^d$ consisting of a finite amount of unique points, and $X = \{x_1, x_2, \ldots, x_p\} \subset \Omega$ a subset of cardinality $p = |X| \in \mathbb{N}$ we define the fill distance of $X$ in $\Omega$ as

$$h_{X,\Omega} := \max_{x \in \Omega} \min_{x_j \in X} \|x - x_j\|_2. \tag{10.15}$$

Put differently, we have that any point $x \in \Omega$ has a point $x_j \in X$ not farther away than $h_{X,\Omega}$. Notice that, if $X, \bar{X} \subset \Omega$ with $p = |X| = |\bar{X}|$ and $h_{X,\Omega} < h_{\bar{X},\Omega}$ then $X$ consists of data points that are more widely distributed in $\Omega$, thus more diverse, than those in $\bar{X}$.

Fixing the number of points $p \in \mathbb{N}$ we want to select from $\Omega$, we aim to find $X \subset \Omega$ such that

$$X = \underset{\bar{X} \subset \Omega, |\bar{X}| = p}{\arg\min} \; h_{\bar{X},\Omega}. \tag{10.16}$$

The naive approach to solve the minimization problem in (10.16) would first require computing the fill distance for all possible sets $X \subset \Omega$ with $|X| = p$ and then choosing one of those sets where the minimum of the fill distance is attained. Unfortunately, such an approach is very time consuming and computationally intractable. Therefore, as an alternative approach we use the FPS algorithm [13]. FPS is a greedy selection method, which means that the points are progressively selected starting from an initial a-priori chosen point, i.e., given a set of selected points $X^s = \{x_1, x_2, \ldots, x_s\} \subset \Omega$ with cardinality $|X^s| = s < p$, the next chosen point is

$$x_{s+1} = \arg\max_{x \in \Omega} \min_{x_j \in X^s} \|x - x_j\|. \tag{10.17}$$

$x_{s+1}$ is the point which is farthest away from the points in $X^s$ and it is the point where the fill distance $h_{X^s, \Omega}$ is attained. In other words, the next selected sample is the center of the largest empty ball in the dataset.

#### 10.3.5.2   Representativeness

We say that data points are representatively selected for the entire dataset, when the distribution of properties in the selected subset are as close as possible to the corresponding distribution in the whole dataset. To this aim, we divide $\Omega$ into clusters and select data points from them so that the distribution of the clusters in the subset resembles that of the whole dataset. For example, if we divide $\Omega$ into two clusters, each containing 50% of the data points, we aim to select a subset consisting of data points which are also equidistributed in the two clusters. The clustering can be performed by clustering algorithms or be based on properties and criteria stemming from domain knowledge, i.e., in the sense of [50] the training data is selected based on scientific knowledge. Furthermore, data points within each cluster are selected using the farthest point sampling, which ensures that in the various clusters a set of diverse data points is chosen. We call this approach cluster-based farthest point sampling (C-FPS).

## 10.4   Numerical Experiments

### 10.4.1   QM9 Dataset

In this chapter, we analyze the publicly available QM9 dataset [47, 49] containing a diverse set of organic molecules. Precisely, the QM9 consists of 133,885 organic molecules in equilibrium with up to 9 heavy atoms of four different types: C, O, N and F. The dataset provides the SMILES [59] representation of the relaxed molecules, their geometric configurations and 19 physical and chemical properties. To guarantee a consistent dataset, we remove all 3054 molecules that failed the consistency test proposed by [47]. Moreover, we remove the 612 compounds for which the RDKit package [33] can not interpret the SMILES. After this preprocessing procedure, we obtain at a smaller version of the QM9 dataset consisting of 130,219 molecules.

#### 10.4.1.1   Knowledge Based Molecular Representation

The domain knowledge based molecular representation we employ is based on Mordred [42], a publicly available library that exploits the molecules' topological information encoded in the SMILES strings to provide 1826 physical and chemical features. Such molecular features are defined as the "final result of a logical and mathematical procedure, which transforms chemical information encoded within a symbolic representation of a molecule into a useful number or the result of some standardized experiment" [57] and encode scientific knowledge reflecting algebraic equations, logic rules, or invariances [50]. Using the Mordred library, we represent each molecule in the QM9 dataset with a high-dimensional vector where each vector's entry is associated with a distinct feature.

To work with a more compact representation, after generating the Mordred vectors, we use the CUR [41] approach to select a subset of relevant features. The CUR algorithm takes as input the Mordred vector representation of each of the molecules in the analyzed dataset and ranks the significance of the features by associating them with an importance score. We select the first 59 top-ranked features and normalize their values in the range (0,1) using the "MinMaxScaler" function provided by the scikit-learn python library [44]. Moreover, to ensure the uniqueness of the representation, we consider an additional set of features representing the atom type distribution within each molecule. Specifically, for each data point, we add five features, each expressing the amount of atoms of a particular type within the molecule, in percentage. The possible atom types are H, C, O, N and F. In conclusion, the Mordred based representation we employ to sample the QM9 dataset consists of 64-dimensional vectors.

### 10.4.1.2  Diverse and Representative Sets of Molecules

The knowledge related to the molecules in the QM9 enables us to employ the data sampling strategies introduced in Sect. 10.3.5 to create diverse and representative sets.

Diverse sets are constructed using the FPS algorithm on the Mordred-based vector representations of the molecules in the QM9. The Mordred vectors allow the representation molecules as points in $\mathbb{R}^d$, $d \in \mathbb{N}$, and the definition of a distance between the molecules, the Euclidean distance. Thus, we represent the QM9 as a finite set $\Omega \subset \mathbb{R}^d$ and use the FPS to sample from $\Omega$ a sub-optimal minimizer of the fill distance.

Representative sets are constructed using the procedure introduced in Sect. 10.3.5 consisting of segmenting the QM9 in clusters and then sampling from each cluster so that the distribution of the chosen molecules resembles that of the whole dataset. The segmentation procedure is based on the molecules' topological information and considers their size, atom types and bond types, which following [50], reflects scientific knowledge in the selection of training data so that selected molecular properties are invariant per cluster. Specifically, we define the clusters through a process consisting of three main steps. In the first step, we split the molecules according to their sizes. As a result of this first step, we divided the QM9 dataset into 26 sets. After that, we separate each cluster obtained in the first step into subclusters defined by the different heavy atom types within the molecules. Overall, molecules in the QM9 consist of 4 heavy atom types. Thus, each molecule could consist of 15 different combinations of such atom types, e.g., a molecule can contain up to four distinct heavy atoms, and for each amount of distinct heavy atom types, various combinations are possible. After this second step, each of the initial 26 clusters is divided into 15 subclusters. The third and final step is further splitting the data points in each subcluster into different sets according to the various bond types present in each molecule. We consider four different bond types: single, double, triple and aromatic bonds. Thus, each of the subclusters is further divided into 15 distinct sets. As a result of this clustering procedure, we divide the QM9 in 5850 different clusters that account for molecular size, atom types and bond types. Molecules within the clusters are selected using the farthest point sampling, which ensures that in the various clusters, a set of diverse molecules is chosen.

### 10.4.1.3    Sampling the QM9 Dataset

For the experiments, we select training sets of different sizes and according to different strategies from the entire preprocessed QM9 dataset. After that, we test each trained model's predictive accuracy on all the molecules that have not been selected to train it. We construct training sets consisting of 100, 250, 500, 1000 and 5000 samples. Such sets are created following three different selection criteria: random sampling (RDM), as a benchmark, and the two selection strategies introduced in Sect. 10.3.5, namely, diversity sampling (FPS) and representative sampling (C-FPS). For each sampling strategy and training set size, we run the training set selection process independently five times. For RDM, at each run the points are independently and uniformly selected, while in the case of FPS and C-FPS the initial point to initialize the FPS algorithm is independently selected at random at each run. Thus, for each selection strategy and training set size, each of the analyzed models is trained and tested five times, independently. The test results that follow are averaged over the five runs.

We want to point out that sampling the training data non-randomly will lead to a shift between the training and test distribution, as showcased in Fig. 10.1, where we compare FPS with a random selection. It is not obvious how such a bias effects the different models. Note that for Fig. 10.1 we performed the selection twice, with different initialization for FPS and different seeds for the random selection, respectively. We find that changing the initialization for FPS does not lead to a significant change in the distribution, for different seeds in the random selection we make the same observation.

### 10.4.1.4    Measuring the Error

We evaluate the performances of the employed machine learning methods using three different metrics. Specifically, we consider the mean absolute error (MAE), the root mean squared error (RMSE) and the worst-case error. The mean absolute

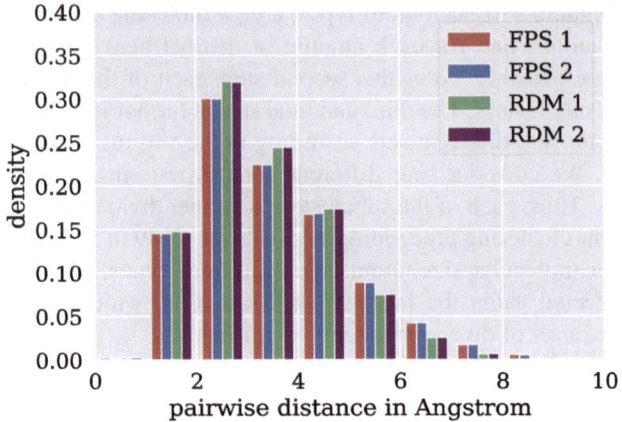

**Fig. 10.1** The distributions of pairwise interatomic distances within a molecule for 5000 molecules sampled with either FPS or randomly (two different splits each) differ

error (MAE) computes the arithmetic average of the absolute errors between the predicted values $\{\tilde{y}_i\}_{i=1}^{N}$ and the ground truths $\{y_i\}_{i=1}^{N}$, that is,

$$\text{MAE} := \sum_{i=1}^{N} |y_i - \tilde{y}_i|, \tag{10.18}$$

where $N \in \mathbb{N}$ is the number of data points in the test set used to evaluate the models. The root mean squared error (RMSE) computes the root of the mean squared error, which is the arithmetic average of the squared errors. It is a measure of how spread out the errors are and it is represented by the formula

$$\text{RMSE} := \sqrt{\sum_{i=1}^{N} (y_i - \tilde{y}_i)^2}. \tag{10.19}$$

The worst-case error calculates the maximum absolute error between the predicted values and the ground truths. It is an indicator of the robustness of a model's predictions, and it is defined as

$$\text{worst-case error} := \max_{1 \le i \le N} |y_i - \tilde{y}_i|. \tag{10.20}$$

### 10.4.2   SchNet

In order to get experimental insights on FPS also for a different class of informed predictive models, we train the publicly available implementation of SchNet from Pytorch Geometric [15] on the defined subsets. In this chapter, we choose a cutoff radius of 4 Å while keeping the other hyperparameters to be the default ones suggested by the Pytorch Geometric implementation (version 2.0.4). Besides the test set that is used for final evaluation, we use random 20% from the training set for evaluation during training and refer to it as validation set. We minimize the $L2$-loss function with respect to the model parameters with the Adam optimizer using mini-batches of 32 molecules per iteration and a learning rate of $7 \cdot 10^{-4}$. The learning rate is decayed by a factor of 0.8 if the validation error has not improved for 50 epochs. After each epoch a checkpoint is saved if the model has achieved a smaller validation loss than the current best model. The training process is stopped after the model has not improved for 200 epochs (early stopping). The best model is then used for assessing the model performance on the test set.

The first thing we observe is that SchNet does not seem to profit from FPS-based sampling strategies when examining the MAE and RMSE (Fig. 10.2a, b) alone. Random sampling consistently leads to approximately equal or smaller measurements for the MAE and RMSE for all investigated training set sizes. However, for 100, 250 and 500 training samples, the worst case error is reduced by at least 0.5 eV when employing FPS for the training set selection (Fig. 10.2b).

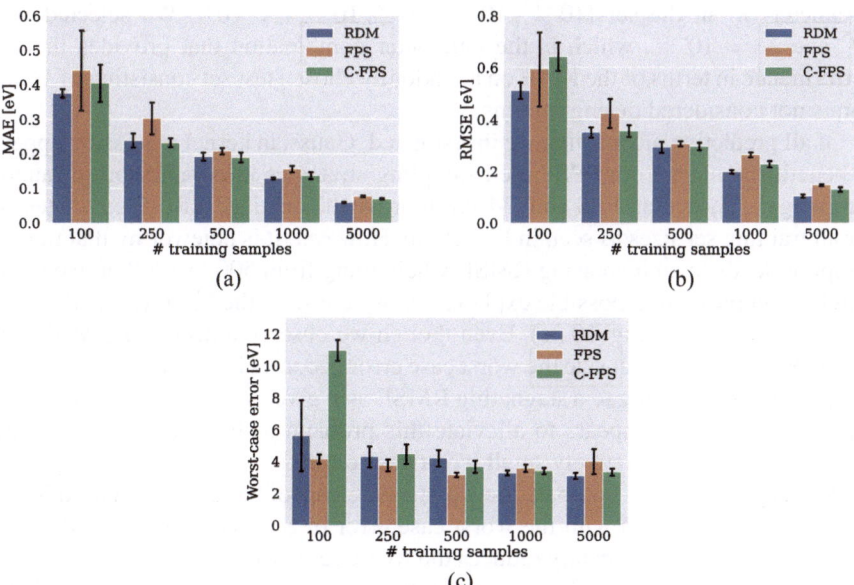

**Fig. 10.2** Results for SchNet. (**a**) MAE. (**b**) RMSE. (**c**) Worst-case error

Considering the comparatively small error bars, we expect FPS to be a reliable technique to reduce the worst case error for small (i.e. $\leq 500$ data points) training sets of QM9. For larger training sets however, this effect vanishes and FPS leads to worse results in the sense of larger worst-case errors. This is possibly due to the fact that FPS is based on Mordred features which yield a rather global description of a molecule. In this sense, FPS selects samples that are maximally far away with respect to those global features. On the contrary, GNNs strongly exploit local information and we believe this discrepancy to be a possible explanation for the merely small effect induced by FPS. However, in absolute numbers, SchNet yields the lowest error metrics of all tested methods. This meets our expectations since it is the only method incorporating geometric information. In fact, the nuclear positions were obtained through DFT calculations and hence the coordinates already encode highly relevant information for predicting the atomization energy. One could argue that SchNet's input is already part of the solution to the problem and view the use of features derived by ab initio methods as some form of information leakage [16], thus making the learning problem easier.

## 10.4.3   Kernel Ridge Regression

The kernel and regression hyperparameters were optimized in a grid search for each of the randomly selected training sets of 1000 points. Specifically, we varied the kernel parameter '$\nu$' in the set $\{10^{-1}, 10^0, 10, 10^2, \ldots, 10^7\}$ and the regularization parameter '$\lambda$' in the set $\{10^{-12}, 10^{-10}, 10^{-8}, 10^{-6}, \ldots, 10^0\}$. We selected $\nu = 10^5$ and $\lambda = 10^{-12}$, which is the parameter combination that provides the best performance in terms of the MAE on a randomly chosen test set consisting of 10,000 points not considered during training.

Of all predictive models that we investigated, Gaussian kernel regression appears to benefit the most from FPS-based sampling strategies in comparison to random sampling. In particular, FPS and C-FPS improve the obtained RMSE on the test set for all training set sizes as seen in Fig. 10.3b. However, it is noteworthy that random sampling leads to an increasing RMSE when going from 500 to 1000 or even 5000 training samples. For a possible explanation, we consider the MAE (Fig. 10.3a) and the worst case error (Fig. 10.3c). Even though we observe a decreasing MAE with an increasing training set size the worst case error becomes larger with more training samples as well, leading to a stagnating RMSE as it gives a higher weight to outliers than the MAE. FPS appears to alleviate this problem as becomes apparent when considering the comparatively small worst case errors.

At this point, we can compare the worst case errors of the SMILES-based Kernel ridge regression (KRR) with the worst case error of SchNet. From Fig. 10.3c it is apparent that FPS significantly reduces the worst-case error of KRR by one order of magnitude compared to random sampling. In order to contextualize this effect better we consider Fig. 10.4 that shows the worst-case errors of SchNet and KRR side by side for different numbers of training samples. We find KRR to approach the values

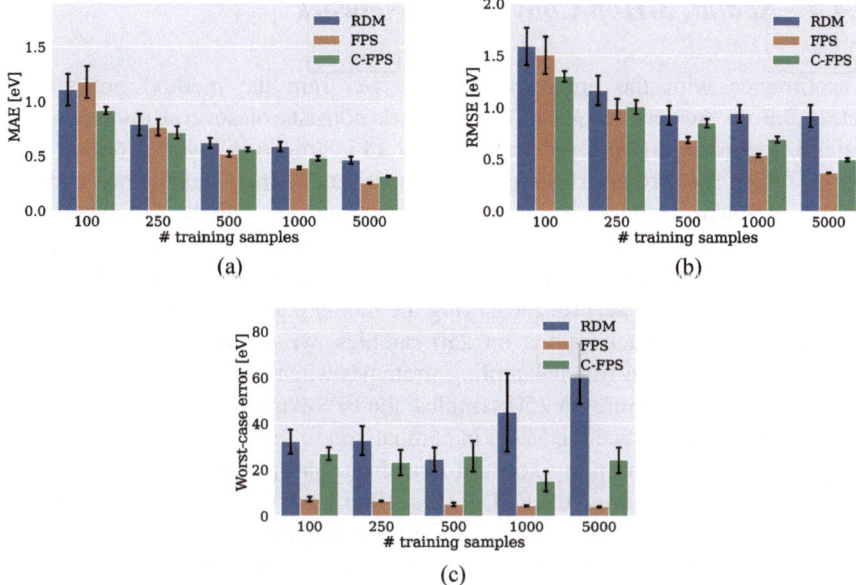

(a)                                                    (b)

(c)

**Fig. 10.3**  Results for Gaussian kernel regression. (**a**) MAE. (**b**) RMSE. (**c**) Worst-case error

**Fig. 10.4**  The worst-case
error of KRR can be reduced
by FPS such that the order of
magnitude is comparable to
SchNet

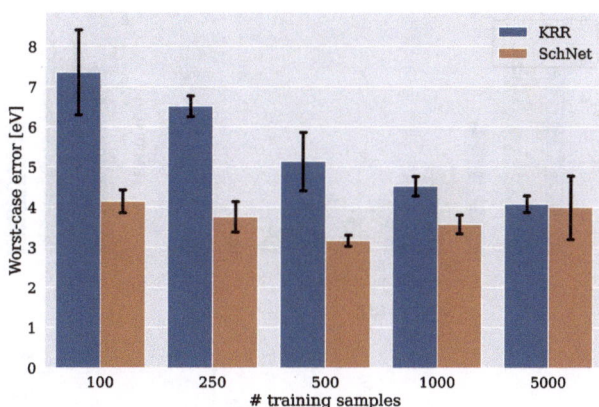

of SchNet with an increasing number of training samples. In particular, we observe
the errors to have the same order of magnitude. This is noteworthy as the KRR
only exploits topological information while SchNet requires the atom coordinates
obtained from DFT as input.

### *10.4.4   Spatial 3-Hop Convolution Network*

In accordance with the previous sections, we train the method presented in
Sect. 10.3.3 on subsets of QM9. The network consists of several updates by the
spatial 3-hop convolution layer, followed by an aggregation layer to obtain graph
features which are further processed by linear layers. During training we minimize
the L1-loss function with respect to the models parameters with the Adam optimizer,
use a learning rate of $2 \cdot 10^{-4}$ and a batch size of 32 molecules per iteration. We train
each model for 500 epochs and choose the best model with respect to a validation
set (20% of the training set) for measuring the model performance on a test set.

Apart from the model trained on 250 samples, we do not observe significant
differences among the different sampling strategies when considering only the MAE
(Fig. 10.5a). When training on 250 samples, the FPS-based methods appear to lead
to an advantage and reduce the MAE in comparison to random sampling. For larger
training sets random sampling seems to catch up and perform on par with FPS-
based sampling. Considering the RMSE (Fig. 10.5b), our observations are somewhat
different. In particular, we find FPS and C-FPS to outperform random sampling for
most sizes of the training set. In line with the other methods, FPS reduces the worst-
case error in comparison to random sampling (Fig. 10.5c).

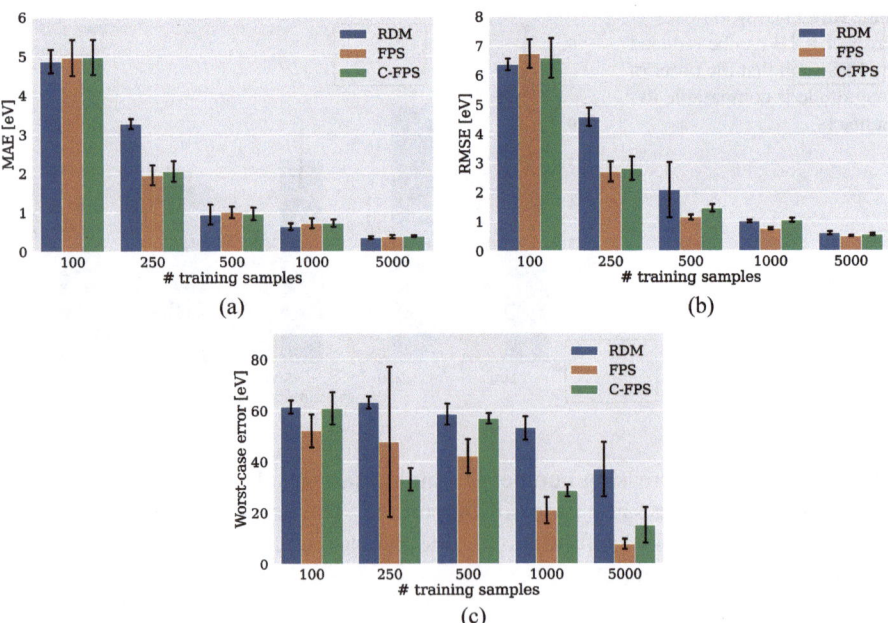

**Fig. 10.5** Results for spatial 3-hop convolution network. (**a**) MAE. (**b**) RMSE. (**c**) Worst-case
error

We observe comparatively large values for all metrics, especially for small training set sizes. For example, the MAE for 100 training samples obtained with FPS amounts to approximately 5 eV. This is around 4 times larger than what we measure for KRR and more than 10 times larger than the value of SchNet. This was to be expected, since both KRR and SchNet are relatively data efficient. We note that the good performance of SchNet is to be expected as it uses more features than the spatial 3-hop convolution network and especially uses the positions of the atoms (a powerful information which allows to compute the atomization energy explicitly). The KRR has the advantage of being a kernel method which has empirically shown to be effective in the realm of small datasets [45]. However, for larger training sets the relative difference between the methods becomes smaller: For 500 training samples, KRR yields only a two times smaller and SchNet only a 5 times smaller MAE. Moreover, the spatial 3-hop convolution network can benefit the most from larger datasets, i.e. we observe a significant improvement in performance whenever the size of the dataset is increased.

## 10.4.5 Explanation

With GRDE framework from Sect. 10.3.4 we now investigate the domain knowledge learned by the spatial 3-hop convolution network from Sect. 10.3.3 using the sampling strategies from Sect. 10.3.5.

### 10.4.5.1 Setup of the Experiments

For the experiments, we utilize the spatial 3-hop convolution network from Sect. 10.3.3 that has been pre-trained using the sampling strategies from Sect. 10.3.5. More specifically, we compare explanations on the pre-trained model for the cases of random sampling (RDM) and diversity sampling (FPS) of 5000 samples as the training dataset. We fix the distortion measure $d$ as the $L^2$ distance for a regression task, and randomly initialize masks $S$. Furthermore, given the sparsity of the data, we also set a low value on $\lambda_A, \lambda_X = 20$ (which corresponds to choosing 10–15% of the non-zero elements in the respective masks) and set $(\nu_{S_A}, \nu_{S_X})$ to null. Since the QM9 dataset possesses edge features, we optimize $S_A = [S_{A_1}, S_{A_2}, ..., S_{A_h}]$ where $A_i$ is the adjacency matrix with respect to the edge feature $i$ $\forall i \in \{1, 2, ..., h\}$; $h$ being the number of edge features. The results that follow are obtained as an average over 3 independent runs on 100 graphs randomly sampled from the respective test datasets. Since the setup produces positive relevance masks, i.e., the masks only obfuscate features that exist for each node/edge, and do not show the relevance of the lack of a feature for a node/edge), we aggregate and average the node- and edge-wise scores to obtain feature-wise scores. Furthermore, we offset the imbalance in the scores by weighting them with respect to the frequency of their occurrence over the sampled data. Our explanation

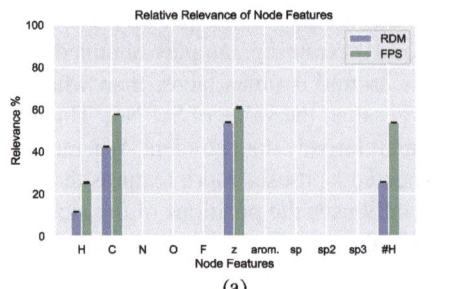
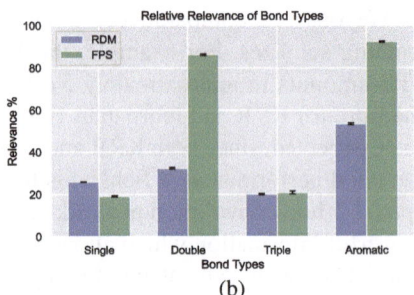

(a)                                                                          (b)

**Fig. 10.6** Results for feature-level GRDE. (**a**) Node feature explanations. (**b**) Edge feature explanations

query is as follows: *For a randomly selected graph unseen by the pre-trained model, which features does the model consider important for its prediction?*

#### 10.4.5.2    Results

From Fig. 10.6a, b, we see that the model trained using FPS places a stronger importance in both edge and node features than in the case of RDM, especially in the case of atomic numbers (Z), double bonds and aromatic rings. In comparison, single bonds, though the most frequent of the bond types in the sampled data are not considered as important. Furthermore, Fig. 10.6a shows that the model requires certain nodes to be specific atom types and have a certain type of neighborhood for its predictions, as can be seen from the atomic number (Z) as well as the importance of carbon (C) and the number of hydrogen atoms (#H) surrounding the node in the case of FPS. In the case of RDM, we also have a clear indication that edge features are not as important as the node features, whereas this is significantly more balanced in the case of FPS. Finally, we find that, though there exist non-zero values for some node features in the sampled graphs such as in the case of Nitrogen (N) and Oxygen (O), GRDE does not attribute any importance to them. This implies that the model treats these features as noise and ignores them regardless of the sampling strategy used.

### 10.5    Conclusion

In this chapter, we employed three informed ML models to predict the atomization energy of molecules in the QM9 dataset. We used KRR with a kernel obtained from molecular topological features, a geometry-based GNN (SchNet) and a topology-based GNN. We saw that maximizing molecular diversity in the training set selection process improves the accuracy and robustness of those methods. Our main finding is that by training topology-based ML models with sets of diverse molecules, we can significantly reduce their test maximum absolute error, thus increasing their robustness to distribution shifts. For SchNet, this effect was still observable but only for small training sets. Moreover, by maximizing diversity in the training sets, we could substantially reduce the gap between the maximum absolute errors of a topology-based regression method as KRR and the SchNet, which is a geometry-based GNN. This chapter proposes only an empirical investigation, in the field of molecular property prediction, on the effects of maximizing molecular diversity in the training set selection. Ongoing research seeks to provide a theoretical foundation for the observed empirical results.

We believe that reducing the worst-case error is of great importance for applications that require a high degree of robustness but have limited budget for data generation. One example would be the application of Machine Learning Interatomic Potentials (MLIPs) for molecular dynamics simulations. In this scenario, the predictions of the model are used to integrate the equations of motion and to compute particle trajectories. Thus, large errors in the predictions could potentially lead to a failure of the simulation and techniques to prevent this are needed. Investigating this scenario in particular, could be a direction of future research.

**Acknowledgments** This work was supported in part by the BMBF-project 05M20 MaGriDo (Mathematics for Machine Learning Methods for Graph-Based Data with Integrated Domain Knowledge) and in part by the Fraunhofer Cluster of Excellence "Cognitive Internet Technologies".

# References

1. Agarwal, C., Queen, O., Lakkaraju, H., Zitnik, M.: Evaluating explainability for graph neural networks. Scientific Data **10**(1), 144 (2023)
2. Agarwal, P.K., Har-Peled, S., Varadarajan, K.R.: Geometric approximation via coresets. In: Combinatorial and Computational Geometry, vol. 52, pp. 1–30. Cambridge University Press (2005)
3. Barker, J., Berg, L.S., Hamaekers, J., Maass, A.: Rapid prescreening of organic compounds for redox flow batteries: A graph convolutional network for predicting reaction enthalpies from SMILES. Batteries & Supercaps **4**(9), 1482–1490 (2021). https://doi.org/10.1002/batt. 202100059
4. Batzner, S., Musaelian, A., Sun, L., Geiger, M., Mailoa, J.P., Kornbluth, M., Molinari, N., Smidt, T.E., Kozinsky, B.: E(3)-equivariant graph neural networks for data-efficient and accurate interatomic potentials. Nature Communications **13**(1) (2022). https://doi.org/10.1038/s41467-022-29939-5
5. Bochkarev, A., Lysogorskiy, Y., Menon, S., Qamar, M., Mrovec, M., Drautz, R.: Efficient parametrization of the atomic cluster expansion. Phys. Rev. Materials **6**, 013804 (2022). https://doi.org/10.1103/PhysRevMaterials.6.013804
6. Brandstetter, J., Hesselink, R., van der Pol, E., Bekkers, E.J., Welling, M.: Geometric and physical quantities improve e(3) equivariant message passing. In: International Conference on Learning Representations (2022)
7. Braverman, V., Feldman, D., Lang, H.: New frameworks for offline and streaming coreset constructions. ArXiv **abs/1612.00889** (2016)

8. Cai, S., Wang, Z., Wang, S., Perdikaris, P., Karniadakis, G.E.: Physics-informed neural networks for heat transfer problems. Journal of Heat Transfer **143**(6) (2021)

9. Cuomo, S., Di Cola, V.S., Giampaolo, F., Rozza, G., Raissi, M., Piccialli, F.: Scientific machine learning through physics–informed neural networks: where we are and what's next. Journal of Scientific Computing **92**(3), 88 (2022)

10. Defferrard, M., Bresson, X., Vandergheynst, P.: Convolutional neural networks on graphs with fast localized spectral filtering. In: Proceedings of the 30th International Conference on Neural Information Processing Systems, NIPS'16, p. 3844–3852. Curran Associates Inc., Red Hook, NY, USA (2016)

11. Deringer, V.L., Bartók, A.P., Bernstein, N., Wilkins, D.M., Ceriotti, M., Csányi, G.: Gaussian process regression for materials and molecules. Chemical Reviews **121**(16), 10073–10141 (2021). https://doi.org/10.1021/acs.chemrev.1c00022

12. Duval, A., Malliaros, F.D.: Graphsvx: Shapley value explanations for graph neural networks. In: Joint European Conference on Machine Learning and Knowledge Discovery in Databases, pp. 302–318. Springer (2021)

13. Eldar, Y., Lindenbaum, M., Porat, M., Zeevi, Y.: The farthest point strategy for progressive image sampling. IEEE Transactions on Image Processing **6**(9), 1305–1315 (1997). https://doi.org/10.1109/83.623193

14. Feldman, D.: Core-sets: Updated survey. In: Sampling Techniques for Supervised or Unsupervised Tasks, pp. 23–44. Springer International Publishing (2019). https://doi.org/10.1007/978-3-030-29349-9_2

15. Fey, M., Lenssen, J.E.: Fast graph representation learning with PyTorch Geometric. In: ICLR Workshop on Representation Learning on Graphs and Manifolds (2019). URL https://github.com/pyg-team/pytorch_geometric

16. Gasteiger, J., Yeshwanth, C., Günnemann, S.: Directional message passing on molecular graphs via synthetic coordinates. In: Conference on Neural Information Processing Systems (NeurIPS) (2021)

17. Gilmer, J., Schoenholz, S.S., Riley, P.F., Vinyals, O., Dahl, G.E.: Neural message passing for quantum chemistry. In: Proceedings of the 34th International Conference on Machine Learning - Volume 70, ICML'17, p. 1263–1272. JMLR.org (2017)

18. Gori, M., Monfardini, G., Scarselli, F.: A new model for learning in graph domains. In: Proceedings. 2005 IEEE International Joint Conference on Neural Networks, vol. 2, pp. 729–734 vol. 2 (2005). https://doi.org/10.1109/IJCNN.2005.1555942

19. Gould, T., Dale, S.G.: Poisoning density functional theory with benchmark sets of difficult systems. Phys. Chem. Chem. Phys. **24**, 6398–6403 (2022). https://doi.org/10.1039/D2CP00268J

20. Hernández, Q., Badías, A., Chinesta, F., Cueto, E.: Thermodynamics-informed graph neural networks. arXiv preprint arXiv:2203.01874 (2022)

21. Huang, Q., Yamada, M., Tian, Y., Singh, D., Chang, Y.: Graphlime: Local interpretable model explanations for graph neural networks. IEEE Transactions on Knowledge and Data Engineering (2022)

22. Jang, E., Gu, S., Poole, B.: Categorical reparameterization with Gumbel-softmax. In: International Conference on Learning Representations (2017)

23. Joerding, W.H., Meador, J.L.: Encoding a priori information in feedforward networks. Neural Networks **4**(6), 847–856 (1991)

24. Jørgensen, P., Jacobsen, K., Schmidt, M.: Neural message passing with edge updates for predicting properties of molecules and materials. In: 32nd Conference on Neural Information Processing Systems, NIPS 2018 (2018)

25. Kim, Y., Cho, J., Naser, N., Kumar, S., Jeong, K., McCormick, R.L., John, P.C.S., Kim, S.: Physics-informed graph neural networks for predicting cetane number with systematic data quality analysis. Proceedings of the Combustion Institute (2022)

26. Kipf, T.N., Welling, M.: Semi-supervised classification with graph convolutional networks. In: 5th International Conference on Learning Representations, ICLR 2017. OpenReview.net (2017). https://openreview.net/forum?id=SJU4ayYgl

27. Kolek, S., Nguyen, D.A., Levie, R., Bruna, J., Kutyniok, G.: A rate-distortion framework for explaining black-box model decisions. xxAI - Beyond Explainable AI p. 91–115 (2022). https://doi.org/10.1007/978-3-031-04083-2_6

28. Kolek, S., Windesheim, R., Loarca, H.A., Kutyniok, G., Levie, R.: Explaining image classifiers with multiscale directional image representation. arXiv preprint arXiv:2211.12857 (2022)

29. Kramer, M.A., Thompson, M.L., Bhagat, P.M.: Embedding theoretical models in neural networks. In: 1992 American Control Conference, pp. 475–479. IEEE (1992)

30. Krause, A., Golovin, D.: Submodular function maximization. Tractability **3**, 71–104 (2014)

31. Kulik, H.J., Hammerschmidt, T., Schmidt, J., Botti, S., Marques, M.A.L., Boley, M., Scheffler, M., Todorović, M., Rinke, P., Oses, C., Smolyanyuk, A., Curtarolo, S., Tkatchenko, A., Bartók, A.P., Manzhos, S., Ihara, M., Carrington, T., Behler, J., Isayev, O., Veit, M., Grisafi, A., Nigam, J., Ceriotti, M., Schütt, K.T., Westermayr, J., Gastegger, M., Maurer, R.J., Kalita, B., Burke, K., Nagai, R., Akashi, R., Sugino, O., Hermann, J., Noé, F., Pilati, S., Draxl, C., Kuban, M., Rigamonti, S., Scheidgen, M., Esters, M., Hicks, D., Toher, C., Balachandran, P.V., Tamblyn, I., Whitelam, S., Bellinger, C., Ghiringhelli, L.M.: Roadmap on machine learning in electronic structure. Electronic Structure **4**(2), 023004 (2022). https://doi.org/10.1088/2516-1075/ac572f

32. Kung, S.Y.: Kernel Methods and Machine Learning. Cambridge University Press (2014). https://doi.org/10.1017/cbo9781139176224

33. Landrum, G.: Rdkit:. Open-source cheminformatics (2012). URL http://www.rdkit.org

34. Li, Y., Tarlow, D., Brockschmidt, M., Zemel, R.: Gated graph sequence neural networks. In: Y. Bengio, Y. LeCun (eds.) 4th International Conference on Learning Representations, ICLR 2016, San Juan, Puerto Rico, May 2–4, 2016, Conference Track Proceedings (2016). URL http://arxiv.org/abs/1511.05493

35. Liu, Z., Zhou, J.: Introduction to graph neural networks. Synthesis Lectures on Artificial Intelligence and Machine Learning **14**(2), 1–127 (2020)

36. Luo, D., Cheng, W., Xu, D., Yu, W., Zong, B., Chen, H., Zhang, X.: Parameterized explainer for graph neural network. Advances in neural information processing systems **33**, 19620–19631 (2020)

37. Ma, J., Sheridan, R.P., Liaw, A., Dahl, G.E., Svetnik, V.: Deep neural nets as a method for quantitative structure–activity relationships. Journal of Chemical Information and Modeling **55**(2), 263–274 (2015). https://doi.org/10.1021/ci500747n

38. MacDonald, J., Wäldchen, S., Hauch, S., Kutyniok, G.: A rate-distortion framework for explaining neural network decisions. arXiv preprint arXiv:1905.11092 (2019)

39. Maddison, C.J., Mnih, A., Teh, Y.W.: The concrete distribution: A continuous relaxation of discrete random variables. In: International Conference on Learning Representations (2017)

40. Magister, L.C., Kazhdan, D., Singh, V., Liò, P.: GCExplainer: Human-in-the-loop concept-based explanations for graph neural networks. In: 3rd ICML Workshop on Human in the Loop Learning (2021). ArXiv preprint arXiv:2107.11889

41. Mahoney, M.W., Drineas, P.: CUR matrix decompositions for improved data analysis. Proceedings of the National Academy of Sciences **106**(3), 697–702 (2009). https://doi.org/10.1073/pnas.0803205106

42. Moriwaki, H., Tian, Y.S., Kawashita, N., Takagi, T.: Mordred: A molecular descriptor calculator. Journal of Cheminformatics **10**(1) (2018). https://doi.org/10.1186/s13321-018-0258-y

43. Mueller, T., Kusne, A.G., Ramprasad, R.: Machine learning in materials science. In: Reviews in Computational Chemistry, pp. 186–273. John Wiley & Sons, Inc (2016). https://doi.org/10.1002/9781119148739.ch4

44. Pedregosa, F., Varoquaux, G., Gramfort, A., Michel, V., Thirion, B., Grisel, O., Blondel, M., Prettenhofer, P., Weiss, R., Dubourg, V., et al.: Scikit-learn: Machine learning in python. Journal of Machine Learning Research **12**(Oct), 2825–2830 (2011)

45. Pinheiro, M., Ge, F., Ferré, N., Dral, P.O., Barbatti, M.: Choosing the right molecular machine learning potential. Chem. Sci. **12**, 14396–14413 (2021). https://doi.org/10.1039/D1SC03564A

46. Pope, P.E., Kolouri, S., Rostami, M., Martin, C.E., Hoffmann, H.: Explainability methods for graph convolutional neural networks. In: 2019 IEEE/CVF Conference on Computer Vision and

Pattern Recognition (CVPR), pp. 10764–10773 (2019). https://doi.org/10.1109/CVPR.2019.01103

47. Ramakrishnan, R., Dral, P.O., Rupp, M., von Lilienfeld, O.A.: Quantum chemistry structures and properties of 134 kilo molecules. Scientific Data **1** (2014)

48. Roscher, R., Bohn, B., Duarte, M.F., Garcke, J.: Explainable Machine Learning for Scientific Insights and Discoveries. IEEE Access **8**(1), 42200–42216 (2020). https://doi.org/10.1109/ACCESS.2020.2976199

49. Ruddigkeit, L., van Deursen, R., Blum, L.C., Reymond, J.L.: Enumeration of 166 billion organic small molecules in the chemical universe database gdb-17. Journal of Chemical Information and Modeling **52**(11), 2864–2875 (2012). https://doi.org/10.1021/ci300415d. PMID: 23088335

50. von Rueden, L., Mayer, S., Beckh, K., Georgiev, B., Giesselbach, S., Heese, R., Kirsch, B., Pfrommer, J., Pick, A., Ramamurthy, R., Walczak, M., Garcke, J., Bauckhage, C., Schuecker, J.: Informed Machine Learning - A Taxonomy and Survey of Integrating Knowledge into Learning Systems. IEEE Transactions on Knowledge and Data Engineering **35**(1), 614–633 (2023). https://doi.org/10.1109/TKDE.2021.3079836

51. Scarselli, F., Gori, M., Tsoi, A.C., Hagenbuchner, M., Monfardini, G.: The graph neural network model. IEEE Transactions on Neural Networks **20**(1), 61–80 (2009). https://doi.org/10.1109/TNN.2008.2005605

52. Schlichtkrull, M.S., Cao, N.D., Titov, I.: Interpreting graph neural networks for {nlp} with differentiable edge masking. In: International Conference on Learning Representations (2021)

53. Schütt, K.T., Kindermans, P.J., Sauceda, H.E., Chmiela, S., Tkatchenko, A., Müller, K.R.: Schnet: A continuous-filter convolutional neural network for modeling quantum interactions. In: Proceedings of the 31st International Conference on Neural Information Processing Systems, NIPS'17, p. 992–1002. Curran Associates Inc., Red Hook, NY, USA (2017)

54. Shukla, K., Di Leoni, P.C., Blackshire, J., Sparkman, D., Karniadakis, G.E.: Physics-informed neural network for ultrasound nondestructive quantification of surface breaking cracks. Journal of Nondestructive Evaluation **39**, 1–20 (2020)

55. Smilkov, D., Thorat, N., Kim, B., Viégas, F., Wattenberg, M.: Smoothgrad: removing noise by adding noise. arXiv preprint arXiv:1706.03825 (2017)

56. Sutton, C., Boley, M., Ghiringhelli, L.M., Rupp, M., Vreeken, J., Scheffler, M.: Identifying domains of applicability of machine learning models for materials science. Nature Communications **11**(1), 4428 (2020). https://doi.org/10.1038/s41467-020-17112-9

57. Todeschini, R., Consonni, V.: Molecular Descriptors for Chemoinformatics. Wiley-VCH (2009)

58. Veličković, P., Cucurull, G., Casanova, A., Romero, A., Lio, P., Bengio, Y.: Graph attention networks. In: International Conference on Learning Representations (2018)

59. Weininger, D.: SMILES, a chemical language and information system. 1. Introduction to methodology and encoding rules. Journal of Chemical Information and Modeling **28**(1), 31–36 (1988). https://doi.org/10.1021/ci00057a005

60. Ying, Z., Bourgeois, D., You, J., Zitnik, M., Leskovec, J.: Gnnexplainer: Generating explanations for graph neural networks. Advances in neural information processing systems **32** (2019)

61. Yu, K., Bi, J., Tresp, V.: Active learning via transductive experimental design. In: Proceedings of the 23rd International Conference on Machine Learning, ICML '06, p. 1081–1088. Association for Computing Machinery, New York, NY, USA (2006). https://doi.org/10.1145/1143844.1143980

62. Yuan, H., Tang, J., Hu, X., Ji, S.: Xgnn: Towards model-level explanations of graph neural networks. In: Proceedings of the 26th ACM SIGKDD International Conference on Knowledge Discovery & Data Mining, pp. 430–438 (2020)

63. Yuan, H., Yu, H., Gui, S., Ji, S.: Explainability in graph neural networks: A taxonomic survey. IEEE Transactions on Pattern Analysis and Machine Intelligence (2022)

64. Yuan, H., Yu, H., Wang, J., Li, K., Ji, S.: On explainability of graph neural networks via subgraph explorations. In: International Conference on Machine Learning, pp. 12241–12252. PMLR (2021)
65. Zaverkin, V., Holzmüller, D., Steinwart, I., Kästner, J.: Exploring chemical and conformational spaces by batch mode deep active learning. Digital Discovery (2022). https://doi.org/10.1039/D2DD00034B

# Chapter 11
# Informed Machine Learning Aspects for the Multi-Agent Neural Rewriter

Nathalie Paul, Tim Wirtz, Stefan Wrobel, and Alexander Kister

**Abstract** We regard the multi-vehicle routing problem as a cooperative multi-agent system where agents (vehicles) seek to determine the team-optimal agent routes with minimal total cost. Each agent can hereby observe solely its own cost information. Our multi-agent reinforcement learning approach builds on an existing method for solving a single-vehicle routing problem by iteratively rewriting solutions. We define new rewriting rules to enable agents to act and interact in a parallel conflict-free manner. We use a form of Informed Machine Learning to integrate knowledge about the underlying cost distribution into the learning process of the agent policy. It enables the (solely own cost observing) agent to act globally optimal within a representative team. Empirical results on simulated data of small problem sizes show that our approach competes with a well-performing heuristic which has also only imperfect cost knowledge.

## 11.1 Introduction

The logistics market is mostly dominated by competing players. Yet in the view of environmental as well as economical benefits, there exists the ambition to promote cooperation [18]. We consider the setup of collaborating logistics companies which want to jointly serve a given set of customers in a globally optimal way, i.e., with minimal total cost. Each company, however, has a justified interest in not revealing its cost structure to its collaborators as it represents business-internal and -sensitive information. Yet knowing solely its own costs, a company cannot behave optimally for the consortium: Assume it has only one collaborator and there is a

N. Paul (✉) · T. Wirtz · S. Wrobel
Fraunhofer IAIS, Sankt Augustin, Germany
e-mail: nathalie.paul@iais.fraunhofer.de; tim.wirtz@iais.fraunhofer.de; stefan.wrobel@iais.fraunhofer.de

A. Kister
Federal Institute for Materials Research and Testing, Berlin, Germany
e-mail: alexander.kister@bam.de

© The Author(s) 2025  
D. Schulz, C. Bauckhage (eds.), *Informed Machine Learning*,
Cognitive Technologies, https://doi.org/10.1007/978-3-031-83097-6_11

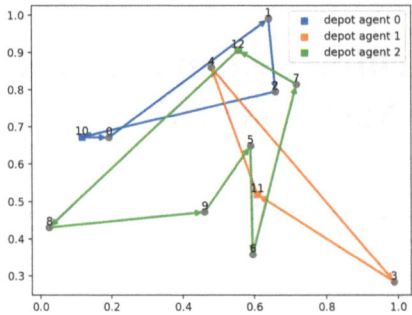

**Fig. 11.1** Exemplifying solution of a multi-vehicle routing problem with 10 customers and 3 vehicles (agents): An agent's route is given by a sequence of visited customer nodes, starting and ending at its depot. Each customer node must be visited exactly once in total. Each edge induces an agent-specific cost

single customer to serve. It is impossible for the company to decide whether it should serve the customer or leave it for its collaborator. The company's optimal decision depends on the collaborator's cost. Some sort of global cost knowledge is thus indispensable for a company's optimal behaviour.

We view the setup as a cooperative multi-agent system with companies corresponding to agents. The team task is given by a multi-vehicle routing problem [12]: It consists of determining which agent visits which customers and in which order such that the total cost is minimized (cf. Fig. 11.1). The agents hereby seek to keep their own cost information as private as possible.

Throughout this chapter, we take on the perspective of a single agent: The agent is a team player without any self-interested goals. It must be aware that in a (feasible) solution all customer nodes have to be visited, i.e., it needs an understanding of the problem. Moreover, it has only limited access to its teammates' costs. In particular, we ultimately require the agent to make decisions while observing solely its own cost information. This raises the following questions: How to inform the agent about the problem formulation? And given limited teammate cost knowledge, how to inform the agent about how to behave as a team player?

The limited cost observability excludes classic routing solvers as a possible solution approach since they require complete knowledge about all cost information. We use reinforcement learning (RL) to tackle the problem which has lately become very attractive for learning heuristics for combinatorial optimization problems [1]. In RL, an agent learns to solve a sequential decision problem by repeatedly interacting with an environment: It chooses an action in the current state, transitions to a subsequent state and observes a corresponding reward. The concept is formalized by a Markov decision process [20]. It can be extended to multiple cooperative agents in terms of a team Markov game [25] which consists of a (global) state space, a joint action space defined by the agent's own possible actions and a single common reward signal. To design a team Markov game for our multi-agent setup, we build on an existing RL based approach for solving a routing problem with a single agent: The so-called neural rewriter [2] takes a given routing solution and iteratively rewrites it by swapping two nodes in the visit order.

The proposed multi-agent neural rewriter (MANR) extends the concept of rewriting routing solutions to the setup of multiple agents. Preliminary results were presented on a workshop [17]. The multi-agent setup requires to define new rewriting rules for the game which guarantee that the self-determined agents cannot directly interfere in a teammate's route but at the same time allow for exchanging customers within the team. A key aspect in the game design is the introduction of a pool set which serves as a temporary storage for customer nodes where an agent can drop-off and integrate customers from. Teaching the agent team that all nodes from the pool must be ultimately reintegrated in some agent's route is realized in the reward definition and can be viewed as a form of Informed Machine Learning (ML) [24].

For implementing our game, we make the assumption that the individual agent costs originate from the same underlying distribution. This allows to learn a single agent policy, used by each agent individually. The agent policy is only allowed to process the agent's own cost explicitly, but as the introductory example demonstrates, it needs to incorporate some kind of global cost knowledge. In the spirit of centralized-learning-decentralized execution [27] we allow the agent costs to be revealed during training to build up global cost knowledge in the shape of a modelled action value function $Q$ which can assess the team benefit in terms of the expected cumulative future reward when executing a joint agent action in a given state. We employ an actor-critic approach [10] to pass the knowledge of $Q$ (critic) on to the agent policy (actor), i.e., the policy's decisions are judged by the cost-omniscient $Q$ model. It enables the agent to learn about the underlying cost distribution and thus how to act globally optimal in a representative team. Note that sharing costs during training is a valid assumption since training can be performed based on realistic but fictitious agent costs. The key is that during execution, the agent acts solely based on its own cost information.

As an additional investigation in relation to informed ML, we analyze the potential of transferring knowledge between models with different sized agent teams. In particular, we aim at transferring the skill of finding the best visiting order given a set of assigned customers. This skill must be always learned by the agent policy independent of the given team constellation.

We give an overview of the related work in Sect. 11.2. In particular, we put our approach into context of the informed ML taxonomy presented in [24]. In Sect. 11.3 we define the considered vehicle routing problem and discuss our solution approach in terms of the problem's design as a team Markov game and its implementation using variants of informed ML. Empirical evaluations of our approach are performed on simulated data in Sect. 11.4. We compare the performance of our MANR to a heuristic which has also imperfect cost knowledge as well as to a cost-omniscient classic routing solver. Finally, the section discusses the potential for further informed ML in terms of applying two transfer learning approaches. Section 11.5 summarizes the results and outlines plans for future work.

## 11.2   Related Work

RL has become popular for learning heuristics to solve NP-hard combinatorial optimization problems like vehicle routing problems. It has the potential to outperform
expert-designed heuristics and generate solutions faster compared to mathematical
optimization methods [1]. There are only few RL approaches for solving vehicle
routing problems with multiple vehicles. In order to avoid conflicting agent actions
there, the agents typically choose actions sequentially [3, 26]. We aim for a multi-
agent system which allows agents to execute actions in parallel. The case of a
single vehicle is on the other hand considered in a variety of RL approaches.
The typical RL design there is to build the routing solution step by step, i.e., to
visit one customer more in each state of the episode [11, 16]. Instead, we follow
the idea of [2] and model the solution process as a rewriting procedure of given
routing solutions. It implements the concept of local search, a widely used heuristic
for solving combinatorial optimization problems [8], as a RL task. The rewriting
problem design avoids the challenge of a sparse reward function as the success of
a rewriting action can be directly evaluated in the subsequent state (by computing
the total team cost). It also allows us to easier avoid conflicts between agents when
extending to a multi-agent setup.

Producing optimal solutions in multi-agent systems while minimizing the
amount of shared agent information is an active research topic for both competitive
and collaborative setups. The extreme case of independent learning, where no agent
information is revealed at all, is challenging and for cooperative tasks known to be
outperformed by methods involving some sort of information exchange [21]. An
interesting direction is to share only implicit agent information, e.g., their model
parameters, which is pursued in the field of distributed machine learning [15, 28].
As a first step, we employ the well-established scheme of centralized-learning-
decentralized-execution [27]. It allows agent information to be disclosed during
training to build up knowledge which during execution enables an agent to act
decentrally on its own observing solely its local information [5, 14].

To the best of our knowledge, the work we initiated in [17] is the first RL
based approach to solve a multi-vehicle routing problem with simultaneously
acting cooperating agents which can observe solely their own cost information.
We continue the work by modifying and extending our approach: E.g., we use
a more realistic and at the same time more challenging model for simulating
individual agent costs. We also make technical adaptions by introducing new model
components and enable a better evaluation of our approach by developing a new
benchmark which has also imperfect cost knowledge. Moreover, we investigate
transfer learning approaches for improving our model.

## 11.2.1 Informed Machine Learning

Integrating knowledge in RL algorithms is typically achieved by the means of human feedback. For example, actions are assessed by an expert instead of a reward signal or expert demonstrations are used instead of on-the-fly simulated data [24]. In this chapter, we widen the perspective on informed ML as presented in [24]: We take on the view of a single agent which is part of a collaborative multi-agent system and has only limited knowledge about its teammates and thus about the overall optimization problem. How can the agent locally learn to act globally good for the team? What kind of mechanisms can we employ during learning to inform the agent about its teammates? On the one hand, we apply expert knowledge for formulating multi-vehicle routing as a problem that can be solved by RL. On the other hand, we employ an actor-critic approach where a cost-omniscient critic passes on its knowledge to the (initially) cost unknowing actor by criticizing its actions. Integrating these two pieces of knowledge is essential for our setup to allow the agents to solve the team task in a globally optimal way. We additionally investigate two transfer learning approaches where the knowledge of a trained agent policy is leveraged for learning in a new environment. Most importantly, we widen the understanding of informed ML by considering learned knowledge as a valid source of knowledge. In this extended view, the above aspects can be put into context of the informed ML taxonomy presented in [24] as follows:

- *Informing the agent about the problem formulation:* We use expert knowledge to design our team Markov game in such a way, that it allows an individual agent to learn how to collaborate and solve the optimization problem. For each step in this formulation process, we provide a detailed description of the used expert knowledge and also of the way how it was used in Sect. 11.3. To give an example, the agent must know that all customers have to be visited to yield a feasible solution. The introduction of the pool set to the game leads to the fact that not all nodes are visited in each state of the system, as will be explained in more detail in Sect. 11.3.2. The agent must understand that, potentially against first actual plans, has to react by visiting a customer if none of its teammates does. We integrate this expert knowledge about the game design in form of an algebraic equation into the definition of the reward, i.e., into the learning algorithm: The agent team gets penalized if nodes stay unvisited for a too long time.
- *Informing the agent about the global cost distribution:* The agent must be able to assess its teammates' costs for optimal behaviour, as it cannot explicitly observe these due to the limited cost disclosure requirement. We integrate knowledge in terms of a modelled action value function $Q$ which can evaluate a team action with respect to the profit for the team. In the context of informed ML this is a special kind of knowledge, namely learned knowledge. $Q$ is defined by an algebraic equation and inserted into the learning algorithm of the agent policy: Inside the agent policy's loss, $Q$ criticizes the policy's decisions, making it learn about the underlying cost distribution (cf. Sect. 11.3.4.1). In contrast to the classic idea of informed ML, we build up the knowledge $Q$ only during learning, i.e., at

the same time while learning the agent policy of interest. However, $Q$ could have been also a priori learned on simulated data.

- *Informing the agent policy about useful sub-strategies:* We study an additional way of how the agent policy could profit from integrating knowledge via transfer learning. The idea is to use the learned knowledge of an already trained policy in terms of a learned (sub-)strategy which is necessary to be acquired independent of the specific agent team environment. The knowledge in terms of algebraic equations is integrated into the policy's learning algorithm by modified loss functions (cf. Sect. 11.4.3).

## 11.3  Multi-Agent Neural Rewriter (MANR)

The agent team members have to solve the conflict between finding a routing solution with minimal total team cost and not sharing cost information between each other. Since we assume the agent costs to originate from the same underlying distribution, we can learn a single agent policy (for optimal agent behaviour in the team). In this way, we counteract the non-stationarity problem in multi-agent RL: Learning becomes easily unstable if agents learn their own policies, since from one agent's perspective the environment changes dynamically then [7]. Our learned agent policy is used by each agent individually, processing the accessible information out of the respective agent's perspective. The desired game concept needs to fulfill the following requirements:

- An agent action can solely modify the agent's own route and not the ones of its teammates. This is to ensure independence and self-determinedness of each team member.
- An agent action cannot be in conflict with another teammate's action. This requirement allows agents to act in parallel in the multi-agent system.
- The agent policy processes solely the agent's own cost information and not the one of its teammates. This guarantees the limited cost disclosure requirement.

The first two points are realized via the game design, which will be discussed in Sect. 11.3.2. The third point is a matter of the game implementation, cf. Sect. 11.3.4. The desired agent policy will be used by each agent for decentral decision-making during execution. As the policy does not explicitly process the teammates' costs, we have to impart global cost knowledge to it during learning. This knowledge is built up simultaneously during training. To facilitate its learning process, the game workflow during training differs from the one during execution as described in Sect. 11.3.3. The integration of the global cost knowledge into the learning algorithm of the agent policy is discussed in Sect. 11.3.4.1. In Sect. 11.3.1, we start by formalizing the problem definition.

### 11.3.1 Problem Definition

We consider a multi-vehicle routing problem in which $n$ vehicles (agents) cooperate to serve a set of customers $V$. Each agent is characterised by an individual cost function and depot. We call a solution a feasible solution if each agent's route starts and ends at its depot and each customer node $v \in V$ is visited exactly once in total by any of the agents. The quality of a feasible solution is assessed with the team average cost, i.e., the average over all agent route costs. The goal is to find an optimal solution which is a feasible solution and has minimal team average costs.

### 11.3.2 Game Design

For designing our game we build on the rewriting approach of the neural rewriter [2]. It regards the solution process of a vehicle routing problem with a single agent as a Markov decision process: States represent feasible routing solutions, actions rewrite a solution by swapping two nodes and rewards express the resulting improvements in the solution quality. We extend this concept to a collaborative multi-agent setup where agents perform rewriting actions on their routes simultaneously. Our extension, the multi-agent neural rewriter (MANR), has already been introduced in our previous work [17]. An important novelty in the multi-agent setup is that the initial assignment of nodes to agents might not be optimal and hence the agents have to exchange nodes and not just rewrite their own routes. Enabling node exchange by allowing an agent to directly swap any nodes in the routing problem would however contradict its teammates' self-determinedness and also create room for conflicts. Thus, we need to define new rewriting rules for our game. We introduce a so-called pool set to the system where an agent can drop-off and pick-up customer nodes. The pool is the only possibility for an agent to (indirectly) interact with its teammates and exchange nodes. In this way we guarantee self-determined team members and conflict-free agent actions. Yet the pool introduction comes with its own challenge: An agent giving a customer node to the pool produces an infeasible solution in the resulting state as not all customers are visited by some agent then. The agent team must understand and act accordingly that nodes in the pool must be reintegrated into some agent's route. We use a form of informed ML to insert this expert knowledge into the learning algorithm: We shape the reward function such that the team gets penalized if nodes stay in the pool for a too long time. In the following, we formalize the rewriting approach to multi-vehicle routing as a team Markov game.

**Team State** A team state $s_t = (s_t^1, s_t^2, ..., s_t^n, \mathbb{P}_t)$ at time $t$ is given by the agent states $(s_t^i)_i$ and the corresponding current state of the pool set $\mathbb{P}_t$. The state $s_t^i$ of agent $i$ at time $t$ is defined by the agent's route, i.e., the sequence of its visited nodes starting and ending at the agent's depot. The pool state $\mathbb{P}_t$ is the set of unvisited nodes at time $t$. Only if the pool state is empty, the team state represents a feasible solution. See Fig. 11.2 for exemplary team states.

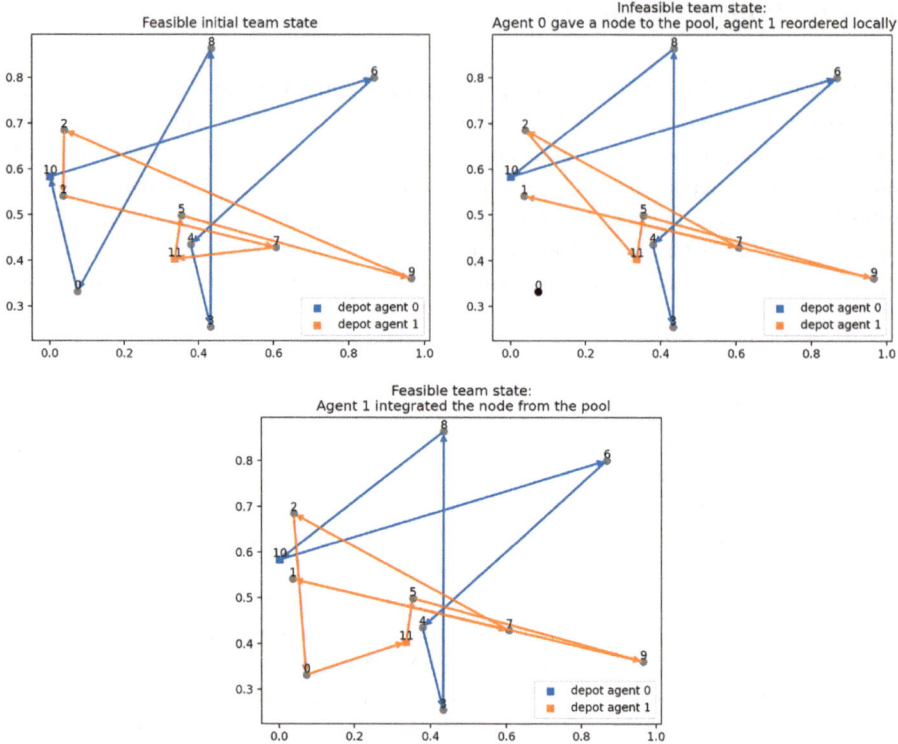

**Fig. 11.2** Exemplifying rewriting sequence of three team states with corresponding semantic agent actions (left, right, bottom)

**Team Action** A team action $a_t = (a_t^1, a_t^2, ..., a_t^n)$ at time $t$ is given by all agent actions at time $t$. An agent's action consists of either giving one of its visited nodes to the pool, integrating a new node from the pool in its route, reorder a node within its route or do nothing. The set of allowed agent actions is hereby dependent on the pool state: If the pool is empty, an agent can either reorder its route locally, give a node to the pool or do nothing. If the pool is filled, an agent is only allowed to either integrate a node from the pool or do nothing. This shall encourage the frequent generation of feasible solutions. Technically, the action $a_t^i = (w_t^i, u_t^i)$ of agent $i$ at time $t$ consists of making two consecutive decisions: First, a region node $w_t^i$ is selected which will be placed after the subsequently selected rule node $u_t^i$ to invoke a rewriting step. If the pool is empty, the region is a node from the agent's state $w_t^i \in \{s_t^i\}$. It can be locally reordered by choosing the rule also from its state or given to the pool by choosing an artificially introduced pool node $p$, i.e., $u_t^i \in \{s_t^i\} \cup \{p\}$. The action of doing nothing simply translates to choosing the region's predecessor node in the agent's state as a rule. For a filled pool, the region node $w_t^i \in \mathbb{P}_t$ is from the pool. An agent can integrate it by choosing a rule node from its state or keep its route unchanged by choosing the artificial pool node (and thus deny the offer of integrating it), i.e., as above, $u_t^i \in \{s_t^i\} \cup \{p\}$. See Fig. 11.2 for exemplary agent actions.

**Team Reward** In order to learn the team behaviour via RL, we have to define a reward function that expresses how beneficial it is to perform a team action in any possible team state: If the system is in a feasible solution, we compute the improvement in the team average cost in comparison to the last feasible solution. In case of an infeasible solution, we penalize the team if infeasible solutions were created consecutively for a too long time, and until then assess an infeasible state neutrally with a reward of 0. The designed team reward thus informs each agent about the problem formulation.

$$
r_t = \begin{cases} c(s_{prev_f(t)}) - c(s_t) & \text{if } s_t \text{ is feasible,} \\ -10 & \text{if } s_t \text{ is infeasible and the last } m \text{ team} \\ & \text{states were infeasible,} \\ 0 & \text{otherwise,} \end{cases} \tag{11.1}
$$

where $c(s_t) = \frac{1}{n} \sum_{i=1}^{n} c^i(s_t^i)$ denotes the team average cost of the team state $s_t$ given the agent-individual costs of their current routes $c^i(s_t^i)$ and $s_{prev_f(t)}$ the last feasible team state before time $t$. Choosing the hyperparameter $m > 1$ results in a non-Markovian reward which could be explicitly handled in the team state representation but is currently left to the agents to learn about.

**Episode** The game's rewriting episode is given by a sequence of team states, team actions and team rewards and is observed by each agent. It requires an initial feasible team state $s_0$ to start with and is limited by a fixed number of rewriting steps $T$:

$$(s_0, a_0, r_1, s_1, a_1, r_2, s_2, ..., s_{T-1}, a_{T-1}, r_T, s_T).$$

The result of our game is defined as the last feasible team state in the episode.

### 11.3.3 Game Workflow

As described above, the game consists of agents performing rewriting actions on their routes to find a team-optimal solution. The procedure to yield a team action is set up differently during training than during execution since the focus in these phases is a different one: During training we must build up global cost knowledge and pass it on to the agent policy. During execution we have an informed agent policy ready to be used by each agent for decision making. In the following, we describe how a team action is determined in the respective phase.

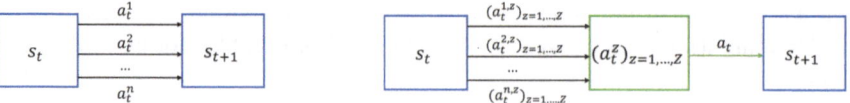

**Fig. 11.3** Team action determination during execution (left) and during training (right). During execution each agent acts decentrally via the agent policy. During training agent actions are centrally coordinated by choosing a team action out of a produced candidate set with the team action scorer $Q$. The setup follows the common approach of centralized-learning-decentralized-execution

**During Execution** Each agent acts decentrally and chooses its own action via the learned agent policy observing solely its own cost. The agents' chosen actions then together automatically determine the team action, cf. Fig. 11.3.

**During Training** A central decision is made about which team action to choose out of some candidates. To generate $Z$ candidate team actions, each agent individually produces $Z$ actions with the agent policy which are centrally zipped together. A team action then gets chosen with the so-called team action scorer, an action value function model $Q$ which learns to judge a team action with respect to its expected team benefit, cf. Fig. 11.3. In the light of informed ML it represents the knowledge about the global cost distribution each agent must get informed about. To facilitate the learning process of $Q$ we explicitly involve it in decision-making instead of a posteriori showing it decentrally determined team actions. $Q$ is hereby used in an epsilon-greedy strategy which allows for a good trade-off between exploration and exploitation. This is critical given the high-dimensional team action space.

### 11.3.4   Game Implementation

In the following, we describe the models used to implement our game design and workflow.[1] We discuss the different models for encoding nodes of the routing problem (node encoders), which produce node representations that are processed by the remaining models: The agent policy which determines one agent's action as well as the team action scorer which during training builds up global cost knowledge and passes it on to the agent policy.

**Node Encoders** Each agent makes use of three node encoder models for representing different types of nodes: Nodes in its own state, nodes in a teammate's state and nodes in the pool state. The multiple encoders are necessary due to the different semantic structure of the states as well as the different kind of information availability.

---

[1] https://github.com/fraunhofer-iais/MANR.

- Agent state encoder: An LSTM-based model encodes nodes in the agent's own route to incorporate the sequential relation between the visited nodes. It generally processes the nodes' coordinates as well as the agent's travelling cost information. More precisely, the model is given by a bidirectional LSTM with one hidden layer of dimension 256. It processes each node $v$ in the agent's state as a 5-dimensional vector $(v_x, v_y, v_x^p, v_y^p, c^i(v, v^p))$ containing x,y coordinates of the node itself and of its current predecessor $v^p$ as well as the agent's cost to travel between them. The resulting node encoding is of size 512 as the LSTM outputs from both directions are concatenated.
- Teammate state encoder: Analogously to the agent state encoder, an LSTM-based model encodes the nodes in a teammate's state. It is also given by a bidirectional LSTM with one hidden layer of size 256. It only differs from the agent state encoder that it does not process the teammate's cost information since it is not accessible by the agent. I.e., each node $v$ in a teammate's state is processed as a 4-dimensional vector $(v_x, v_y, v_x^p, v_y^p)$ containing the x,y coordinates of the node itself and of its current predecessor $v^p$.
- Pool state encoder: Since there is no semantic sequential order in the pool state, a node is encoded by an attention-based model. It processes each node $v$ in the pool as a 2-dimensional vector $(v_x, v_y)$ solely containing its coordinates. More precisely, the model consists of three MLPs for generating a query, keys and values respectively [23]. The query and keys networks have the same architecture and are defined by two hidden layers of size 256 and an (attention) output size of 16. They are used to compute the compatibility of a node in the pool with the other nodes in the pool (self-attention). Based on the compatibilities, the values produced by the values network are aggregated to determine the node encoding. The values network is given by two hidden layers of size 1024 and outputs a vector of size 512. All networks use ReLU activation in the hidden layers.

**Agent Policy**   The agent policy chooses an agent action. By game design, it involves selecting a region and a corresponding rule node. Defining the policy over the tuples of region and rule nodes would lead to a sample space which is quadratic in the problem size. Following [2], we reduce the modelling complexity by handling the selection process of a region and a rule separately. In the current implementation, only the rule selection is learned while the region selection is performed randomly considering some rules.

- Region selector: By game design, the allowed choices for a region node depend on the pool state. If the pool is empty, the region node to be moved is one of the agent's currently visited nodes. The agent chooses it by sampling from a uniform distribution over its nodes in the current state. If the pool is filled, the region node must be from the pool state. To avoid conflicts, the pool is equipped with an automatic mechanism which distributes all nodes in the pool to agents. An agent is thus automatically provided with a region node in case of a filled pool. The distribution mechanism of the pool does not operate completely random but respects some rules for smarter behaviour: A node is always first offered to an agent which did not drop the node there. Moreover, an agent's denial of integrating a node is respected in the sense of not giving it the same node as a region again as long as there are other teammates which haven't declined the integration yet. Note that the randomness in the region node selection cannot harm the course of the game in terms of the overall solution quality since an agent can always decide to do nothing via the subsequently chosen rule node. It can only harm the game in terms of slowing the solution process down.

- Rule selector: Given a region node, the agent selects a rule node with an attention-based model $\pi_u$ which predicts a probability distribution over all potential rule nodes. The model bases its prediction on information about the current agent state and pool state as well as some context information which aggregates the info about which teammate visits which node and in which order in the current team state. Technically, the model consists of two MLPs with two hidden layers of size 256 with ReLU activation and an output size of 16 for producing a query and keys. The query network processes the node encoding of the selected region node and the average of the teammates' nodes encodings as context info. The keys network processes the information about the possible rule nodes in terms of their node encodings as well as how the choice of a respective rule would affect the agent state. The compatibilities between the (region) query and (rule) keys are used to define a probability distribution via the softmax function. During training, the rule is sampled from this distribution while during inference we choose the rule with the highest probability. Note that through the described inputted node encodings, the model processes solely the cost of the agent itself and not of its teammates.

**Team Action Scorer** The team action scorer $Q$ is given by a MLP and quantifies the expected team benefit when rewriting a team state with a team action. It processes the node encodings of the involved region and rule nodes produced by the respective agent state encoder and thus observes the cost information of all agents. The employed architecture depends on the number of agents. Since an agent action (choosing region and rule) is two-dimensional, the network's input is of size $2 \cdot 512 \cdot$ number of agents. The output is one-dimensional. We successively introduce hidden layers with ReLU activation by halving the hidden dimension as long as the current size is bigger than 200.

Combining the discussed models with the above presented game workflow, we obtain the following summarized procedure for generating a rewriting episode for a given routing instance:

---

**Algorithm 1** Rewriting episode of one routing problem

---

**Data:** Number of rewriting steps $T$, number of candidate actions $Z$, probability $\epsilon$ for choosing a
  random team action, Boolean training_flag to indicate training (or inference) process.

generate initial solution for the routing problem;             /* Discussed in Sect. 11.4.1 */
rewriting step = 0
**while** *rewriting step* $< T$ **do**
  each agent updates its node encodings using the agent, teammate and pool state encoder
  **if** *training_flag* **then**
    each agent perceives $Z$ candidate regions from the region selector
    each agent samples $Z$ corresponding rules from the rule selector
    all agents' candidate actions are centrally zipped together to $Z$ team actions
    a team action is $\epsilon$-greedily determined with the team action scorer $Q$
  **else**
    each agent perceives a region from the region selector
    each agent selects the corresponding best rule from the rule selector
  **end**
  each agent executes its action and perceives the resulting team state
  rewriting step = rewriting step + 1
**end**

---

### 11.3.4.1    Loss Functions

As discussed above, the agent uses only the agent policy for making a rewriting
decision during execution. The policy processes solely the own agent's cost
information. To allow for globally optimal team behaviour, it must be provided with
some sort of knowledge about its teammates' costs during training.

In the spirit of centralized-learning-decentralized execution [27] we allow the
teammates' costs to be revealed during training. It allows to compute the team
reward within observed rewriting episodes and use it to learn the team action scorer
$Q$ as

$$L_a(\theta, \tilde{\psi}) = \frac{1}{T} \sum_{t=0}^{T-1} \left( \sum_{t'=t}^{T-1} \gamma^{t'-t} r_{t'+1} - Q(s_t, a_t; \theta, \tilde{\psi}) \right)^2,$$

where $\theta$ denote the parameters of the $Q$-model, $\tilde{\psi} = [\psi_1, \psi_2]$ the parameters of the
agent state and pool state encoder and $\gamma < 1$ the discount factor. Unlike [2], we fit
$Q$ to the *cumulative* discounted observed team reward within the rewriting episode
instead of the *maximal* one. This is necessary to guarantee the limited disclosure
requirement during execution: We cannot evaluate all generated team states within
the rewriting rollout to identify the best one. Our final result has to be the last
(feasible) team state within the rollout, requiring it to be the optimized one.

We integrate the knowledge of the team action scorer $Q$ into the learning process
of the agent policy in the form of an actor-critic approach: The agent policy's
(actor's) decisions are criticized by the cost-omniscient $Q$-model (critic). It enables
the agent policy to learn about the underlying team member cost distribution and

thus how to behave in a representative team. More precisely, the rule selector $\pi_u$ of the agent policy is fitted as

$$L_u(\phi, \psi) = \frac{1}{n} \sum_{i=1}^{n} \left( -\sum_{t=0}^{T-1} A(s_t, u_t^i, a_t^{-i}, w_t^i) \log \pi_u(u_t^i | w_t^i, s_t^i, \mathbb{P}_t; \phi, \psi) \right),$$

where $\phi$ denote the parameters of the rule selector, $\psi = [\tilde{\psi}, \psi_3] = [\psi_1, \psi_2, \psi_3]$ the parameters of the agent state encoder, the pool state encoder and the teammate state encoder and $A(s_t, u_t^i, a_t^{-i}, w_t^i)$ the advantage of choosing rule node $u_t^i$ in the team state $s_t$ given the agent's region node $w_t^i$ and all other teammates' actions $a_t^{-i} = (a_t^1, ..., a_t^{i-1}, a_t^{i+1}, ..., a_t^n)$. The advantage of an agent's rule node choice is defined as

$$\begin{aligned}
&A(s_t, u_t^i, a_t^{-i}, w_t^i) \\
&= Q(s_t, u_t^i, a_t^{-i}, w_t^i; \theta, \tilde{\psi}) \\
&\quad - \sum_{\substack{\text{all candidate} \\ \text{rules } \tilde{u}^i}} \pi(\tilde{u}^i | w_t^i, s_t^i, \mathbb{P}_t; \phi, \psi) Q(s_t, \tilde{u}^i, a_t^{-i}, w_t^i; \theta, \tilde{\psi}).
\end{aligned}$$

It compares the by $Q$ predicted team benefit when choosing the agent's rule node to the expected one over all possible rule nodes under the agent policy (while keeping the teammates' actions fixed). In this way, $Q$ informs and guides the learning process of the agent policy. Note that it is arguably correct to evaluate the advantage of an agent action in the context of fixed teammate actions since the optimal decision of the agent at a time step is independent of its teammates' actions at the same time step.

The presented models (agent state encoder, teammate state encoder, pool state encoder, agent policy and the team action scorer) are trained centrally and simultaneously with a combined loss function as

$$L(\theta, \phi, \psi) = L_a(\theta, \tilde{\psi}) + \alpha L_u(\phi, \psi), \tag{11.2}$$

where $\alpha \in (0, 1)$ downscales the loss of the agent policy's rule selector. This gives the team action scorer $Q$ a head start to perform well first to be a good critic when assessing the agent policy's decisions. We use the Adam optimizer [9] for minimizing the combined loss.

## 11.4 Empirical Evaluation

We empirically evaluate the presented MANR on simulated data for the setups of 10 and 20 customers with team sizes of 2, 3 and 5 agents respectively. We outline the procedure for data simulation in Sect. 11.4.1 and assess the performance of our approach by comparing it to two benchmarks in the subsequent Sect. 11.4.2. The potential for additional informed ML is discussed in Sect. 11.4.3 by introducing and evaluating two transfer learning approaches.

### *11.4.1 Data Generation*

In the following we describe our method to generate vehicle routing instances in terms of node topologies and individual agent costs. We also discuss the way of creating an initial feasible team state which is needed to start the rewriting episode.

**Vehicle Routing Topologies** We are interested in scenarios in which it is optimal that all agents contribute to the solution. Therefore we design the random assignment of depot and customer locations such that it is very unlikely that one agent serving all customers is optimal. Depot nodes as well as a random fraction of customer nodes are sampled uniformly in the unit square $[0, 1]^2$. The remaining customer nodes are assigned as equally as possible to the agents and are drawn close to the respective agent's depot. This complex sampling procedure is necessary to encourage the participation of multiple agents since we observed that when distributing all nodes uniformly, it is often cheapest if a single agent visits them all. To accomplish the closeness of the remaining customers to the respective agent's depot, they are drawn from a bivariate truncated normal distribution in the unit square $\mathcal{N}(d^i, 0.1^2)$ centered at the corresponding agent's depot $d^i$ with a standard deviation of 0.1. Exemplary topologies are shown in Fig. 11.4.

**Agent Cost Model** We use the Euclidean distance to compute the distance between two nodes $v, z \in [0, 1]^2$. Agent-specific travel costs are then achieved by introducing an agent-specific speed for each edge. To obtain agents that are generally faster or generally slower we model the final speeds as the product of the agent-edge-specific velocities and a general velocity. More precisely, for agent $i$ we draw its general velocity $\eta^i$ uniformly from $[0.5, 1]$ allowing an agent to be at most 50% faster in general than its teammates. Moreover, we make the actual agent's speed also dependent on the specific travelling edge by sampling its edge-specific velocity $\eta^i_{v,z}$ for the edge between node $v$ and $z$ uniformly from $[0.5, 1.5]$. An agent's speed can thus be decreased or increased on an edge by at most 50%. In summary, the cost of agent $i$ to travel between two nodes $v, z \in [0, 1]^2$ is given as

$$c^i(v, z) = \frac{1}{\eta^i * \eta^i_{v,z}} \|v - z\|_2.$$

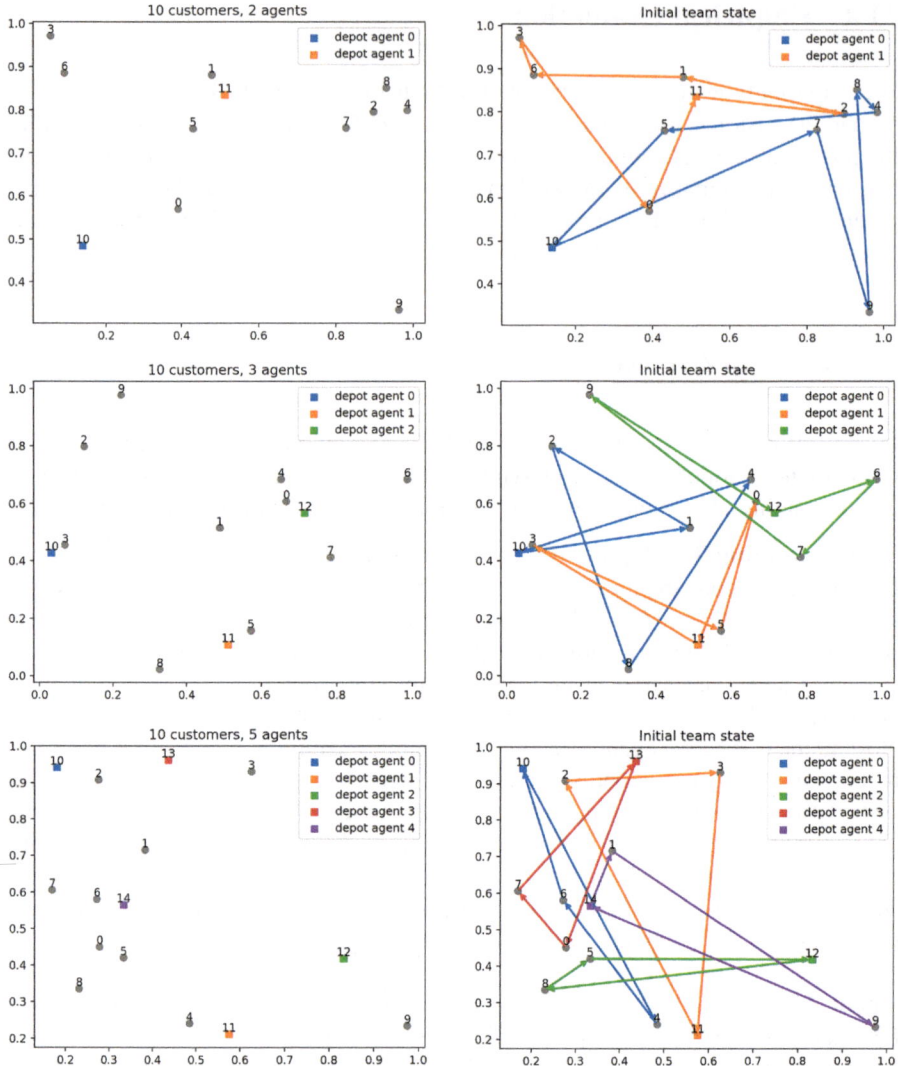

**Fig. 11.4** Exemplary generated routing instances with corresponding initial solutions for 10 customer nodes and a varying amount of agents

**Initial Solution** The rewriting procedure of our game requires an initial feasible solution as a starting point. We generate it by randomly and as equally as possible assigning customer nodes to agents. Each agent then uses the nearest-neighbour-heuristic [19] to create a route through its assigned nodes. Exemplary initial solutions are depicted in Fig. 11.4.

## 11.4.2 Experiment Results for the MANR

For empirical evaluation we consider the vehicle routing setups of 10 and 20 customer nodes with 2 ,3 and 5 agents respectively. For each setup we generate 6250 routing instances split into three parts of 80%–10%–10% for training, validating and testing. In the training set we include the same topologies multiple times with different cost matrix samples to better enable an agent to learn about the underlying cost distribution. More precisely, the training set with 5000 routing instances contains 100 different topologies with 50 cost matrices each. In the validation and test set all topologies are different. In the following, we give details on our model's training as well as inference process. We describe the employed benchmarks and the considered performance metric. Based thereon, we discuss the evaluation results of our approach.

**Hyperparameters** Hyperparameter tuning was performed on the validation set with Tune [13]. Some values could be set equally throughout setups with a fixed amount of customers $k \in \{10, 20\}$ or a fixed agent team size $n \in \{2, 3, 5\}$, see Table 11.1 for a detailed overview. To mention those which are most essential for the

**Table 11.1** Overview of selected hyperparameter values for all experiments. The first seven rows contain the hyperparameters which were set equally throughout all setups

| Hyperparameter | 10 nodes 2 agents | 10 nodes 3 agents | 10 nodes 5 agents | 20 nodes 2 agents | 20 nodes 3 agents | 20 nodes 5 agents |
|---|---|---|---|---|---|---|
| Num trained epochs | 30 | 30 | 30 | 30 | 30 | 30 |
| Discount factor $\gamma$ | 0.5 | 0.5 | 0.5 | 0.5 | 0.5 | 0.5 |
| $Q$ learning rate | 5e−04 | 5e−04 | 5e−04 | 5e−04 | 5e−04 | 5e−04 |
| Learning rate decay rate | 0.9 | 0.9 | 0.9 | 0.9 | 0.9 | 0.9 |
| Learning rate decay steps | 200 | 200 | 200 | 200 | 200 | 200 |
| Epsilon-greedy $\epsilon$ | 0.15 | 0.15 | 0.15 | 0.15 | 0.15 | 0.15 |
| Gradient clip | 0.05 | 0.05 | 0.05 | 0.05 | 0.05 | 0.05 |
| Rewriting steps $T$ | 30 | 30 | 30 | 40 | 40 | 40 |
| Num candidates actions $Z$ | 5 | 5 | 5 | 10 | 10 | 10 |
| Max num pool filled $m$ | 3 | 4 | 6 | 3 | 4 | 6 |
| Policy learning rate factor $\alpha$ | 1e−06 | 1e−06 | 1e−05 | 5e−07 | 5e−07 | 1e−06 |

game workflow: For a fixed agent team size, we set the value of $m$ which regulates the time a penalty is assigned to the team (cf. (11.1)) to $m = n + 1$. The idea is that all team members, which are asked in turn for integrating a specific node from the pool, are allowed to deny once. The team is penalized as soon as an agent declines for the second time. For a fixed amount of 10 resp. 20 customers, we set the number of rewriting steps in an episode to $T = 30$, resp. $T = 40$ and the number of centrally produced candidate team actions during training to $Z = 5$, resp. $Z = 10$.

**Inference** While the generation of multiple candidate actions has to some extent compensated the stochasticity in the region selection during training (since the actions most probably contained different region nodes), this is not given anymore during execution. For execution, the agent policy is used once decentrally by each agent to determine the final agent action. Thus to counteract the region stochasticity during inference, we increase the rewriting steps to $T = 100$ for all considered vehicle routing setups. Moreover, we execute the MANR multiple times to assess its performance: We make 20 inference runs and report the results when considering

- the best run for each routing instance ("MANR best"),
- the average performance over all runs for each routing instance ("MANR").

**Benchmarks** We compare our model to two benchmarks: A cost-omniscient one given by a classic routing solver and a self-developed heuristic which has also imperfect knowledge regarding the agent costs. The cost-omniscient benchmark is an upper bound on the performance, it can not be surpassed by any approach that respects our assumptions on the limited disclosure of the agent costs. The heuristic with imperfect cost knowledge is a lower bound on the performance which we aim to surpass or at least compete with.

More precisely, we consider the routing solver from OR-Tools which is based on local search and specifically tuned for vehicle routing.[2] The solver is configured with the default search parameters and starts from the same initial solutions as our model. It needs to be provided with the cost information of all agents for optimization. Note that by using OR-Tools we might only find an approximation to the optimal solution, but for our purposes this one will usually be good enough to still serve as an upper bound.

Our imperfect cost knowledge benchmark PrivAssign relies on the idea that we make a decision about which agent visits which nodes without knowing the true agent costs. Instead, we only use knowledge about the cost distribution built up from observed cost matrix samples. After the decision is made, we access the true costs and use a routing solver to find the best visiting order of the assigned nodes for each agent to evaluate the solution. To be exact, for a given routing instance we generate 100 random agent-node-assignments as possible solutions. To assess one node assignment, we sample costs for all agents in total 50 times. For each

---

[2] https://developers.google.com/optimization/routing/vrp.

cost-setup we compute optimized agent routes with a routing solver,[3] i.e., we solve one travelling salesman problem per agent, and evaluate the whole solution with the team average cost. One node assignment is then assessed by the median team average cost (over the 50 cost setups). For the given routing instance we choose the node assignment with the minimal median team average cost out of the 100 options as the final one. For this node assignment we consider the true agent costs and, analogously to above, compute optimized agent routes yielding the final result.

**Performance Metric**  The solution of a routing instance $r$ generated by model $m$ is evaluated with the team average cost $c_{\text{team},r}^{m} = \frac{1}{n} \sum_{i=1}^{n} c_r^{i,m}$, where $c_r^{i,m}$ denotes the cost of the route of agent $i$. We denote the mean test set performance for model $m$ by $c_{\text{team}}^{m} = \frac{1}{R} \sum_{r=1}^{R} c_{\text{team},r}^{m}$. For all models we report the final performances in terms of percentage cost reductions relative to the initial solution which is given by

$$c_{\text{gap}} = \frac{c_{\text{team}}^{\text{init}} - c_{\text{team}}^{m}}{c_{\text{team}}^{\text{init}}} \qquad (11.3)$$

with $c_{\text{team}}^{\text{init}}$ denoting the mean test set team average costs of the initial solutions and model $m \in \{\text{MANR, MANR best, OR-Tools, PrivAssign}\}$. We note that PrivAssign is the only model which does not start from an initial solution for optimization. Nevertheless we can take the quality of the initial solutions used in the other models as a reference for evaluating the performance of PrivAssign.

**Results**  The performance gaps in Tables 11.2 and 11.3 show that the MANR significantly improves over the initial solutions for all considered setups of 10 and 20 customer nodes with 2, 3 and 5 agents respectively. The more agents, the higher the gap which can be explained with the fact that the generated initial solutions get naturally worse with an increasing amount of agents: The more agents, the higher the probability that a node gets assigned to the wrong agent initially. We see a pronounced difference in the performance gaps of MANR and MANR best. This is not only related to the stochasticity in the region selection, but also to the nature

**Table 11.2**  Experimental results for 10 customer nodes in terms of $c_{\text{gap}}$

| 10 nodes | PrivAssign | MANR | MANR best | OR-Tools |
|----------|-----------|------|-----------|----------|
| 2 agents | 43% | 31% | 43% | 52% |
| 3 agents | 55% | 48% | 60% | 65% |
| 5 agents | 67% | 60% | 72% | 76% |

**Table 11.3**  Experimental results for 20 customer nodes in terms of $c_{\text{gap}}$

| 20 nodes | PrivAssign | MANR | MANR best | OR-Tools |
|----------|-----------|------|-----------|----------|
| 2 agents | 43% | 28% | 43% | 50% |
| 3 agents | 56% | 40% | 57% | 63% |
| 5 agents | 67% | 50% | 65% | 75% |

---

[3] https://developers.google.com/optimization/routing/tsp.

**Table 11.4** Execution times in seconds for solving one routing instance on average, reported for the models with imperfect cost knowledge. Note that MANR is exactly 20 times faster as MANR best as MANR evaluates a problem instance once and MANR best 20 times

| 10 nodes | PrivAssign | MANR | MANR best | 20 nodes | PrivAssign | MANR | MANR best |
|----------|-----------|------|-----------|----------|-----------|------|-----------|
| 2 agents | 33.15 | 0.48 | 9.61 | 2 agents | 109.2 | 0.57 | 11.43 |
| 3 agents | 41.50 | 0.55 | 10.92 | 3 agents | 108.92 | 0.67 | 13.48 |
| 5 agents | 46.51 | 0.77 | 15.4 | 5 agents | 132.29 | 0.81 | 16.28 |

of the modelled local search procedure which happens to get stuck in local minima. It can also indicate that some team states in the large state space have not been well explored and that there is thus room for improvement during training.

As expected, the results of the cost-omniscient OR-Tools benchmark cannot be reached by our approach. Yet taken into the account the limited cost observability, MANR best gets reasonably close. Comparing our model results to the one of the PrivAssign benchmark which also uses solely knowledge about the underlying cost distribution instead of the actual costs, we see that MANR best indeed manages to compete. In the setups of 10 nodes with 3 and 5 agents the performance of PrivAssign is even notably surpassed. Only in case of 20 nodes and 5 agents MANR best is outperformed by PrivAssign. It illustrates that learning becomes more difficult for larger problem and team sizes. Yet our approach gives the better trade-off between solution quality and execution time: For the setup of 10 customers, MANR best is three times faster than PrivAssign and in case of 20 customer nodes even eight times, cf. Table 11.4.

## 11.4.3 Transfer Learning Investigations

Learning in setups with larger teams becomes more challenging: The team action scorer $Q$ has to learn about the joint action space whose dimension scales with the number of agents. Can we further improve our models by integrating additional information during training?

We note that the agent policy must acquire two different skills during training. On the one hand, it needs to be aware of the best visiting order given a set of assigned nodes and on the other hand, it has to know how and when to use the pool for exchanging nodes. The latter pool usage skill greatly depends on the number of teammates: Whether the agent should exclude or integrate a node depends on the given team constellation. Yet the local reordering skill is basically independent of the team, it affects solely the agent itself. Only the typical agent route length will vary for different team sizes. What if we could transfer the skill of local reordering between different models of varying team sizes to facilitate learning? An agent already having an idea about how to locally reorder could then focus on the interaction with the pool, the part it must always newly learn about given a new team setup.

**Fig. 11.5** Distribution of the local reordering performance, i.e., the cost gap to the TSP solver, over the 5ag-test set instances for the 5ag- and 2ag-model

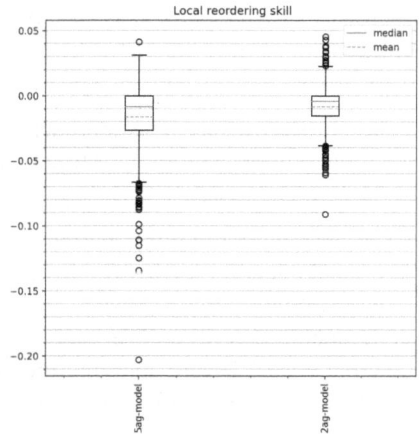

On that account we analyzed the local reordering skill of the above presented two-agent (2ag) and five-agent (5ag) model trained for 10 nodes. Both models were evaluated on the 5ag-test set. Note that as the agent policy does technically not depend on the team size, it can be directly used for execution in all team setups. To assess the local reordering performance of a model, we compare each generated agent route in the final results to the corresponding one optimized by a TSP solver (given the respective same node assignment). More precisely, for each agent route we quantitatively measure the skill as the cost gap to the TSP solver, normalized by the respective route length

$$\text{gap}_k^{i,\text{TSP}} = \frac{c_k^{i,\text{tsp}} - c_k^{i,\text{MANR}}}{\text{agent } i\text{'s route length in test instance } k},$$  (11.4)

where $c_k^{i,\text{MANR}}$ denotes the cost of agent $i$ in the final routing solution of test instance $k$ produced by MANR best and $c_k^{i,\text{tsp}}$ the resulting agent cost when computing the route through agent $i$'s nodes by a TSP solver. The higher the gap, the better our model: A positive gap indicates that the model outperforms the TSP solver at local reordering. Figure 11.5 shows the distribution of the computed gap over the 5ag-test set for the 5ag- and 2ag-model. We see that the 2ag-model is clearly better at local reordering than the 5ag-model as it has the higher mean and median values. Also the distribution is more concentrated indicating that the skill was acquired in a more stable and reliable way.

This raises the question: Can we use the 2ag-model skill of local reordering as prior knowledge when learning in the more challenging 5-agent environment?

To this end we evaluate two transfer learning approaches. The first approach transfers the local reordering skill by initializing the agent policy with a pre-trained one [22]. The second approach is a policy distillation technique [4] which distills the local reordering skill from a knowledgeable teacher during learning. In both approaches we adapt the loss term of the agent policy in the combined loss in (11.2) to focus on learning the interactions with the pool after the local reordering skill was transferred.

**Policy Initialization** The approach reuses pre-trained model components and fine-tunes them in the new setup. More precisely, it implements three changes compared to the original model:

- Initialization: We don't initialize all model components randomly, but use the trained agent policy as well as all the three trained encoders from the 2ag-setup as a starting point.
- Temperature scaling: We scale the logits of the agent policy with a temperature value $T > 1$ before applying the softmax function. This pushes the distribution towards a more uniform distribution [6]. This is necessary as the initial policy is already too decided about which action to choose which prevents the policy from learning. We adaptively decrease the temperature over time such that the mechanism loses its effect (with $T = 1$) after 500 model updates.
- Agent policy loss: We modify the loss term of the agent policy's rule selector by introducing a weighing factor $\lambda$. It puts more weight on actions involving the pool, i.e., actions which give a node to the pool or take a node out of the pool. More precisely, we define the new loss term as

$$L_{u,\text{new}}(\phi, \psi) = \frac{1}{n} \sum_{i=1}^{n} \left( - \sum_{t=0}^{T-1} \lambda(a_t^i) \, A(s_t, u_t^i, a_t^{-i}, w_t^i) \right.$$

$$\left. \log \pi_u(u_t^i | w_t^i, s_t^i, P_t; \phi, \psi) \right)$$

with

$$\begin{cases} \lambda(a_t^i) \gg 1 & \text{if } a_t^i \text{ contains pool usage,} \\ \lambda(a_t^i) = 1 & \text{else.} \end{cases}$$

**Policy Distillation** The idea of policy distillation is to train a so-called student policy by transferring knowledge from a given teacher policy. We use a student-driven distillation approach called expected entropy regularised distillation presented in [4]. The idea is to align the student policy $\tilde{\pi}_u$ (5ag-policy) with the teacher policy $\tilde{\pi}_{\text{teacher}}$ (2ag-policy) by minimizing the cross entropy between their distributions and additionally maximizing the agreement of the teacher with the student's ultimately chosen actions. We restrict these mechanisms to solely local reordering actions to transfer the local reordering skill. We integrate the policy distillation loss as an additional weighted term in the agent policy loss as follows:

$$L_{u,\text{new}}(\phi, \psi) = L_u(\phi, \psi) + \beta L_{u,\text{transfer}}(\phi, \pi_{\text{teacher}})$$

with

$$L_{u,\text{transfer}}(\phi, \pi_{\text{teacher}})$$

$$= \frac{1}{n} \sum_{i=1}^{n} \left( -\sum_{t=0}^{T-1} \hat{R}_t^i \, \log \tilde{\pi}_u(u_t^i | w_t^i, s_t^i, P_t; \phi, \psi) \mathbb{1}_{(w_t^i, u_t^i) = locReord} \right.$$

$$\left. + H\left( \tilde{\pi}_u(s_t^i, w_t^i; \phi) || \tilde{\pi}_{\text{teacher}}(s_t^i, w_t^i) \right) \mathbb{1}_{(w_t^i, u_t^i) = locReord} \right),$$

where $\tilde{\pi}_m$ with $m \in \{u, \text{teacher}\}$ is the respective conditional policy distribution defined over all nodes except the pool node $p$:

$$\tilde{\pi}(\cdot | s_t^i, w_t^i) = \pi(\cdot | u_t^i \neq p, s_t^i, w_t^i),$$

as we only want to align the behaviour regarding local reordering actions.

In the first term of the transfer loss the teacher assesses the student's decision regarding a chosen local reordering action at time $t$ by an aggregated agreement regarding the student's chosen local reordering actions after time $t$:

$$\hat{R}_t^i = \sum_{k=t}^{T} \log \tilde{\pi}_{\text{teacher}}(u_{k+1}^i | w_{k+1}^i, s_{k+1}^i, P_{k+1}) \mathbb{1}_{(w_{k+1}^i, u_{k+1}^i) = locReord}.$$

In the second term, the similarity of the teacher's and student's distribution over potential local reordering actions at time $t$ is computed in terms of their entropy as

$$H\left( \tilde{\pi}_u(s_t^i, w_t^i; \phi) || \tilde{\pi}_{\text{teacher}}(s_t^i, w_t^i) \right)$$

$$= -\sum_{v \in V : v \neq p} \tilde{\pi}_u(u_t^i = v | s_t^i, w_t^i; \phi) \log \tilde{\pi}_{\text{teacher}}(u_t^i = v | s_t^i, w_t^i).$$

**Benchmarks** For the considered setup of 5 agents and 10 nodes, we compare the presented transfer learning approaches to two baselines. First, to the 5ag-model itself which we ultimately want to improve. Second, to the 2ag-model which is used as the teacher for transferring the local reordering skill and can be directly applied in the 5ag-setup.

**Table 11.5** Transfer learning results (TL distill, TL init) for 10 customer nodes in terms of $c_{gap}$, i.e., the improvement over the initial solutions. For interpreting the scale of the values, we recall that PrivAssign reached a gap of 67% and OR-Tools of 76%

| 10 nodes | 5ag | TL distill | TL init | 2ag |
|---|---|---|---|---|
| MANR | 60% | 61% | 62% | 63% |
| MANR best | 72% | 73% | 73% | 74% |

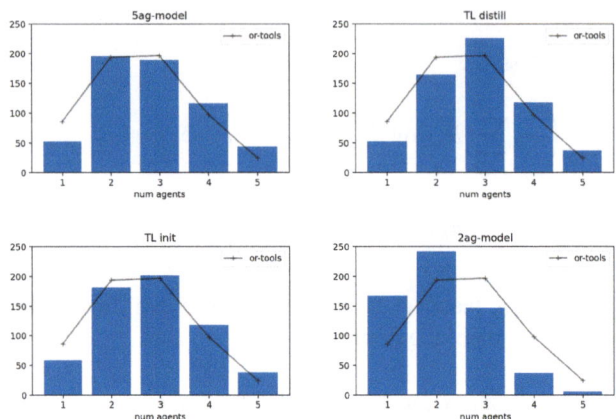

**Fig. 11.6** Number of participating agents in the respective model solutions for the 5ag-test set

**Results** Both transfer learning approaches lead to an improvement over the 5ag-model (cf. Table 11.5). While considering the best found solution per routing instance (MANR best) leads to the same performance of the policy initialization (TL init) and the distillation (TL distill) approach, TL init leads to better results more consequently throughout multiple inference runs (MANR). However, the direct application of the 2ag-model which is the only model not trained in the 5ag-environment actually outperforms the other approaches. We trace it back to the fact that the 2ag-model prefers solutions in which only one or two agents participate (cf. Fig. 11.6), as it has learned it during training. Such solutions represent a local minimum. The amount of collaboration in the OR-Tools solution (cf. black line in Fig. 11.6) shows that generally more collaboration leads to even better results. Even though all models trained in the 5ag-setup learned to collaborate more, they didn't learn to do it in a proper way. Since the blind transfer of the 2ag-model works better than the elaborated approaches, the experiments demonstrate the necessity for additional and different kind of knowledge integration during training. More knowledge about the problem structure must be incorporated to improve learning in larger teams.

## 11.5  Conclusion

The presented multi-agent neural rewriter (MANR) models and implements collaborative vehicle routing as a team Markov game with partially observable costs. The learned agent policy chooses an agent action solely based on the own's agent cost. The idea of rewriting solutions has been extended from the (single-agent) neural rewriter by introducing new rewriting rules. Most importantly, they allow agents to exchange customer nodes without causing conflicts and thus act simultaneously in the multi-agent system. We use variants of informed ML to enable an agent to behave optimally within and for its team: First, the agent is informed about the necessity to create feasible solutions by shaping the reward function appropriately. To produce not only feasible solutions but such with minimal total team cost, the solely local cost observing agent is provided with global knowledge about the underlying cost distribution. This knowledge is represented in terms of a modelled $Q$ function and is acquired during training by processing exemplary cost samples. It is inserted into the agent policy's learning algorithm through an actor-critic approach. It enables the agent to assess its teammates' costs and thus to act for the sake of its team. This is confirmed by first empirical evaluations on small problem sizes: The MANR competes with a well-performing heuristic with the same limited cost knowledge. In particular, our approach yields a significantly better trade-off between solution quality and execution time. We have observed the further potential of informed ML through transfer learning (TL), specifically for models trained in larger team setups. Yet the TL experiments have also demonstrated that there is room for improving our modelling approach. In the future, we want to investigate the integration of additional and different kind of knowledge about the problem structure.

**Acknowledgments** This contribution was supported by the Fraunhofer Cluster of Excellence "Cognitive Internet Technologies".

## References

1. Bengio, Y., Lodi, A., Prouvost, A.: Machine learning for combinatorial optimization: a methodological tour d'horizon. European Journal of Operational Research. **290**, 405–421 (2021)
2. Chen, X., Tian, Y.: Learning to Perform Local Rewriting for Combinatorial Optimization. Advances in Neural Information Processing Systems. **32**, 6281–6292 (2019)
3. Correll, R., Weinberg, S. J., Sanches, F., Ide, T., Suzuki, T.: Reinforcement Learning for Multi-Truck Vehicle Routing Problems. arXiv preprint arXiv:2211.17078 (2022)
4. Czarnecki, W., Pascanu, R., Osindero, S., Jayakumar, S., Swirszcz, G., Jaderberg, M.: Distilling policy distillation. The 22nd International Conference on Artificial Intelligence and Statistics. 1331–1340 (2019)

5. Foerster, J., Farquhar, G., Afouras, T., Nardelli, N., Whiteson, S.: Counterfactual multi-agent policy gradients. Proceedings of the AAAI conference on artificial intelligence. **32** (2018)
6. Guo, C., Pleiss, G., Sun, Y., Weinberger, K.: On calibration of modern neural networks. International conference on machine learning. 1321–1330 (2017)
7. Gupta, J. K., Egorov, M., Kochenderfer, M.: Cooperative multi-agent control using deep reinforcement learning. International conference on autonomous agents and multiagent systems. 66–83 (2017)
8. Hromkovič, J.: Algorithmics for hard problems: introduction to combinatorial optimization, randomization, approximation, and heuristics. Springer Science & Business Media (2013)
9. Kingma, D. P., Ba, J.: Adam: A method for stochastic optimization. International Conference on Learning Representations (2015)
10. Konda, V., Tsitsiklis, J.: Actor-critic algorithms. Advances in neural information processing systems. **12** (1999)
11. Kool, W., van Hoof, H., Welling, M.: Attention! learn to solve routing problems! International Conference on Learning Representations. (2019)
12. Laporte, G.: The vehicle routing problem: An overview of exact and approximate algorithms. European journal of operational research. **59**, 345–358 (1992)
13. Liaw, R., Liang, E., Nishihara, R., Moritz, P., Gonzalez, J. E., Stoica, I.: Tune: A Research Platform for Distributed Model Selection and Training. arXiv preprint arXiv:1807.05118 (2018)
14. Lowe, R., Wu, Y., Tamar, A., Harb, J., Pieter, A., Mordatch, I.: Multi-agent actor-critic for mixed cooperative-competitive environments. Advances in neural information processing systems. **30** (2017)
15. McMahan, B., Moore, E., Ramage, D., Hampson, S., y Arcas, B. A.: Communication-efficient learning of deep networks from decentralized data. Artificial intelligence and statistics. 1273–1282 (2017)
16. Nazari, M., Oroojlooy, A., Snyder, L., Takác, M.: Reinforcement learning for solving the vehicle routing problem. Advances in neural information processing systems. **31** (2018)
17. Paul, N., Wirtz, T., Wrobel, S., Kister, A.: Multi-Agent Neural Rewriter for Vehicle Routing with Limited Disclosure of Costs. Presented at the Gamification and Multiagent Solutions Workshop within Tenth International Conference on Learning Representations, ICLR. (2022)
18. Pomponi, F., Fratocchi, L., Tafuri, S., Palumbo, M.: Horizontal collaboration in logistics: a comprehensive framework. Research in Logistics & Production. **3** (2013)
19. Rosenkrantz, D. J., Stearns, R. E., Lewis, II, P. M.: An analysis of several heuristics for the traveling salesman problem. SIAM journal on computing. **6**, 563–581 (1977)
20. Sutton, R. S., Barto, A. G.: Reinforcement learning: An introduction. MIT press. (2018)
21. Tan, M.: Multi-agent reinforcement learning: Independent vs. cooperative agents. Proceedings of the tenth international conference on machine learning. 330–337 (1993)
22. Tan, C., Sun, F., Kong, T., Zhang, W., Yang, C., Liu, C.: A survey on deep transfer learning. International conference on artificial neural networks. 270–279 (2018)
23. Vaswani, A., Shazeer, N., Parmar, N., Uszkoreit, J., Jones, L., Gomez, A. N., Kaiser, Ł., Polosukhin, I.: Attention is all you need. Advances in neural information processing systems. **30** (2017)
24. von Rueden, L., Mayer, S., Beckh, K., Georgiev, B., Giesselbach, S., Heese, R., Kirsch, B., Walczak, M., Pfrommer, J., Pick, A., Ramamurthy, R., Garcke, J., Bauckhage, C., Schuecker, J.: Informed machine learning - a taxonomy and survey of integrating prior knowledge into learning systems. IEEE Transactions on Knowledge and Data Engineering. **35**, 614–633 (2021)
25. Wang, X., Sandholm, T.: Reinforcement learning to play an optimal Nash equilibrium in team Markov games. Advances in neural information processing systems. **15** (2002)
26. Zhang, K., He, F., Zhang, Z., Lin, X., Li, M.: Multi-vehicle routing problems with soft time windows: A multi-agent reinforcement learning approach. Transportation Research Part C: Emerging Technologies. **121** (2020)

27. Zhang, K., Yang, Z., Başar, T.: Multi-agent reinforcement learning: A selective overview of theories and algorithms. Handbook of Reinforcement Learning and Control. 321–384 (2021)
28. Zhang, K., Yang, Z., Liu, H., Zhang, T., Basar, T.: Fully decentralized multi-agent reinforcement learning with networked agents. International Conference on Machine Learning. 5872–5881 (2018)

# Part IV
# Hybrid Methods

# Chapter 12
# Training Support Vector Machines by Solving Differential Equations

**Christian Bauckhage and Rafet Sifa**

**Abstract** The increasingly popular idea of Physics Informed Machine Learning uses trained machine learning models as tools for differential equation solving. Here, we turn this idea upside down and consider differential equation solving as a tool for training machine learning models. We focus on support vector machines for binary classification and explore the merits of training them by means of solving gradient flows. We thus assume a continuous time perspective on a fundamental machine learning problem which, in the mid- to long term, may inform implementations on (re)emerging hardware platforms such as analog- or quantum computers.

## 12.1 Introduction

The term *Informed Machine Learning* refers to the idea of designing computational intelligence systems which combine data- and knowledge-driven models and algorithms [35]. Reasons for pursuing this idea are manifold and we briefly review some of them.

For instance, modern end-to-end learning with deep neural networks of up to several hundred billion parameters is undeniably successful across a wide variety of tasks. Yet, in order to generalize well, such models must typically be trained with vast amounts of training data. With respect to practical use cases in industry, this may not be a problem for fully digitized businesses, however, even in the age of big data, more traditional industries still struggle with acquiring sufficient amounts of appropriate training data. Here, incorporating business- or process specific prior knowledge into model building and learning algorithms may reduce model complexity (measured, say, in terms of the number of adjustable model parameters) and can thus circumvent the need for massive data [34, 49].

C. Bauckhage (✉) · R. Sifa
Fraunhofer IAIS, Sankt Augustin, Germany
e-mail: christian.bauckhage@iais.fraunhofer.de; rafet.sifa@iais.fraunhofer.de

© The Author(s) 2025
D. Schulz, C. Bauckhage (eds.), *Informed Machine Learning*,
Cognitive Technologies, https://doi.org/10.1007/978-3-031-83097-6_12

Other reasons for informed learning are improved explainability or accountability. If the structure or parameter ranges of a model reflect known facts about its intended application domain, training processes are less likely to overfit or to learn spurious correlations. Outputs of such models better hold up to scrutiny and are easier to understand or retrace by human experts. Exemplary approaches include knowledge-based restrictions on model parameters or on learning objectives [2, 25], problem specific neural network topologies [27, 28], and physics informed learning. The latter recently rose to prominence and works with models that reflect physical background knowledge, usually in form of differential equations, and uses this knowledge to guide training processes [19, 32].

The methods we introduce below can be seen as a variant of physics informed approaches in that we apply differential equations as a tool for model training. However, what distinguishes our approach is that the models we are training do not represent specific physical phenomena or processes but are general purpose machine learning models.

To be more specific, we show how to train support vector machines (SVMs) for binary classification by means of solving systems of ordinary differential equations.To accomplish this, we devise continuous time gradient flows over the feasible sets of corresponding dual training problems. These flows are known to converge to asymptotically stable stationary points from which we can compute the parameters of the sought after classifier.

Our motivations for considering continuous time models for SVM training are at least fourfold: First of all, there is renewed interest in SVMs because their underlying theory has recently been connected to deep neural networks [4, 9, 18]. Gradient flows play an important role in establishing this connection and, since the kind of flows we consider below differ from those considered before, they may inform further analysis.

Second of all, systems of ordinary differential equations (ODEs) occur in all of the hard sciences. Since they often do not have closed form solutions, numerical ODE solving has a long and venerable history and there exists a plethora of methods and corresponding software packages. Hence, by setting up the SVM training problem as a problem of solving continuous time gradient flows, we gain access to domain-agnostic learning algorithms as well as to versatile computational paradigms which can be implemented on various kinds of hardware platforms.

Speaking of hardware platforms, there is, third of all, growing concern as to the environmental sustainability of (deep) learning on modern GPU or TPU clusters [43, 46]. The enormous energy demands of operating such high performance infrastructures have prompted researchers to (re)consider more energy efficient analog computing [16, 37]. Indeed, analog circuits composed of resistors, capacitors, inductors, and operational amplifiers are known to allow for differential equation solving [44]. This, in turn, suggests that it may be possible to train SVMs on low cost, low power hardware.

Forth of all and somewhat orthogonal to our previous point, there also is increasing interest in Quantum Machine Learning. Since it has recently been shown that quantum computing algorithms can solve differential equations [23, 24, 40, 48], our continuous time perspective on SVM training could therefore also lead to novel quantum learning solutions.

### 12.1.1  Overview

In what follows, we particularly focus on the problem of training $L_2$ SVMs. This choice is admittedly informed by personal preferences but does not constitute any loss of generality. Seen from the point of view of mathematical intricacy, our focus rather marks a middle ground: On the one hand, gradient flows for $L_2$ SVM training are easier to set up than those for conventional SVMs. On the other hand, they are not as trivial as those for least squares SVMs.

Since $L_2$ SVMs seem to be less well known than other kinds of SVMs, Sect. 12.2 briefly summarizes the underlying concepts. In Sect. 12.3, we then discuss $L_2$ SVM training via solving gradient flows. Section 12.4 presents and discusses several didactic baseline experiments which allow for illustrating practical performances. Finally, in Sect. 12.5, we summarize our contributions and findings and provide an outlook to auspicious future work.

### 12.1.2  Mathematical Notation

Throughout, we will be working with vectors and matrices over the field of real numbers. Vectors will be written using bold lower case letters ($\mathbf{v}$) and matrices using bold upper case letters ($\mathbf{M}$).

The operator "$\mathsf{T}$" denotes vector- and matrix transposition. Using this operator, inner- and outer products of two vectors will be written as $\mathbf{u}^\mathsf{T}\mathbf{v}$ and $\mathbf{u}\mathbf{v}^\mathsf{T}$, respectively.

The symbol "$\odot$" denotes the Hadamard or element-wise product of vectors or of matrices. That is, $\mathbf{u} = \mathbf{v} \odot \mathbf{w} \Leftrightarrow u_i = v_i \cdot w_i$ and $\mathbf{M} = \mathbf{N} \odot \mathbf{O} \Leftrightarrow M_{ij} = N_{ij} \cdot O_{ij}$.

## 12.2  Setting the Stage

This section briefly recalls the basic theory behind $L_2$ SVM training and application. Readers familiar with this topic may safely skip ahead.

Since we address the problem of binary classifier training, we assume that we have access to labeled samples of training data $\left\{(\mathbf{x}_i, y_i)\right\}_{i=1}^{n}$ where the data vectors $\mathbf{x}_i \in \mathbb{R}^m$ represent entities from two classes and the label values $y_i \in \{-1, +1\}$ indicate class membership.

Below, we will write very compact mathematical expressions. To be able to do so, we gather the given training data points in an $m \times n$ matrix and their labels in an $n$ vector, namely

$$
\mathbf{X} = \begin{bmatrix} | & | & & | \\ \mathbf{x}_1 & \mathbf{x}_2 & \cdots & \mathbf{x}_n \\ | & | & & | \end{bmatrix} \tag{12.1}
$$

$$
\mathbf{y} = \begin{bmatrix} y_1 & y_2 & \cdots & y_n \end{bmatrix}^\mathsf{T}. \tag{12.2}
$$

Training a binary classifier for data like these means to estimate the parameters of a function $y : \mathbb{R}^m \to \{-1, +1\}$. A common, simple, and generic ansatz for such a function is a linear classifier

$$
y(\mathbf{x}) = \text{sign}(\mathbf{x}^\mathsf{T}\mathbf{w} - \theta) \tag{12.3}
$$

whose parameters are a weight vector $\mathbf{w} \in \mathbb{R}^m$ and a threshold value $\theta \in \mathbb{R}$.

What turns the generic model in (12.3) into an SVM is the idea of estimating its weight vector $\mathbf{w}$ such that projections $\mathbf{x}_i^\mathsf{T}\mathbf{w}$ of the training data from both classes are maximally separated or, equivalently, such that the margin $\rho \in \mathbb{R}$ between the projected training data is as large as possible.

As there exist various loss functions for formalizing max-margin criteria, SVMs come in many different flavors. Well known variants are the standard SVMs introduced in pioneering work by Cortes and Vapnik [11]. Other popular flavors include least squares SVMs due to Suykens and Vanderwalle [42] or $\nu$-SVMs as proposed by Schölkopf et al. [36].

Another variant are the so called $L_2$ SVMs whose origins can be traced back to work by Frieß and Harrison [13] or by Mangasarian and Musicant [26]. These are of practical interest as they are surprisingly easy to train [3, 39, 47]. Next, we corroborate this claim and sketch how $L_2$ SVM training can be accomplished.

### 12.2.1  $L_2$ Support Vector Machines

A (linear) support vector machine for binary classification determines the max-margin hyperplane between the training data for two given classes. Should these classes not be linearly separable, one typically incorporates additional slack variables $\xi_i \geq 0 \in \mathbb{R}$ gathered in a vector $\boldsymbol{\xi} \in \mathbb{R}^n$ whose influence on the training procedure is controlled by a parameter $C \geq 0 \in \mathbb{R}$.

When training an $L_2$ SVM, slack variables enter the primal objective in form of a sum of squares which differs from conventional SVMs where they appear in a simple sum. Similar to conventional SVMs, the primal problem of training an $L_2$ SVM comes with inequality constraints which differs from least squares SVMs where there are none.

In short, one can show [3, 39] that the *primal problem of training an $L_2$ SVM* consists in solving the following constrained optimization problem

$$\underset{\mathbf{w},\theta,\rho,\boldsymbol{\xi}}{\text{argmin}} \ \tfrac{1}{2}\mathbf{w}^\mathsf{T}\mathbf{w} + \tfrac{1}{2}\theta^2 - \rho + \tfrac{C}{2}\boldsymbol{\xi}^\mathsf{T}\boldsymbol{\xi}$$

$$\text{s.t.} \quad \left[\mathbf{Z}^\mathsf{T}\mathbf{w} - \theta \cdot \mathbf{y}\right] - \rho \cdot \mathbf{1} + \boldsymbol{\xi} \geq \mathbf{0},$$

(12.4)

where $\mathbf{0}, \mathbf{1} \in \mathbb{R}^n$ denote the vectors of all zeros and all ones, respectively and matrix

$$\mathbf{Z} = \begin{bmatrix} | & | & & | \\ \mathbf{z}_1 & \mathbf{z}_2 & \cdots & \mathbf{z}_n \\ | & | & & | \end{bmatrix}$$

is a matrix of size $m \times n$ whose individual columns are given by

$$\mathbf{z}_i = y_i \cdot \mathbf{x}_i.$$

In the appendix, we show that evaluating the Karush-Kuhn-Tucker conditions of optimality leads to the following *dual problem of training an $L_2$ SVM*

$$\underset{\boldsymbol{\mu}\in\mathbb{R}^n}{\text{argmin}} \ \tfrac{1}{2}\boldsymbol{\mu}^\mathsf{T}\left[\mathbf{X}^\mathsf{T}\mathbf{X} \odot \mathbf{y}\mathbf{y}^\mathsf{T} + \mathbf{y}\mathbf{y}^\mathsf{T} + \tfrac{1}{C}\mathbf{I}\right]\boldsymbol{\mu}$$

$$\text{s.t.} \quad \begin{aligned} \mathbf{1}^\mathsf{T}\boldsymbol{\mu} &= 1 \\ \boldsymbol{\mu} &\geq \mathbf{0}. \end{aligned}$$

(12.5)

Here, $\mathbf{I}$ denotes the $n \times n$ identity matrix, $\boldsymbol{\mu} \in \mathbb{R}^n$ is a vector of $n$ Lagrange multipliers $\mu_i$, and $\odot$ denotes the Hadamard product, i.e. the element-wise product of vectors or matrices.

Once the minimizer $\boldsymbol{\mu}$ of the problem in (12.5) has been found, those entries $\mu_s$ of $\boldsymbol{\mu}$ which exceed zero identify which training data points support the sought after hyperplane. This, in turn, allows for computing the model parameters

$$\mathbf{w} = \quad \mathbf{X}\left[\mathbf{y} \odot \boldsymbol{\mu}\right]$$

(12.6)

$$\theta = -\mathbf{1}^\mathsf{T}\left[\mathbf{y} \odot \boldsymbol{\mu}\right].$$

(12.7)

Given these, the sought after classifier in (12.3) becomes

$$y(\mathbf{x}) = \text{sign}\left(\mathbf{x}^\mathsf{T}\mathbf{X}\left[\mathbf{y} \odot \boldsymbol{\mu}\right] + \mathbf{1}^\mathsf{T}\left[\mathbf{y} \odot \boldsymbol{\mu}\right]\right)$$

(12.8)

$$= \text{sign}\left(\left[\mathbf{x}^\mathsf{T}\mathbf{X} + \mathbf{1}^\mathsf{T}\right]\left[\mathbf{y} \odot \boldsymbol{\mu}\right]\right).$$

(12.9)

### 12.2.2 Invoking the Kernel Trick

Note that, during training (12.5) as well as during application (12.9) of an $L_2$ SVM, data vectors exclusively occur within inner products, namely $\mathbf{X}^\mathsf{T}\mathbf{X}$ and $\mathbf{x}^\mathsf{T}\mathbf{X}$, respectively. This allows for invoking the kernel trick and therefore for dealing with non-linear settings.

In other words, considering a Mercer kernel $k : \mathbb{R}^m \times \mathbb{R}^m \to \mathbb{R}$, a non-linear binary classifier can be trained by replacing the Gram matrix $\mathbf{X}^\mathsf{T}\mathbf{X}$ in (12.5) with an $n \times n$ kernel matrix $\mathbf{K}$ whose elements are given by

$$K_{ij} = k(\mathbf{x}_i, \mathbf{x}_j). \tag{12.10}$$

By the same token, the trained classifier in (12.9) can be rewritten as

$$y(\mathbf{x}) = \mathrm{sign}\left([\mathbf{k}(\mathbf{x}) + \mathbf{1}]^\mathsf{T}[\mathbf{y} \odot \boldsymbol{\mu}]\right), \tag{12.11}$$

where the elements of the $n$-dimensional kernel vector $\mathbf{k}(\mathbf{x})$ amount to

$$k_i(\mathbf{x}) = k(\mathbf{x}, \mathbf{x}_i). \tag{12.12}$$

### 12.2.3 A Baseline Training Algorithm

Observing that the feasible set of the dual $L_2$-SVM training problem in (12.5) is the standard simplex

$$\Delta^{n-1} = \left\{\boldsymbol{\mu} \in \mathbb{R}^n \mid \boldsymbol{\mu} \succeq \mathbf{0} \wedge \mathbf{1}^\mathsf{T}\boldsymbol{\mu} = 1\right\} \tag{12.13}$$

and, introducing the following two shorthands for brevity

$$\mathbf{H} \equiv \mathbf{K} \odot \mathbf{y}\mathbf{y}^\mathsf{T} + \mathbf{y}\mathbf{y}^\mathsf{T} + \tfrac{1}{C}\mathbf{I} \tag{12.14}$$

$$f(\boldsymbol{\mu}) \equiv \tfrac{1}{2}\boldsymbol{\mu}^\mathsf{T}\mathbf{H}\boldsymbol{\mu}, \tag{12.15}$$

we find that the parameter estimation problem in (12.5) can be written more succinctly, namely

$$\underset{\boldsymbol{\mu}\in\Delta^{n-1}}{\mathrm{argmin}} f(\boldsymbol{\mu}). \tag{12.16}$$

Written like this, our classifier training problem is now clearly revealed as a quadratic minimization problem over a compact convex set. Since it can therefore rather easily be solved using the Frank-Wolfe algorithm [12], our practical examples presented and discussed below will consider iterative Frank-Wolfe optimization as baseline method for $L_2$ SVM training; the algorithm's favorable characteristics for this purpose have previously been discussed in [3, 39].

## 12.3 Gradient Flows for $L_2$ SVM Training

In this section, we devise a gradient flow tuned to the peculiarities of the optimization problem in (12.16). Solutions to this flow represent parameter space trajectories which approach an optimal choice of the Lagrange multipliers which constitute our decision variables. Given the optimized Lagrange multipliers, we can then compute $L_2$ SVM weight and bias parameters.

Since the term *gradient flow* is now often re-appropriated to mean the backward flow of gradient information during neural network training, we emphasize that we understand it in its classical sense, namely

**Definition** Given a finite dimensional Euclidean vector space $\mathbb{R}^n$ and a smooth function $f : \mathbb{R}^n \to \mathbb{R}$, a *gradient flow* is a smooth curve $\boldsymbol{\mu} : \mathbb{R} \to \mathbb{R}^n$, $t \mapsto \boldsymbol{\mu}(t)$ such that $\dot{\boldsymbol{\mu}}(t) = -\nabla f(\boldsymbol{\mu}(t))$.

Given the wide range of possible behaviors of dynamical systems in general, the dynamics of finite dimensional gradient flows are rather simple. One can show [50] that they either converge to a stationary point $\boldsymbol{\mu}_*$ of $f$ where $\nabla f(\boldsymbol{\mu}_*) = \mathbf{0}$ or diverge as $t \to \infty$. One can further show [1, 17] the following: if $\boldsymbol{\mu}_*$ is an isolated stationary point and a local minimum of $f$, then $\boldsymbol{\mu}_*$ is an asymptotically stable equilibrium of the system $\dot{\boldsymbol{\mu}}(t) = -\nabla f(\boldsymbol{\mu}(t))$.

Dropping the dependence on time $t$ for readability, the Euclidean gradient of the objective function of the optimization problem in (12.16) is given by

$$\nabla f(\boldsymbol{\mu}) = \mathbf{H}\boldsymbol{\mu}. \tag{12.17}$$

However, a crucial difference between the gradient flow in the above definition and the $L_2$ SVM training problem in (12.16) is that the former consider flows in $\mathbb{R}^n$ whereas solutions to the latter are confined to the standard simplex $\Delta^{n-1}$.

We therefore note that the standard simplex $\Delta^{n-1}$ is but a specific instance of a convex polytope

$$\mathcal{P} = \left\{ \boldsymbol{\mu} \in \mathbb{R}^n \mid \boldsymbol{\mu} \succeq \mathbf{0}, \mathbf{A}\boldsymbol{\mu} = \mathbf{b} \right\} \tag{12.18}$$

and resort to a crucial result by Helmke and Moore [17]. They have shown that and how an *open* convex polytope

$$\overset{\circ}{\mathcal{P}} = \left\{ \boldsymbol{\mu} \in \mathbb{R}^n \mid \boldsymbol{\mu} \succ \mathbf{0}, \mathbf{A}\boldsymbol{\mu} = \mathbf{b} \right\} \tag{12.19}$$

can be endowed with a Riemannian metric. Gradients with respect to this metric, i.e. Riemannian gradients on an open convex polytope, are written as "grad" and the gradient flow

$$\dot{\mu} = - \operatorname{grad} f(\mu) \tag{12.20}$$

of a function $f : \overset{\circ}{\mathcal{P}} \to \mathbb{R}$ with respect to this Riemannian metric is given by

$$\dot{\mu} = -\left[\mathbf{I} - \mathbf{D}\mathbf{A}^{\mathsf{T}}[\mathbf{A}\mathbf{D}\mathbf{A}^{\mathsf{T}}]^{-1}\mathbf{A}\right]\mathbf{D}\,\nabla f(\mu) , \tag{12.21}$$

where $\nabla$ again denotes the conventional Euclidean gradient and matrix $\mathbf{D} = \operatorname{diag}(\mu)$ is an $n \times n$ diagonal matrix.

Comparing the definition of the standard simplex $\Delta^{n-1}$ in (12.13) to the definition of a general polytope $\mathcal{P}$ in (12.18), we note the following: For the standard simplex, we have $\mathbf{A} = \mathbf{1}^{\mathsf{T}}$. Regarding the problem in (12.16), we further note that $\mathbf{D}\mathbf{1} = \mu$ as well as $\mathbf{1}^{\mathsf{T}}\mu = 1$ so that $\mathbf{1}^{\mathsf{T}}\mathbf{D}\mathbf{1} = 1$. Restricted to the open simplex $\overset{\circ}{\Delta}{}^{n-1}$ and using $\nabla f(\mu) = \mathbf{H}\,\mu$, a gradient flow for (12.16) is therefore given by

$$\dot{\mu} = -\left[\mathbf{I} - \mathbf{D}\mathbf{1}[\mathbf{1}^{\mathsf{T}}\mathbf{D}\mathbf{1}]^{-1}\mathbf{1}^{\mathsf{T}}\right]\mathbf{D}\mathbf{H}\,\mu \tag{12.22}$$

$$= -\left[\mathbf{I} - \mu\mathbf{1}^{\mathsf{T}}\right]\mathbf{D}\mathbf{H}\,\mu \tag{12.23}$$

$$= -\mathbf{D}\mathbf{H}\,\mu + \mu\mu^{\mathsf{T}}\mathbf{H}\,\mu. \tag{12.24}$$

With respect to practical implementations of numerical computing code, we further note that it can be more efficient to work with the equivalent expression

$$\dot{\mu} = -\left[\mathbf{H}\,\mu - \mu^{\mathsf{T}}\mathbf{H}\,\mu \cdot \mathbf{1}\right] \odot \mu. \tag{12.25}$$

Using software for numerical integration, this ordinary dynamical system can be practically solved for $\mu(t)$.

For instance, in our practical performance evaluations which we discuss below, we worked with the odeint functionalities provided in the *Python* scientific computing libraries *NumPy/SciPy* [29].

A general introduction to numerical differential equation solving and its extensive underlying theory is far beyond the scope of this chapter but respective details can, for instance, be found in a comprehensive textbook by Stuart and Humphries [41].

## 12.4 Practical Examples

In this section, we present and discuss practical examples which illustrate how $L_2$ SVM training by means of solving the gradient flow in (12.25) compares to $L_2$ SVM training by using the Frank-Wolfe algorithm to solve the problem in (12.16).

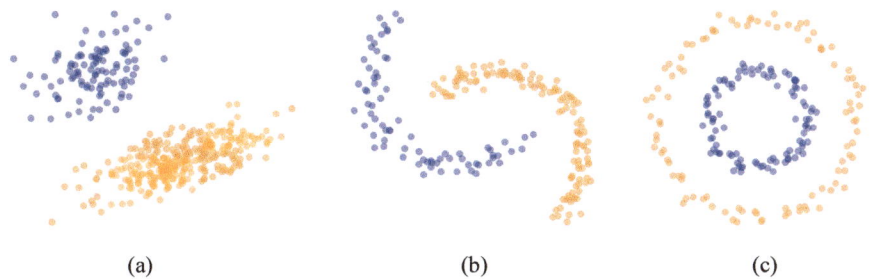

**Fig. 12.1** Didactic training sets of 2D data points on which to train $L_2$ SVMs for binary classification. While the two classes represented by the data in (a) are linearly separable, the classes in (b) and (c) are not. For the setting in (a) we may thus simply train a linear SVM; for the settings in (b) and (c), however, we should work with non-linear kernel SVMs. (**a**) Two Gaussian blobs. (**b**) Two moons. (**c**) Two nested circles

For ease of visualization as well as for ease of discussion, we consider three simple and deliberately didactic binary classification scenarios. Each of these scenarios only involves two-dimensional data points and the corresponding training data sets are shown in Fig. 12.1. Looking at the three panels in this figure, we realize that the two classes in Fig. 12.1a are linearly separable whereas the classes in Fig. 12.1b and c are not.

For the former, we therefore trained linear SVMs or, equivalently, kernel SVMs with linear kernels of the form

$$k\big(\mathbf{x}_i, \mathbf{x}_j\big) = \mathbf{x}_i^\mathsf{T} \mathbf{x}_j . \tag{12.26}$$

For the two latter cases, on the other hand, we trained kernel SVMs with non-linear kernels. To be specific, we considered the following 3rd order polynomial kernel

$$k\big(\mathbf{x}_i, \mathbf{x}_j\big) = \big(\mathbf{x}_i^\mathsf{T} \mathbf{x}_j + 1\big)^3 \tag{12.27}$$

for the "two moons" data in Fig. 12.1b and the following Gaussian kernel function

$$k\big(\mathbf{x}_i, \mathbf{x}_j\big) = \exp\Big(-\tfrac{1}{2 \cdot 0.75^2} \, \big\| \mathbf{x}_i - \mathbf{x}_j \big\|^2\Big) \tag{12.28}$$

for the "two nested circles" data in Fig. 12.1c.

The SVM slack parameters were determined manually and set to $C = 500$, $C = 1$, and $C = 500$ for the settings in Fig. 12.1a, b, and c, respectively.

Figure 12.2 illustrates training processes and results for the "two Gaussian blobs" data in Fig. 12.1a.

The panels in the upper row show the situation when using the Frank-Wolfe algorithm for iteratively solving the dual $L_2$ SVM training problem in (12.16). Here, we started the process with an initial value of $\boldsymbol{\mu}_0 = \tfrac{1}{n}\mathbf{1}$ and considered a total of

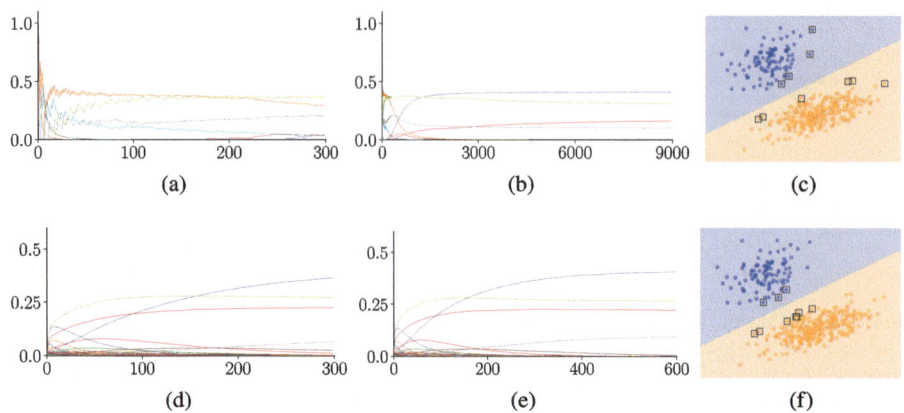

**Fig. 12.2** Examples of training kernel $L_2$ SVM classifiers on the "two Gaussian blobs" data in Fig. 12.1a. The upper row visualizes the training process and its result when using Frank-Wolfe optimization; the lower row visualizes the training process and its result when solving the gradient flow in (12.25). The kernel function in both cases is a simple linear kernel. From a practical point of view, the resulting classifiers are virtually indistinguishable; yet, the (feature space) separating hyperplane found via the Frank-Wolfe algorithm is supported by notably different support vectors than the hyperplane that results from solving (12.25). (**a**) Discrete Frank-Wolfe iterates $\mu_t$. (**b**) Discrete Frank-Wolfe iterates $\mu_t$. (**c**) Resulting classifier. (**d**) Continuous gradient flow $\mu(t)$. (**e**) Continuous gradient flow $\mu(t)$. (**f**) Resulting classifier

6000 iterations. For visual clarity, the evolution of the Lagrange multipliers over the first 300 of these iterations is shown in Fig. 12.2a and their evolution over the whole iterative process can be seen in Fig. 12.2b.

In Fig. 12.2a we clearly recognize the typical jittering behavior of the Frank-Wolfe iterates $\mu_t$ which occurs when running plain vanilla versions of the algorithm [14]. Figure 12.2b illustrates the convergence behavior of the overall optimization process. Figure 12.2c visualizes the decision boundary of the resulting classifier and highlights those training data points which support the (feature space) separating hyperplane by means of black squares.

The panels in the lower row of Fig. 12.2 show the situation when numerically solving the continuous time gradient flow in (12.25). Here, we considered an initial value of $\mu(0) = \frac{1}{n}\mathbf{1}$ and solved the dynamical systems on an equally spaced grid of time points $t = 0.0, 0.5, 1.0, \ldots, 600.0$. Again for visual clarity, the evolution of the Lagrange multipliers from time $t = 0$ to time $t = 300$ is shown in Fig. 12.2d and their evolution over the whole period can be seen in Fig. 12.2e.

Both these panels reveal the values $\mu(t)$ to vary smoothly; Fig. 12.2b illustrates the convergence behavior of the underlying gradient flow. Figure 12.2f visualizes the decision boundary of the resulting classifier and highlights those training data points which support the (feature space) separating hyperplane by means of black squares. Note that, to obtain the latter, we rounded the entries of $\mu(600)$ to five decimal places and then renormalized them (to sum to unity) because the flow in (12.25)

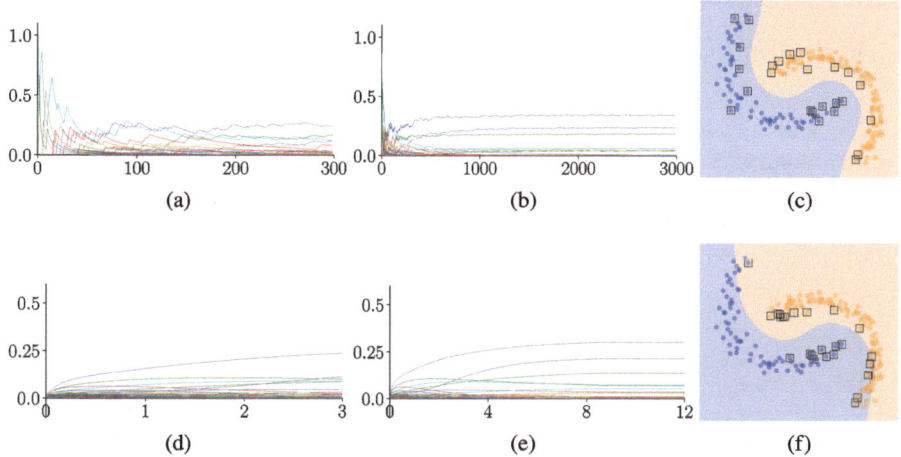

**Fig. 12.3** Examples of training kernel $L_2$ SVM classifiers on the "two moons" data in Fig. 12.1b. The upper row visualizes the training process and its result when using Frank-Wolfe optimization; the lower row visualizes the training process and its result when solving the gradient flow in (12.25). The kernel function in both cases is a third order polynomial. The resulting classifiers are virtually indistinguishable; yet, the (feature space) separating hyperplane found via Frank-Wolfe iterations is supported by fewer and different support vectors than the one found from solving (12.25). (**a**) Discrete Frank-Wolfe iterates $\mu_t$. (**b**) Discrete Frank-Wolfe iterates $\mu_t$. (**c**) Resulting classifier. (**d**) Continuous gradient flow $\mu(t)$. (**e**) Continuous gradient flow $\mu(t)$. (**f**) Resulting classifier

evolves in the *open* simplex $\mathring{\Delta}^{n-1}$ so that no entry of $\mu(t)$ can ever truly drop to zero.

Looking at Fig. 12.2c and f, it appears that, from a practitioners point of view, the decision boundaries of both $L_2$ SVM classifiers are virtually indistinguishable. However, we note that the Frank-Wolfe- and the gradient flow solutions for $\mu$ lead to quite different support vectors. We further point out that, in this example, the gradient flow converged faster (in terms of fewer overall iterations or computational steps) than the Frank-Wolfe algorithm.

Figure 12.3 illustrates training processes and results for the "two moons" data in Fig. 12.1b. The panels in the upper row again reflect the situation when using Frank-Wolfe optimization and the panels in the lower row show the situation when solving the corresponding gradient flow. In both settings, we again considered feasible initial values of $\mu_0 = \mu(0) = \frac{1}{n}\mathbf{1}$. Frank-Wolfe optimization was run for a total of 3000 iterations and the gradient flow was numerically solved on a grid of time points $t = 0.0, 0.5, 1.0, \ldots, 12.0$.

Looking at this figure, it appears that all the above observations apply again: whereas the Frank-Wolfe iterates jitter considerably, the flow evolves smoothly. It also converges faster than the Frank-Wolfe algorithm, and the decision boundaries of the resulting classifiers are again virtually indistinguishable although the corresponding (feature space) separating hyperplanes are supported by different support

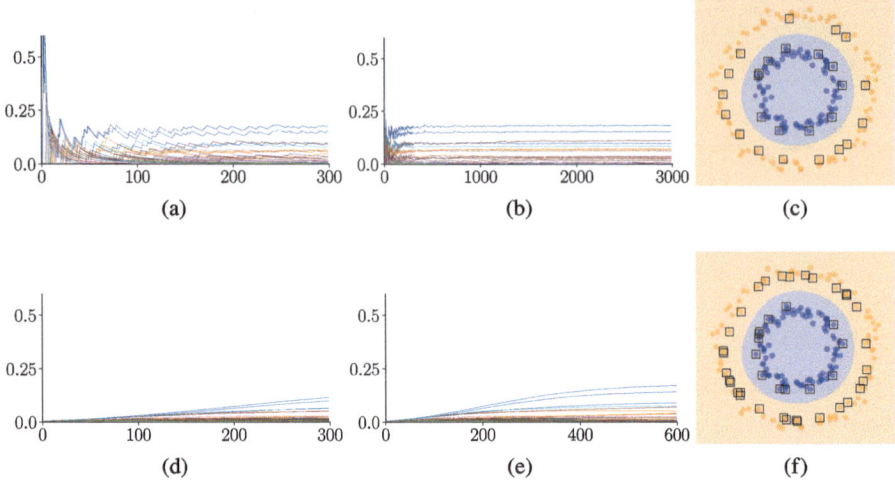

**Fig. 12.4** Examples of training kernel $L_2$ SVM classifiers on the "two nested circles" data in Fig. 12.1c. The upper row visualizes the training process and its result when using Frank-Wolfe optimization; the lower row visualizes the training process and its result when solving the gradient flow in (12.25). The kernel function in both cases is a Gaussian. The resulting classifiers are virtually indistinguishable; yet, the (feature space) separating hyperplane found via Frank-Wolfe iterations is supported by fewer and different support vectors than the one found from solving (12.25). (**a**) Discrete Frank-Wolfe iterates $\mu_t$. (**b**) Discrete Frank-Wolfe iterates $\mu_t$. (**c**) Resulting classifier. (**d**) Continuous gradient flow $\mu(t)$. (**e**) Continuous gradient flow $\mu(t)$. (**f**) Resulting classifier

vectors. However, we also note that training via Frank-Wolfe optimization resulted in a solution with slightly fewer support vectors.

Figure 12.4 illustrates training processes and results for the "two nested circles" in Fig. 12.1c. Again, the content of this figure is structured as in the previous two examples. Moreover, we once again set the starting points of the two training processes to $\mu_0 = \mu(0) = \frac{1}{n}\mathbf{1}$. Frank-Wolfe optimization was again run for a total of 3000 iterations and the gradient flow was numerically solved at time points $t = 0.0, 0.5, 1.0, \ldots, 600.0$.

Looking at this figure, it appears that all the observation we made above apply once more. Most notably, the decision boundaries of the two $L_2$ SVM classifiers resulting from the two training methods are again virtually indistinguishable although the corresponding (feature space) separating hyperplanes are supported by different support vectors. Regarding the latter, the solution produced by Frank-Wolfe optimization is a gain sparser than the one resulting from solving the corresponding gradient flow; in other words, the $L_2$ SVM trained via Frank-Wolfe optimization comes with fewer support vectors than the one trained via solving a gradient flow.

Overall, these three examples empirically corroborate the theoretical expectation that the training of SVMs for binary classification can be accomplished via solving gradient flows, i.e. via solving systems of ordinary differential equations.

Practical results and performance characteristics observed for the above examples are typical. This is to say that we also observed the above behavior of gradient flows for SVM training when working with other kinds of data sets and different types of kernel functions. An interesting empirical observation in this regard is that polynomial kernels seem to always entail rapidly converging gradient flows whereas the arguably more popular Gaussian kernels seem to cause rather slow convergence to an equilibrium of the corresponding flow. This points out an auspicious direction for future work: a rigorous mathematical analysis of the convergence behavior of SVM gradient flows under different kernels is currently under way and results will be reported soon.

Another open question at this point pertains to our empirical observation that SVMs trained via Frank-Wolfe optimization tend to require fewer support vectors than those trained via solving gradient flows. Again, a mathematical analysis of this phenomenon is currently under way.

## 12.5   Conclusion

In this contribution, we turned the idea of physics informed learning where machine learning models are used for differential equation solving on its head and considered differential equation solving as a tool for training machine learning models.

Focusing on the basic but fundamental learning task of binary classifier training, we considered the use of $L_2$ SVMs which are nowadays commonly trained by means of running the Frank-Wolfe algorithm to iteratively solve the Lagrangian dual of the primal training problem. Our main and novel contribution in this chapter was to show that $L_2$ SVM training can also be accomplished in terms of solving continuous time gradient flows or, equivalently, in terms of solving systems of ordinary differential equations. This required us to consider Riemannian gradients on the open simplex but, using a result due to Helmke and Moore, corresponding expressions we easily formulated.

Since the problem of differential equation solving arises in all of the hard sciences, numerous mathematical techniques and software tools have been developed for its solution. Hence, using our formulation of the model training problem, we gain access to a wide variety of domain-agnostic learning algorithms or versatile computational paradigms. This ties in with the idea of Informed Machine Learning as it allows for the (re-)use of well established procedural knowledge when training complex machine learning models. Moreover, applying these methods to the task of training machine learning models may be of interest for sustainable computing because they can, in principle, be implemented on various kinds of hardware platforms.

Indeed, the methods and results reported here are part of ongoing research efforts in which we investigate what kind of (classical) machine learning techniques lend themselves to implementations on energy efficient platforms or on low power and low cost hardware. A crucial observation in this regard is that gradient flow formulations of machine learning tasks are not restricted to the ($L_2$) SVM classifiers we considered in this chapter. Similar continuous time dynamical systems for least squares classifiers or linear discriminant classifiers are easy to come by and can also be devised for applications in the context of regression and forecasting [7]. The classic textbook by Helmke and Moore [17] provides a detailed account of the underlying theory and may serve as a source of inspiration for further developments in this direction.

Important practical goals behind our efforts are to develop light weight solutions for more sustainable Machine Learning on the one hand and for robust edge computing or on-sensor data analysis and decision making on the other. The latter are in high demand, for instance, in environmental- or agricultural applications which address sustainability challenges in food production [6, 15, 21]. In settings such as these, continuous time gradient flow formulations of machine learning objectives appear to be auspicious because simple analog circuits can, again in principle, solve differential equations while only requiring Milliwatts of energy [44].

However, implementations on corresponding hardware still face practical challenges such as, say, limited reliable numerical resolution on off-the-shelve analog devices. In other words, the challenges with respect to practical applications on transistor-less edge devices are first and foremost technical rather than theoretical. Nevertheless, in line with currently increasing engineering efforts towards designing analog circuitry specifically for Machine Learning [16, 37], we recently began more long-term efforts to investigate possible analog implementations of the framework presented here.

Finally, we recall yet another motivation for Machine Learning based on gradient flow formulations mentioned in the introduction, namely quantum computing. Given that working quantum computers have by now become a technical reality, research efforts on their use at different stages of the machine learning pipeline are noticeably increasing [8, 45]. Indeed, there already exist proposals for quantum support vector machines [33] and prototype implementations on existing quantum gate computers have been successfully realized [31].

In this context, it is therefore interesting to see that classical SVM training can be accomplished by means of solving continuous time gradient flows. This is because, recently, several quantum algorithms for differential equation solving have been proposed [23, 24, 40, 48] which therefore suggests that further novel approaches towards quantum SVMs or quantum SVM training might be possible.

# Appendix

In the main text of this contribution, we stated that the *primal problem of training an $L_2$ support vector machine* consist in solving

$$\underset{\mathbf{w}, \theta, \rho, \boldsymbol{\xi}}{\operatorname{argmin}} \ \tfrac{1}{2}\mathbf{w}^\mathsf{T}\mathbf{w} + \tfrac{1}{2}\theta^2 - \rho + \tfrac{C}{2}\boldsymbol{\xi}^\mathsf{T}\boldsymbol{\xi}$$

$$\text{s.t.} \quad \left[\mathbf{Z}^\mathsf{I}\mathbf{w} - \theta \cdot \mathbf{y}\right] - \rho \cdot \mathbf{1} + \boldsymbol{\xi} \succeq \mathbf{0} \tag{12.29}$$

and also presented the corresponding *dual problem of training an $L_2$ support vector machine*, namely

$$\underset{\boldsymbol{\mu}}{\operatorname{argmax}} \ -\tfrac{1}{2}\boldsymbol{\mu}^\mathsf{T}\left[\mathbf{Z}^\mathsf{T}\mathbf{Z} + \mathbf{y}\mathbf{y}^\mathsf{T} + \tfrac{1}{C}\mathbf{I}\right]\boldsymbol{\mu}$$

$$\mathbf{1}^\mathsf{T}\boldsymbol{\mu} = 1 \tag{12.30}$$

$$\text{s.t.}$$

$$\boldsymbol{\mu} \succeq \mathbf{0}.$$

**Note** While the dual problem in (12.30) seems to subtly differ from the one we presented in (12.5), we note that $\mathbf{Z}^\mathsf{T}\mathbf{Z} = \mathbf{X}^\mathsf{T}\mathbf{X} \odot \mathbf{y}\mathbf{y}^\mathsf{T}$ so that (12.5) and (12.30) are actually in perfect agreement.

In order for this contribution to be more self-contained, this appendix will show how to obtain the $L_2$ SVM training problem in (12.30) from the one in (12.29).

To derive the dual problem from the primal one, we first of all note that (12.29) constitutes a quadratic minimization problem with a total of $n$ greater-than-or-equal-to zero constraints which are subsumed in a single matrix-vector expression. Hence, we may introduce $n$ Lagrange multipliers $\mu_j$ which we may gather in a vector $\boldsymbol{\mu} \in \mathbb{R}^n$, to obtain the following Lagrangian

$$\mathcal{L}\left(\mathbf{w}, \theta, \boldsymbol{\xi}, \rho, \boldsymbol{\mu}\right) = \tfrac{1}{2}\mathbf{w}^\mathsf{T}\mathbf{w} + \tfrac{1}{2}\theta^2 - \rho + \tfrac{C}{2}\boldsymbol{\xi}^\mathsf{T}\boldsymbol{\xi}$$

$$- \boldsymbol{\mu}^\mathsf{T}\left[\mathbf{Z}^\mathsf{T}\mathbf{w} - \theta \cdot \mathbf{y} - \rho \cdot \mathbf{1} + \boldsymbol{\xi}\right]. \tag{12.31}$$

Next, we recall that the Karush-Kuhn-Tucker (KKT) conditions [20, 22] provide us with a set of criteria any valid solution to our inequality constrained problem must fulfill. For instance, the KKT 1 condition (stationarity) demands that, at a solution, we must have $\nabla\mathcal{L} = \mathbf{0}$.

We therefore partially derive the Lagrangian in (12.31) with respect to its parameters and equate the resulting expressions to zero. Recalling basic rules from multivariate calculus [30], this is easily done and we find

$$\frac{\partial\mathcal{L}}{\partial\mathbf{w}} = \mathbf{w} - \mathbf{Z}\boldsymbol{\mu} \ \overset{!}{=} \ \mathbf{0} \quad \Rightarrow \quad \mathbf{w} = \mathbf{Z}\boldsymbol{\mu} \tag{12.32}$$

$$\frac{\partial \mathcal{L}}{\partial \theta} = \theta + \mathbf{y}^\mathsf{T} \boldsymbol{\mu} \;\overset{!}{=}\; 0 \qquad \Rightarrow \qquad \theta = -\mathbf{y}^\mathsf{T} \boldsymbol{\mu} \tag{12.33}$$

$$\frac{\partial \mathcal{L}}{\partial \boldsymbol{\xi}} = C\boldsymbol{\xi} + \boldsymbol{\mu} \;\overset{!}{=}\; \mathbf{0} \qquad \Rightarrow \qquad \boldsymbol{\xi} = -\frac{1}{C}\boldsymbol{\mu} \tag{12.34}$$

$$\frac{\partial \mathcal{L}}{\partial \rho} = -1 + \mathbf{1}^\mathsf{T}\boldsymbol{\mu} \;\overset{!}{=}\; 0 \qquad \Rightarrow \qquad 1 = \mathbf{1}^\mathsf{T}\boldsymbol{\mu}. \tag{12.35}$$

Plugging the three results in (12.32), (12.33), and (12.34) back into (12.31) eliminates the parameters $\mathbf{w}$, $\theta$, and $\boldsymbol{\xi}$ and the Lagrangian becomes

$$
\begin{aligned}
\mathcal{L}(\boldsymbol{\mu}) = {} & \tfrac{1}{2}\boldsymbol{\mu}^\mathsf{T}\mathbf{Z}^\mathsf{T}\mathbf{Z}\boldsymbol{\mu} + \tfrac{1}{2}\boldsymbol{\mu}^\mathsf{T}\mathbf{y}\mathbf{y}^\mathsf{T}\boldsymbol{\mu} - \rho + \tfrac{1}{2C}\boldsymbol{\mu}^\mathsf{T}\boldsymbol{\mu} \\
& - \boldsymbol{\mu}^\mathsf{T}\mathbf{Z}^\mathsf{T}\mathbf{Z}\boldsymbol{\mu} - \boldsymbol{\mu}^\mathsf{T}\mathbf{y}\mathbf{y}^\mathsf{T}\boldsymbol{\mu} + \rho \cdot \boldsymbol{\mu}^\mathsf{T}\mathbf{1} - \tfrac{1}{C}\boldsymbol{\mu}^\mathsf{T}\boldsymbol{\mu}.
\end{aligned}
\tag{12.36}
$$

Further simplification and another application of the result in (12.35) leads to an even cleaner expression

$$
\begin{aligned}
\mathcal{L}(\rho, \boldsymbol{\mu}) &= -\tfrac{1}{2}\boldsymbol{\mu}^\mathsf{T}\mathbf{Z}^\mathsf{T}\mathbf{Z}\boldsymbol{\mu} - \tfrac{1}{2}\boldsymbol{\mu}^\mathsf{T}\mathbf{y}\mathbf{y}^\mathsf{T}\boldsymbol{\mu} - \tfrac{1}{2C}\boldsymbol{\mu}^\mathsf{T}\boldsymbol{\mu} \\
&= -\tfrac{1}{2}\boldsymbol{\mu}^\mathsf{T}\left[\mathbf{Z}^\mathsf{T}\mathbf{Z} + \mathbf{y}\mathbf{y}^\mathsf{T} + \tfrac{1}{C}\mathbf{I}\right]\boldsymbol{\mu} \\
&\equiv \mathcal{D}(\boldsymbol{\mu}).
\end{aligned}
\tag{12.37}
$$

The function $\mathcal{D}(\boldsymbol{\mu})$ we just introduced is called the Lagrangian dual and we note that it only depends on the Lagrange multipliers $\mu_j$ gathered in vector $\boldsymbol{\mu}$. We also note that $\mathcal{D}(\boldsymbol{\mu})$ is a (definite) quadratic form in $\boldsymbol{\mu}$ or, more specifically, a concave function (due to the scaling factor of $-1/2$) so that it has a unique maximum.

However, if we set out to maximize $\mathcal{D}(\boldsymbol{\mu})$ with respect to $\boldsymbol{\mu}$ we must be careful and actually incorporate two constraints. The first of these constraints is a consequence of (12.35) which demands that $\mathbf{1}^\mathsf{T}\boldsymbol{\mu} = 1$. The second one is due to the KKT 3 condition (dual feasibility) which demands that $\boldsymbol{\mu} \succeq \mathbf{0}$.

Because of Lagrange duality and of the two constraints we just pointed out, we therefore find that the dual problem of training an $L_2$ SVM consists in solving

$$
\begin{aligned}
&\underset{\boldsymbol{\mu}}{\mathrm{argmax}} \quad -\tfrac{1}{2}\boldsymbol{\mu}^\mathsf{T}\left[\mathbf{Z}^\mathsf{T}\mathbf{Z} + \mathbf{y}\mathbf{y}^\mathsf{T} + \tfrac{1}{C}\mathbf{I}\right]\boldsymbol{\mu} \\[2mm]
&\qquad\qquad \mathbf{1}^\mathsf{T}\boldsymbol{\mu} = 1 \\
&\text{s.t.} \\
&\qquad\qquad \boldsymbol{\mu} \succeq \mathbf{0}
\end{aligned}
\tag{12.38}
$$

which is exactly the problem we stated in (12.30).

Finally, to conclude our discussion we note the following: First of all, if we could solve the problem in (12.30) for the optimal vector of Lagrange multipliers $\boldsymbol{\mu}_*$, we can also determine the actually sought after SVM parameters $\mathbf{w}$ and $\theta$. This is thanks to (12.32) and (12.33) which provide us with

$$\mathbf{w} = \mathbf{Z}\,\boldsymbol{\mu}_* \qquad (12.39)$$

$$\theta = -\mathbf{y}^\mathsf{T}\boldsymbol{\mu}_*. \qquad (12.40)$$

Second of all, although we are dealing with a constrained optimization problem, the feasible set in which the optimal solution must reside is just the standard simplex $\Delta^{n-1}$. Moreover, since our objective function is quadratic and concave, maximizing $\mathcal{D}(\boldsymbol{\mu})$ is the same as minimizing $-\mathcal{D}(\boldsymbol{\mu})$. These two observation then imply that we could also write the dual $L_2$ SVM training problem as

$$\operatorname*{argmin}_{\boldsymbol{\mu} \in \Delta^{n-1}} \tfrac{1}{2}\,\boldsymbol{\mu}^\mathsf{T}\left[\mathbf{Z}^\mathsf{T}\mathbf{Z} + \mathbf{y}\mathbf{y}^\mathsf{T} + \tfrac{1}{C}\,\mathbf{I}\right]\boldsymbol{\mu}. \qquad (12.41)$$

This latter observation clearly reveals the dual $L_2$ SVM training problem to be a convex minimization problem over a compact convex set. These are the kind of problems the Frank-Wolfe algorithm was designed for [12] and excels at [3, 10, 38]. Interestingly, while it is also known that problems as in (12.41) can just as well be solved using comparatively simple recurrent neural networks [5, 39], the gradient flow methodology we discussed in the main text seems, to the best of our knowledge, not to have been considered before.

**Acknowledgments** This contribution was supported by the Fraunhofer Cluster of Excellence "Cognitive Internet Technologies".

# References

1. Absil, P.A., Kurdyka, K.: On the Stable Equilibrium Points of Gradient Systems. Systems & Control Letters **55**(7), 573–577 (2006)
2. Agombar, R., Bauckhage, C., Lübbering, M., Sifa, R.: An Optimization for Convolutional Network Layers Using the Viola-Jones Framework and Ternary Weight Networks. In: Proc. LION (2021)
3. Alaiz, C., Suykens, J.: Modified Frank-Wolfe Algorithm for Enhanced Sparsity in Support Vector Machine Classifiers. Neurocomputing **320**(Dec), 47–59 (2018)
4. Arora, S., Du, S., Hu, W., Li, Z., Salakhutdinov, R., Wang, R.: On Exact Computation with an Infinitely Wide Neural Net. In: Proc. NeurIPS (2019)
5. Bauckhage, C.: A Neural Network Implementation of Frank-Wolfe Optimization. In: Proc. ICANN (2017)
6. Bauckhage, C., Kersting, K., Schmidt, A.: Agriculture's Technological Makeover. IEEE Pervasive Computing **11**(2), 4–7 (2012)
7. Bauckhage, C., Sifa, R.: Gradient Flows for Linear Discriminant Analysis. In: Proc. LION (2022)

8. Biamonte, J., Wittek, P., Pancotti, N., Rebentrost, P., Wiebe, N., Lloyd, S.: Quantum Machine Learning. Nature **549**(7671), 195–202 (2017)
9. Chen, Y., Huang, W., Nguyen, L., Weng, T.W.: On the Equivalence between Neural Network and Support Vector Machine. In: Proc. NeurIPS (2021)
10. Clarkson, K.: Coresets, Sparse Greedy Approximation, and the Frank-Wolfe Algorithm. ACM Trans. on Algorithms **6**(4), 63:1–63:30 (2010)
11. Cortes, C., Vapnik, V.: Support Vector Networks. Machine Learning **20**(3), 273–297 (1995)
12. Frank, M., Wolfe, P.: An Algorithm for Quadratic Programming. Naval Research Logistics Quarterly **3**(1–2), 95–110 (1956)
13. Frieß, T., Harrison, R.: The Kernel Adatron With Bias Unit: Analysis of the Algorithm (Part 1). Tech. Rep. ACSE Research Report 729, Dept. of Automatic Control and Systems Engineering, University of Sheffield (1998)
14. GueLat, J., Marcotte, P.: Some Comments on Wolfe's "Away Step". Mathematical Programming **35**, 110–119 (1986)
15. Günder, M., Ispizua Yamati, F., Kierdorf, J., Roscher, R., Mahlein, A.K., Bauckhage, C.: Agricultural Plant Cataloging and Establishment of a Data Framework from UAV-based Crop Images by Computer Vision. GigaScience **11** (2022)
16. Haensch, W., Gokmen, T., Puri, R.: The Next Generation of Deep Learning Hardware: Analog Computing. Proceedings of the IEEE **107**(1), 108–122 (2019)
17. Helmke, U., Moore, J.: Optimization and Dynamical Systems, 4th edn. Springer (1994)
18. Jacot, A., Gabriel, F., Hongler, C.: Neural Tangent Kernel: Convergence and Generalization in Neural Networks. In: Proc. NeurIPS (2018)
19. Karniadakis, G., Kevrekidis, I., Lu, L., Perdikaris, P., Wang, S., Yang, L.: Physics-informed Machine Learning. Nature Reviews Physics **3**, 422–440 (2021)
20. Karush, W.: Minima of Functions of Several Variables with Inqualities as Side Constraints. Master's thesis, University of Chicago (1939)
21. Krause, J., Günder, M., Schulz, D., Gruna, R.: New Active Learning Algorithms for Near-infrared Spectroscopy in Agricultural Applications. at – Automatisierungstechnik **69**(4), 297–306 (2021)
22. Kuhn, H., Tucker, A.: Nonlinear Programming. In: Proc. 2nd Berkley Symposium (1951)
23. Liu, J.P., Kolden, H., Krovi, H., Loureiro, N., Trivisa, K., Childs, A.: Efficient Quantum Algorithm for Dissipative Nonlinear Ddifferential Equations. PNAS **118**(35), e2026805118 (2021)
24. Lloyd, S., De Palma, G., Gokler, C., Kiani, B., Liu, Z.W., Marvian, M., Tennie, F., Palmer, T.: Quantum Algorithm for Nonlinear Differential Equations. arXiv:2011.06571 [quant-ph] (2020)
25. Lübbering, M., Ramamurthy, R., Gebauer, M., Bell, T., Sifa, R., Bauckhage, C.: From Imbalanced Classification to Supervised Outlier Detection. In: Proc. ICANN (2020)
26. Mangasarian, O., Musicant, D.: Lagrangian Support Vector Machines. J. of Machine Learning Research **1**, 161–177 (2001)
27. Ojeda, C., Cvejoski, K., Schuecker, J., Georgiev, B., Bauckhage, C., Sanchez, R.: Learning Deep Generative Models for Queuing Systems. In: Proc. AAAI (2021)
28. Ojeda, C., Georgiev, B., Cvejoski, K., Schuecker, J., Bauckhage, C., Sanchez, R.: Switching Dynamical Systems with Deep Neural Networks. In: Proc. ICPR (2021)
29. Oliphant, T.: Python for Scientific Computing. Computing in Science & Engineering **9**(3), 10–20 (2007)
30. Petersen, K.B., Pedersen, M.S.: The Matrix Cookbook. Technical University of Denmark (2012)
31. Piatkowski, N., Gerlach, T., Hugues, R., Sifa, R., Bauckhage, C., Barbaresco, F.: Towards Bundle Adjustment for Satellite Imaging via Quantum Machine Learning. In: Proc. FUSION (2022)
32. Raissi, M., Perdikaris, P., Karniadakis, G.: Physics-informed Neural Networks: A Deep Learning Framework for Solving Forward and Inverse Problems Involving Nonlinear Partial Differential Equations. Journal of Computational Physics **378**, 686–707 (2019)

33. Rebentrost, P., Mohseni, M., Lloyd, S.: Quantum Support Vector Machine for Big Data Classification. Physical Reviev Letters **113**, 130503 (2014)
34. von Rueden, L., Houben, S., Cvejoski, K., Bauckhage, C., Piatkowski, N.: Informed Pre-Training on Prior Knowledge. arXiv:2205.11433 [cs.LG] (2022)
35. von Rueden, L., Mayer, S., Beckh, K., Georgiev, B., Giesselbach, S., Heese, R., Kirsch, B., Walczak, M., Pfrommer, J., Pick, A., Ramamurthy, R., Garcke, J., Bauckhage, C., Schuecker, J.: Informed Machine Learning – A Taxonomy and Survey of Integrating Prior Knowledge into Learning Systems. IEEE Trans. on Knowledge and Data Engineering **35**(1), 614–633 (2023)
36. Schölkopf, B., Smola, A., Williamson, R., Bartlett, P.: New Support Vector Algorithms. Neural Computation **12**(5), 1207–1245 (2000)
37. Schuman, C., Potok, T., Patton, R., Birdwell, J., Dean, M., Rose, G., Plank, J.: A Survey of Neuromorphic Computing and Neural Networks in Hardware. arXiv:1705.06963 [cs.NE] (2017)
38. Sifa, R.: An Overview of Frank-Wolfe Optimization for Stochasticity Constrained Interpretable Matrix and Tensor Factorization. In: Proc. ICANN (2018)
39. Sifa, R., Paurat, D., Trabold, D., Bauckhage, C.: Simple Recurrent Neural Networks for Support Vector Machine Training. In: Proc. ICANN (2018)
40. Srivastava, S., Sundararaghavan, V.: Box Algorithm for the Solution of Differential Equations on a Quantum Annealer. Physical Review A **99**(5), 052355 (2019)
41. Stuart, A., Humphries, A.: Dynamical Systems and Numerical Analysis. Cambridge University Press (1998)
42. Suykens, J., Venderwalle, J.: Least Squares Support Vector Machine Classifiers. Neural Processing Letters **9**(3), 293–300 (1999)
43. Thompson, N., Greenewald, K., Lee, K., Manso, G.: Deep Learning's Diminishing Returns: The Cost of Improvement is Becoming Unsustainable. IEEE Spectrum **58**(10), 50–55 (2021)
44. Ulmann, B.: Analog Computing. De Gruyter Oldenbourg (2013)
45. Wittek, P.: Quantum Machine Learning. Academic Press (2014)
46. Wolff Anthony, L., Kanding, B., Selvan, R.: Carbontracker: Tracking and Predicting the Carbon Footprint of Training Deep Learning Models. arXiv:2007.03051 [cs.CY] (2020)
47. Wu, Y., Thurau, C., Bauckhage, C.: The Good, the Bad, and the Ugly: Predicting Aesthetic Image Labels. In: Proc. ICPR (2010)
48. Zanger, B., Mendl, C., Schulz, M., Schreiber, M.: Quantum Algorithms for Solving Ordinary Differential Equations via Classical Integration Methods. Quantum **5**(502) (2021)
49. Zhang, S., Bauckhage, C., Cremers, A.: Informed Haar-like Features Improve Pedestrian Detection. In: Proc. CVPR (2014)
50. Zinsl, J.: Systems of Evolution Equations with Gradient Flow Structure. Ph.D. thesis, Technical University Munich (2015)

# Chapter 13
# Informed Machine Learning to Maximize Robustness and Computational Performance of Linear Solvers

Sebastian Gries

**Abstract** It is crucial for the efficiency and robustness of numerical simulations that the linear solver strategy therein is adjusted to the type of simulation in a grey-box manner (Stüben et al., Algebraic multigrid - from academia to industry. In: Scientific computing and algorithms in industrial simulations, 2017). Sophisticated solver methods can then provide a remarkable computational performance along with the required precision in various fields of simulation.

And they still comprise a lot of options for a fine-grained control, where an optimal parameter setting is a highly individual and rather volatile trade-off between robustness and computational efficiency—depending on properties of a particular simulation, computing environment and accuracy requirements.

We apply methods of evolutionary and surrogate machine learning for these remaining optimizations. With the hundreds and thousands of different control options, an uninformed learning approach was practically impossible within a simulation. Instead, along with the general application-tailored solver strategy, a parameter optimization space is provided for the learning methods. These also evaluate data within the simulations.

A deep integration into the solver method allows for accessing all relevant data for decision and learning processes and helps to reduce overhead costs. It also allows for reducing the number of solver setups within a simulation run and guarantees robustness by quickly reacting to convergence break-downs.

We will demonstrate the benefits for simulations from different industrial use cases from fluid dynamics and geological simulations towards structural mechanics and battery aging simulations.

S. Gries (✉)
Fraunhofer SCAI, Sankt Augustin, Germany
e-mail: sebastian.gries@scai.fraunhofer.de

© The Author(s) 2025
D. Schulz, C. Bauckhage (eds.), *Informed Machine Learning*,
Cognitive Technologies, https://doi.org/10.1007/978-3-031-83097-6_13

## 13.1   Introduction

Finding optimal parameter sets for linear solvers is crucial with respect to both computational performance and numerical robustness of numerical simulations in various fields of engineering. We will see how machine learning techniques can provide a significant improvement of such settings over manually adjusted ones. Especially the ability of providing a simulation-specific fine-tuning will allow to get the best performance for a certain simulation. This is important not only in order to speed up engineering workflows but also to reduce the consumption of computational resources therein.

As use cases, we focus on simulations of diffusion-based processes in this chapter. This allows to demonstrate the potential of machine learning techniques in a broad range of industrial applications - from fluid flow to structural mechanics. But with battery aging, we will also consider simulations beyond diffusion-based cases.

In either case, the partial differential equation(s) that describe the physical model can hardly be solved analytically. Instead, they are discretized, for instance, based on some mesh, in order to limit the problem to a finite number of degrees of freedom. In the case of non-linear models, an additional linearization, for instance, via Newton's method, is included. We do not describe the setup of numerical simulations in details here. We rather focus on the linear systems of equations that are to be solved to some pre-defined, application-dependent accuracy as the 'numerical kernel' of the simulations. This linear solution process is required frequently during a simulation: For each time step and, in non-linear cases, each linearization step. Thus, the linear solution process typically accounts for the major portion of computational efforts and covers between 40 and up to even 90% of the computational time. Hence, any optimization of this part of the simulation has a high overall impact.

Before allowing for a machine learning-based control of the concrete parameter settings, however, certain decisions still have to be made in the initial design of the solver strategy. This includes the decision on the basic solver strategy: In applications that are based on diffusion(-like) processes, algebraic multigrid (AMG) approaches [1, 29, 35] are the methods of choice. They provide an almost optimal numerical behavior for the respective linear systems and, thus, are also suited for problems with many degrees of freedom in the order of several millions or even billions. They gain their numerical efficiency from exploiting a hierarchy of linear operators that are constructed completely automatically in a so-called setup phase, independent of any geometric information. Thus, complex underlying grid structures and huge material heterogeneities or jumps in coefficients can be handled.

However, AMG can hardly be applied in a black-box manner in many sophisticated numerical simulations. It exploits certain properties of linear systems that are fully fulfilled in prototypical problems, but may be violated to some extent in many practical applications.

Instead, an application-tailored solution strategy needs to be created, where AMG is combined with further techniques that make it applicable in an overall solution approach for a certain type of application. Limitations of textbook AMG for challenging applications are overcome by specific adjustments of the method, bespoke to the type of application. The problem may also be transferred into an equivalent formulation that is better suited for the application of AMG. However, AMG will still automatically adjust its hierarchy to different underlying grids, material parameters and further environmental settings. Examples for such gray-box AMG approaches are in the fields of material design [14] and reservoir simulations [9]. We refer to [36] for a further overview and more applications.

All linear solver approaches have in common that they still provide several parameters to further adjust the method. This in particular holds for AMG methods, as they in fact combine different other iterative methods with their hierarchy. Most of such control options have in common that one can either turn a solver method to be more robust, at the expense of computational costs—or vice versa. Whether and which adjusted parameter setting is beneficial for a certain simulation is hardly predictable beforehand in many cases. A rather cheap setting may be well-suited for some well-conditioned cases, but not robust enough for more challenging ones. Changes on which parameters to prefer can even occur within a simulation run, as certain environmental settings change, time steps increase or physical settings differ.

Therefore, a manual fine-tuning of linear solver parameters is extremely time-consuming and can be rather difficult. This is where we will apply machine learning techniques to take over the fine-grained control. This will follow an informed machine learning approach [28]: the general solver strategy, i.e., the gray-box solver, will be pre-defined for a certain application. Also the search space where the Machine Learning shall find optimal parameters in will be pre-defined. This allows for a priori excluding parameter combinations that are mathematically meaningless. And it allows for ensuring certain parameter settings that may be required from the gray-box approach. However, within the search space, the setting is found automatically, based on machine learning techniques.

That is, for our approach to Machine Learning for controlling the linear solver, we're mainly exploiting information from a combination of simulation results and human feedback (Secs. 5.3 and 5.8 in [28], respectively). The main source of information is the solver behavior that it evaluates itself within simulations, along with intermediate simulation results. However, to guide the machine learning process, some initial parameter search space is necessary, based on knowledge from experts for a certain type of simulation.

In addition, to some extent, our overall approach will also involve information from algebraic equations (cf. [28]). This concerns the transfer learning with the surrogate model with algebraically evaluated matrix properties from Sect. 13.4 that will enhance our main machine learning approach.

There have been earlier approaches to provide an automatic control of solver methods, also with ideas from the field of Machine Learning. Genetic algorithms have been used in an encompassing parameter optimization feature for the linear solver in reservoir simulations [22]. Another approach in this direction is $\alpha$SAMG [4] that can adaptively decide between AMG and an ILU-ish method. It has mainly been applied to groundwater simulations.

The approach that we are going to present will be more versatile and tied deep into the linear solution process. Moreover, it will not use mere evolutionary learning but will be enhanced by further mechanisms.

In fact, the problem of finding optimal (hyper) parameters applies in various applications. In general, this can be seen as optimizing the parameters of some function with respect to some objective. In our case, we want to optimize linear solver parameters with respect to runtime, where the parameter settings result in a robust, well-converging solver method.

Genetic algorithms are considerable options for automated parameter optimizations [6]. They typically provide reasonable results much faster than randomly testing all imaginable variants [27]. The process can be further improved by combining it with tree-based optimization so that the genetic optimization more quickly focuses on promising sets of parameters [7, 16].

Other options for self-adapting parameter optimization comprise regression-based approaches, such as Bayesian optimization [18, 34]. However, we will not follow the latter approaches for our linear solver control. Our parameters are not necessarily set to values of a continuous interval and parameter effects on the execution time may be highly discontinuous. Identifying a probabilistic model, thus, is not straight-forward with the vast range of possible parameters. Moreover, we will typically measure execution time with a timer function and will be subject to noise. Therefore, we need to fully evaluate the computational performance of each considered parameter setting in any case.

Any standardized control mechanism for parameter optimization can only be placed around the linear solver and consider it as a kind of black-box operation that results in a certain runtime result with a given set of parameters. In contrast, we will tie the parameter optimization deeply into the linear solver. While the main decision machinery of our approach still is located around the linear solver's kernel, it can evaluate more detailed information for its decision and learning process. This allows for exploiting synergies on the one hand, in order to minimize computational work and overhead. On the other hand, it easily allows for a continuous monitoring of the numerical solver behavior in order to interfere as soon as any robustness issues may be observed.

Thus, we can well accept that our parameter optimization may also decide for parameters that yield computational efficiency, but, by themselves, may carry some risk for the robustness. As long as the linear system allows for such less robust settings to be applied without problems, we can exploit the efficiency. Whereas, as soon as robustness is at risk, our optimization approach can reconsider its decision immediately, without ever computing questionable results.

An additional synergy of this integrated monitoring mechanism is that we can also use our control mechanism to handle the reusage of solver setups. This could be multigrid hierarchies as well as incomplete factorizations: as long as they are found to be well-applicable in subsequent Newton and time steps, they will not be recomputed. However, our mechanism ensures to have them recomputed as soon as this is either required or simply beneficial for robustness. This further reduces computational efforts as far as this is possible under the objective of providing robust results from the linear solver. In addition, a surrogate learning model helps to further guide this process and attempts to pre-estimate when to recalculate a solver setup.

The parameter optimization itself for the linear solver will be based on a classical genetic algorithm [6], accelerated with a decision tree [16]. In order to further accelerate the either online or offline training, a surrogate learning model will allow for adjusting the initial likelihoods of search directions based on previous simulation runs.

The search directions themselves result from a user-defined set of parameters and possible values or ranges. This informed machine learning approach [28] allows for exploiting knowledge about promising parameters for optimization. And for excluding parameter sets that are meaningless for a particular application.

This chapter will set up on the recent work on autonomous control of solver parameters in reservoir simulation applications [10]. The initial development was performed with the application of battery aging simulations in mind (cf. Sect. 13.8.7). However, this initial development has been extended further and here we will provide more details on the machine learning-based decisions and we will provide results with further applications. The structure of this chapter is as follows: We will give a very brief overview on linear solvers in numerical simulations. This will be followed by a description of the tree-based genetic optimization approach and the complementing surrogate learning model that we will use for controlling the solver parameters. We will then briefly discuss aspects of reproducibility with respect to decisions based on runtimes and transfer the control ideas to the handling of reusing solver setups. Finally, we will present results of our approach with industrial simulation applications from different fields of engineering.

## 13.2  Short Overview on Linear Solvers in Numerical Simulations

Linear solvers can be seen as the computational kernel of numerical simulations. In the numerical simulations under consideration here, some physical problem is discretized in order to make a simulation possible by only considering a finite number of degrees of freedom. If the problem was non-linear in some way, a linearization process, such as Newton's method, is applied in addition. While the linearization can also be applied before discretizing the problem, or no linearization may be required, the process always results in some system of linear equations

$$Ax = f. \tag{13.1}$$

Here, $A$ is an $n \times n$ matrix that is typically sparse, i.e., only a much smaller number of row entries than $n$ is non-zero. $f$ is a right-hand-side vector of dimension $n$ and $x$ needs to be computed. The problem size $n$ results from the number of discretization cells.

A direct solver could be applied to solve this linear system. While, except for round-off impacts, this was exact, both the computational efforts and the memory requirements do grow cubically with the degrees of freedom $n$. This drastic increase does make it practically impossible to apply direct solvers in numerical simulations, where nowadays millions and hundreds of millions of discretization cells are considered.

Iterative solvers are the methods of choice then. These do not solve the problem in one computationally expensive step. But rather step-wise attempt to further and further improve the solution vector per iteration, based on the previous iteration's result. This proceeds until some given, application-specific accuracy is achieved. That is, while the problem (13.1) does not necessarily need to be solved exactly, each iteration step needs to improve the solution sufficiently enough to reach the target accuracy within a maximal number of iterations. We refer to a robust convergence then.

Typically, Krylov solver methods, such as CG [12] and GMRes [31], are applied in numerical simulations. These solve the linear system by mapping it into a very low-dimensional so-called Krylov subspace. This process can be implemented via basic linear algebra operations that are very well-suited also for application in parallel with HPC architectures.

These Krylov methods are further complemented by iterative methods that serve as preconditioner. Their effect is to build up the Krylov subspace more suited for the particular problem. The better a preconditioning method fits for a certain problem, the better each iteration of the Krylov method does converge in the end. We refer to [30] for an extensive description.

Preconditioning methods range from rather simple relaxation methods over incomplete factorizations [21] or sparse approximate inverses [11] towards hierarchical solver methods. While the other methods do work with the initial linear system, hierarchical methods exploit some sort of hierarchy to further accelerate the solution process.

Multigrid [2, 38] is a wide-spread method of choice for such hierarchical linear systems, at least in applications where the simulated process is diffusion-dominated. The idea is to exploit a hierarchy of coarser and coarser levels. These represent the same problem as the initial linear system does, just at different resolutions. This allows to efficiently solve different Fourier-modes of a current overall solution iterate by exploiting certain assumptions on properties of the linear system that hold in diffusion-based cases. It then leads to an optimal numerical efficiency in the sense that the computational efforts and memory requirements do only increase linearly with the degrees of freedom. A special variant is algebraic multigrid: It computes its hierarchy itself, solely based on the linear system and the coefficients therein. This makes it well-suited for problems that are discretized in complicated geometries in unstructured ways as well as for problems that have strong material heterogeneities and, thus, jumps in coefficients.

Which solver method to choose highly depends on the type of application. The underlying physics lead to different properties of the linear systems. Also the problem size has an impact: for very small cases, a direct solver may be efficient enough to outweigh the setup costs of sophisticated iterative solvers. For large cases, however, only iterative methods can be applied on a given hardware resource. While single-level methods may be sufficient in rather well-conditioned cases, these methods are typically outperformed by multigrid methods in more challenging cases, especially in large ones.

However, textbook multigrid may often enough not be applicable, especially not in a black-box manner. Certain requirements that are to be exploited by multigrid may not be (fully) fulfilled in several applications with complex physics behind them. For instance, diagonal dominance may suffer from constraint-like conditions, or discretization aspects may lead to some lack of symmetry or mixed-sign off-diagonal matrix coefficients. Some adjustments, pre-processings and special multigrid approaches may be required in order to make multigrid exploitable then. Examples for such gray-box AMG applications [36] can be found in reservoir simulations [9] and material design [14].

While certain settings are required for the successful application of AMG in some type of application, the solver methods still offer various other control options that can be used for tuning in a specific simulation. Which, and which ranges, depends on the outer requirements from gray-box AMG and, thus, are pre-defined.

## 13.3    Genetic Optimization of Parameters with Tree Hierarchy

Our objective is to find those values of solver parameters $p_1, \ldots, p_m$ that lead to the lowest runtime of the solver within an application. But that still ensure a robust solution of the linear system, meaning that the iterative solution process properly reaches the target accuracy in the solution of (13.1). In the following, we will simply identify non-robust sets of parameters with an infinite runtime. Thus, runtime is our only target function.

Parameters of linear solvers can be either integer, floating numbers or booleans. Values can be from a list of options (e.g., variant A, B or C) or from a continuous range interval of possible values (e.g., control thresholds). In the latter case, we separate the interval into a finite number of chunks, loosely following the idea of kriging [17]. Hence, each parameter has a finite number of possible states. This simplifies comparisons between sets of parameters in the decisions of the optimization method.

We will employ genetic optimization, as this can deal with both the possibly non-continuous range of parameter options as well as the fact that measured runtime as a target function is subject of statistical noise. Genetic optimization is a well-established method for parameter optimization and we can refer to the literature for a detailed description of such methods [6, 27].

Essentially, the method starts with some initial, typically random state for the parameters $p$. Based on this initial generation, the next generation is created, where

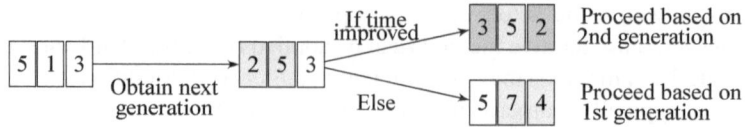

**Fig. 13.1** Visualization of the principle behind genetic optimization algorithms. Parameters are depicted with boxes and their setting by the respective number. Parameter changes are illustrated in gray color

some parameters "mutate" to a different setting. Promising new generations will serve as a base for further generations. Those with no potential for improvement will not be considered further. In this sense, we have an evolutionary process, where the different parameters represent genes that may vary. Which parameters will vary is a randomized decision among those that shall be further considered. Figure 13.1 visualizes this proceeding exemplarily.

The method continues until either no improvement is found any longer or until a pre-described number of generations is reached. The selection of the next generation is limited to those parameter settings that have not yet been considered. Thus, because we have a finite list of options for all parameters, we can safely limit the number of generations to the number of possible parameter combinations. The method stops much earlier in the vast majority of cases because the best set of parameters has already been found, though. More precisely, the method stops because no further options remain within the parameter search space that have the potential for further improvements. That is, the advantage of the genetic algorithm is to rather quickly focus on promising search directions for next test candidates. This drastically reduces the number of evaluations compared to simply testing all possible variants.

We will further strengthen this aspect of finding an optimal setting more quickly with two modifications.

The first is to combine the genetic algorithm with a decision tree [7, 16]. This follows the idea that linear solver methods can be classified into different approaches at high level, such as direct methods, single-level incomplete factorizations, algebraic multigrid, etc. Within each class of approaches, we can further distinguish different types of the respective approaches. Each of them in turn can be controlled by further parameters, some of which may induce the consideration of further options, etc. Our decision tree will represent this rather natural hierarchy of parameter options and categorize the different parameters into different branches and leafs of that tree. Moreover, some parameter settings may only make sense in a certain branch of the higher-level options.

This is more than a mere design advantage: As soon as a higher-level parameter option turns out to be rather non-promising, the entire respective further branch of the decision tree can be either turned to be less likely to be considered further, or to be even completely discarded from the further consideration.

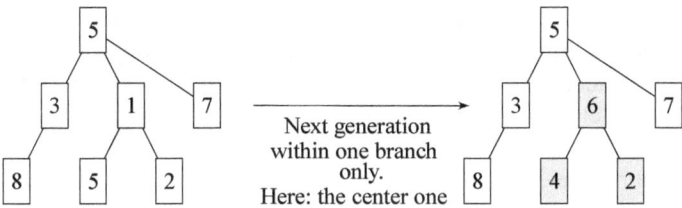

**Fig. 13.2** Visualization of the tree-based genetic algorithm with exemplary settings for parameters (depicted as boxes). Changes in parameter depicted by gray color

Clearly, such decisions should not be based on a single evaluation of a parameter setting from a certain branch. Instead, we require a minimum amount of evaluations of that branch, relative to the possible number of combinations from that branch, in order for such decisions to be taken. If, however, for instance, the direct solver options in the parameter search space have turned out to be significantly slower than multigrid at some stage of the evaluation process, all direct solver options will no longer be considered. This reduces the number of remaining possible parameter combinations and, thus, remaining evaluations. Figure 13.2 visualizes this proceeding.

The second modification of the genetic algorithm changes the likelihood for different parameters to be selected for the next generation. Different linear solver parameters may have different strengths-of-effect on the runtime in different applications. As soon as a parameter has shown to have a rather low impact on the runtime in either direction, we decrease the likelihood for further considering it—independent of the tree-branch it is in. We do so by computing the global Sobol sensitivity indices [33] for each parameter after each evaluation, independent of tree-branches. These are rather cheap to compute and allow to identify parameters with a low impact quickly.

## 13.4   Pre-evolution via Surrogate Learning Model

Linear systems from different simulations, or at later stages of a simulation, in one type of application do often enough feature somewhat comparable properties. Therefore, we intent not to start the genetic optimization process from scratch for each simulation, or when rechecking within the same simulation. We would like to consider, or transfer, the previous results instead and adjust the initial likelihoods for different parameters and settings to be considered.

We use a surrogate learning model for this purpose. While virtually all properties of a linear system do have some impact on the linear solver behavior, some have a more outstanding impact than others. Properties like the number of equations, the number of non-zero coefficients per equation,[1] the amount of unsymmetry[2] and the diagonal dominance[3] are known to have a strong impact. This also holds for the required accuracy, as a less strict target accuracy may allow for different methods than strict ones, and the amount of parallelization, as this may come at the expense of algorithmic compromises in the way a solver is parallelized. All of these information are either directly available, user-defined or cheap to compute, for instance, as average or maximum over the system. These serve as algebraically computed information for guiding the machine learning process (cf. Sec. 5.1 in [28]).

This rather small number of properties is used to characterize a surrogate for a full linear system. Together with the selected parameter settings they are stored in a simple database. Before starting a genetic optimization for a new linear system, the initialization of the genetic process can be taken from this database, based on how close the new linear system is to the previous ones in terms of the surrogate characterizations.

This transfer learning requires computing the comparability of different surrogates. We do so by computing the difference of each surrogate dimension of two data sets. This difference is then expressed relatively to the standard deviation of that dimension across all data sets. Thus, differences of different surrogate dimensions become comparable, although the initial data may be of rather different scales (e.g., solution tolerance vs. size of the linear system compares a value between 0 and 1 vs. a value of millions).

With the result, we can adjust the likelihoods of different parameters and settings to be considered, according to previous results from the same or also other simulations. It is important to note that we are only adjusting likelihoods in most cases and, thus, help the genetic process to focus on promising search directions more quickly. However, we do still consider all search directions in most cases. The thresholds for fully withdrawing search directions are rather high: The surrogate model is a simplification of the huge number of linear system properties to just a few outstanding properties. While this provides some representativity to speed up the genetic optimization, it can hardly be seen as a prediction with certainty.

---

[1] Also referred to as sparsity.

[2] Computed as the difference of two corresponding matrix entries $a_{ij}$ and $a_{ji}$ relative to the average of the respective diagonals $a_{ii}$ and $a_{jj}$.

[3] Computed as the relation of the sum of off-diagonal matrix entries of a matrix row related to the respective diagonal. Values less than one indicate diagonal dominance, one indicates weak diagonal dominance.

## 13.5   Online vs. Offline Training

The genetic optimization will take place during a training phase that can either be in an online or offline mode. In an offline training phase, the optimization is performed for a certain, e.g., the first, linear system. Parameter settings are evaluated for this particular linear system. While this comes at some overhead costs for executing the training phase, results are perfectly comparable.

In contrast, during an online training, the first generation of the genetic optimization is evaluated with the first linear system to solve, the evaluation of the second linear system is performed for the second linear system, etc. This has the advantage that no computed results are wasted: apart from the runtime, each evaluation of the linear solver results in a solution vector for the linear system (13.1) to be solved. During an online training, all of these results will be used to proceed with the actual simulation. The convergence monitoring mechanism that we employ in the linear solver kernel ensures that this only happens if the computed solution vector fulfills the given target accuracy. Otherwise the solver is restarted with another generation of parameters, i.e., two (or more) evaluations are then based on the same system "offline-like".

The monitoring mechanism also stops an online evaluation run if it detects that the convergence is too bad for this parameter set to have a chance to provide a solution with the required accuracy. This is simply done by estimating the final accuracy after reaching the maximal number of iterations, based on the so-far convergence history at the current iteration step.

While it is the advantage of an online training to consider different linear systems to avoid wasting computed results, it has the drawback that solver runtimes for different linear systems need to be compared.

A simple step to improve comparability of different evaluations in an online training is not to compare and store full runtimes but to distinguish between the runtimes for setup and solution phase of a linear solver. Regarding the solution phase, more precisely, we store the runtime per iteration along with the average convergence rate per iteration. This allows to compare results of different solver generations for different linear systems: we can extrapolate how many iterations another parameter setting would have resulted in and how long the solution would have taken then.

This still requires linear systems to be somewhat comparable regarding their properties. If two linear systems differ too drastically, for instance, if we compare results for a really well- and a really ill-conditioned one, the convergence rates of a solver setting for one problem were no longer representative also for the other system.

This, however, is typically not the case in most numerical simulations: The underlying physics, the discretization and even essential material properties remain the same, or at least nearly the same, between different linearization and time steps of a simulation. Therefore, comparability remains to an acceptable level. If not, for instance, due to sudden changes of materials or changes to the physics in certain time steps,[4] falling back to an offline training still is perfectly possible.

We can employ the surrogate model from the pre-evolution for this purpose: As soon as the surrogate representation of the next linear system in the proceeding of a simulation differs too much from the previous ones, the online training automatically falls back to the offline mode.

During an offline training phase, we employ the convergence monitoring mechanism to avoid as much of unnecessary evaluation work as possible. Because we evaluate all parameter generations with the same linear system, we are ensured to obtain a solution vector as long as there is at least one possible parameter setting that results in a robust solver method. Thus, we can safely stop any further evaluation as soon as it is clear that this will not lead to a better runtime than the so-far best parameter setting. Independent of whether the solution has already been achieved with the current parameter generation then, the evolution can be stopped early and the corresponding parameter setting will be discarded. This avoids unnecessary further solver iterations with that setting.

Finally, in the unlikely event that all possible parameter combinations should fail in providing a robust solver method, we will select the method with the best convergence properties. Or one with almost the same convergence properties if it was substantially faster.

This is the best we can do with the provided parameter search space then. In fact, it would be the user's task to provide better information for the machine learning method.

## 13.6   Reproducibility

Our objective is to minimize the runtime of the linear solver that we evaluate with classical timer functions. These are subject of noise due to different workloads on computers, due to timer resolution, memory effects, etc. That is, two runs with the same settings can be expected to have somewhat comparable runtimes. They will hardly have exactly the same runtime, though. Hence, decisions of the genetic algorithm may become different where runtime results are close between two sets of parameters. Generally, this is not a problem in production runs. The convergence monitoring ensures that the solver provides results of the demanded accuracy. Deviations beyond, especially at round-off level can often enough be well-accepted.

In cases where analysis or research purposes require full reproducibility, however, we cannot (solely) base optimization decisions on measured runtimes. We make use of artificial times then (as in the approach in [37]), which are pre-defined for the setup and iterations of different types of solver methods, respectively. This is

---

[4] For instance, switching on and off well-bores in reservoir simulations, cracks in material simulations, sudden changes in boundary conditions, etc.

based on experiences and benchmarks. However, while it still allows for evaluating numerical impacts from parameter changes, e.g. changes of iteration counts, it hardly allows for evaluating fine-grained performance impacts of parameter changes.

Instead, we use a mixed approach for comparing two runtimes $t_1$ and $t_2$, namely

$$\Delta t_{for\,Decision} = \begin{cases} \Delta t_{measured}, & \text{if}\,\frac{\Delta t_{measured}}{t_1 + t_2} \geq thres \\ \xi\,\Delta t_{measured} + (1 - \xi)\Delta t_{artificial}, & \text{else} \end{cases}$$

$$(13.2)$$

where

$$\xi = \left(\frac{\Delta t_{measured}}{(t_1 + t_2)thres}\right)^2.$$

$$(13.3)$$

The choice of $\xi$ in (13.3) ensures that it remains between zero and one if the else-condition in (13.2) applies. It also ensures that $\xi = 1$, where the threshold is exactly met. Thus, if and else condition match continuously at the transition. Using the square puts more emphasis on the measured runtime, if the threshold is only slightly undershot, while the artificial times are emphasized where the measured runtimes $t_1$ and $t_2$ are very close.

While this approach does not exactly guarantee full reproducibility, it drastically reduces non-determinism with small runtime differences but still evaluates measured performances. We call it semi-reproducible.

Another aspect of reproducibility concerns the randomized selection of generations in the genetic optimization process. We make use of the pseudo-random generator of the Fortran runtime environment here. It is initialized with the same seed in each simulation. Thus, reproducibility is guaranteed.

Finally, there may be impacts on the computational performance of a simulation if it does not run exclusively on a computer, but shares the hardware resources with other tasks. This may result in different runtimes, as the simulation process occasionally—and virtually randomly—may need to wait for these other tasks. If a simulation process runs in an environment with other tasks being active, this interference and the resulting limitations for reproducibility are unavoidable. The optimization process can react by re-doing the online training during different stages of the simulation, though. Moreover, the convergence monitoring, based on the runtime per solver iteration, can estimate when such a re-training may be promising. Last but not least, the semi-reproducible approach damps impacts from varying measured runtimes where runtimes of different solver methods are close—and, thus, impacts from process-interference are less likely to lead to different decisions.

## 13.7  Controlling Solver Setup Reusage

Solver methods like incomplete factorizations or AMG do require a setup phase where factorizations and operators are computed that will be applied in the solution iterations. As long as the linear system matrix remains identical, this setup can safely be reused with different linear systems—i.e., different right-hand-sides.

Reusing solver setups, or at least parts thereof, is possible even where the linear system matrix changes. Strictly speaking, the iterative method then is no longer applied to the correct linear system. However, as long as the iterative method is only applied as a preconditioner within a surrounding Krylov method, this may still be acceptable. The Krylov method, as the actual solver method, still is applied to the correct linear system. It is only the preconditioner that may degrade. That is, more iterations of the Krylov method may result, but this increase in iterations may be small enough to be outweighed by avoiding the costs of computing a new solver setup. If, however, the matrix changed too much, reusing the setup can result in the preconditioning quality becoming too bad for the Krylov method to compute an accurate solution within the maximal number of iterations.

Whether the linear system matrix is comparable enough to the one where a setup has initially been computed with is hard to predict before-hand. However, we can use our control mechanisms from the parameter optimization to also control the setup reusage.

The straight-forward transfer is to use the convergence monitoring mechanism to stop the solution process and restart with a fresh setup, as soon as it expects the setup to be beneficial. Just as with the early stops of evaluation runs during training phases, this works with the current convergence speed that is used to estimate the number of iterations to achieve the required solution accuracy. As soon as the additional iterations from a non-fitting reused setup are expected to exceed the costs of a new setup, this will be triggered.

We should note that this can even occur while computing the solution of one linear system: this process can be stopped then and restarted with a fresh setup, based on the so-far computed results.

In addition, we transfer the idea of the surrogate learning model from Sect. 13.4 to easily evaluate changes of linear systems. We compare the current linear system with the one where the initial setup was based on by means of the surrogate model. If the difference exceeds the threshold that was learned from previous setup decisions within this or other simulations, a fresh setup will be computed before even attempting to further reuse the current setup. This does not only avoid the rather small overhead of starting the solution process with an unsuited setup and then stopping it. In the case of setups only being partly reused, this avoids the unnecessary computation of the remaining parts of the setups. In the context of AMG, for instance, instead of reusing the entire setup, also all but the fine-level smoother or all but the AMG operators can be reused. Computing the remainder comes at computational expenses that can be avoided if it is very likely for a full fresh setup to be computed anyway.

## 13.8   Results: Informed Machine Learning for Linear Solver Parameters in Various Practical Applications

We will consider results of the ML-based parameter optimization and setup control with several test cases and simulations from industrial applications in various fields of engineering.

We will consider both full simulations as well as representative single linear systems. With single linear systems, or small sequences, we can demonstrate certain aspects of the parameter optimization. The full simulation results, however, are what matters in the end.

All benchmarks have been performed with Fraunhofer's linear solver library SAMG and the control extension SAMG-ASC both of Release 2022 on compute nodes with two Intel Xeon Gold 6130@2.10 GHz and 192 GB RAM. Nodes had been used exclusively for the performance benchmarks here in order to ensure reproducibility. This also allows to reasonably consider single simulation runs rather than repeated series of simulations. This is the use cases in industrial simulations.

### 13.8.1   Mere Parameter Optimization: Single Reservoir Simulation Problems

We are first considering the achievements of the evolutionary learning for the optimization of solver parameters. We will do so with an offline training for two problems from petroleum reservoir simulation. The underlying SPE10 problem [3] is a well-established reference benchmark in petroleum reservoir simulations that is commonly used to compare simulators and where the input data is publicly available. It is a 3D Black-Oil simulation, i.e., oil, gas and water phases are considered, discretized with 1.1 million grid cells. We are using two linear systems from different time step sizes. They feature different matrix conditions, as they typically arise in different steps of a simulation. While the first is rather easy to solve, the other one is more challenging and requires more solver iterations.

In our informed machine learning approach, we are considering a set of linear solver parameters that, by experience, have a potential for optimizations.[5] At the same time, those parameters that are crucial for a successful AMG application in reservoir simulations remain fixed (cf. [9]).

Figure 13.3 shows the performance results for the optimized parameters with both linear systems. We are not including the training time here, as we are looking for initially optimized parameters with a single linear system. Since we have used

---

[5] Aggressiveness of coarsening, threshold for coupling strength, target coarse level size of the AMG hierarchy (cf. [35] for details) and the floating precision of the incomplete factorization fine-level smoother.

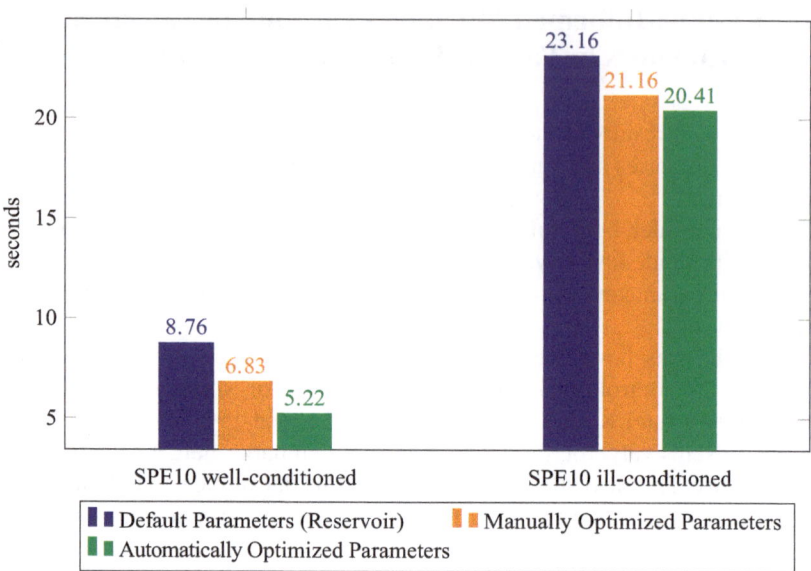

**Fig. 13.3** System-AMG runtimes for single representative linear systems from Black-Oil simulations. Compared are different sets of parameters: Conservative defaults, manually and automatically optimized ones—for each of the two problems. Time of the training phases for both manual and automatic tuning is excluded here

human experience in the definition of the parameter search space, we are also including results of the outcome of a manual optimization of the respective parameters, based on that experience. While this already is a significant improvement towards the standard AMG parameters, the evolutionary learning process can further improve the settings. It saves 40% and 23% of the runtime compared to the default settings for the well-conditioned and the challenging linear system, respectively. And still saves 23% and 4% compared to the manually optimized parameters.

In both cases, the tree-based evolutionary learning approach drastically reduced the number of required evaluations compared to evaluating all combinations. With the provided parameter search space, about 600 parameter combinations would have been possible. Continuous intervals like for thresholds have been discretized using a kriging approach [17] here. However, instead of evaluating all possible options, only about 30 runs have been necessary in the offline training phase.

The comparable background of both linear systems from the same physical model allows to further reduce this. With the proposed surrogate model, we can guide the parameter optimization of the second linear system from a later step of the simulation with results from the first one.

Table 13.1 shows that the surrogate-based initialization of the evolutionary process can significantly help to further reduce the number of evaluations.

The obtained parameter sets for the two linear systems both with and without surrogate learning do target in comparable directions. This indicates that we can

**Table 13.1** Performance of the offline training for two SPE10 linear systems. Either independent training or with the second one being initialized via the surrogate learning approach based on the first one

|  |  | Surrogate learning in second training | Independent training, no surrogate learning |
|---|---|---|---|
| First training | #eval | 30 | 30 |
| (well-cond. problem) | Time | 87.39 sec | 87.39 sec |
| Second training | #eval | 11 | 31 |
| (ill-cond. problem) | Time | 207.10 sec | 570.95 sec |

safely use parameter sets for several time steps and that the surrogate-based initialization does not impose risks for robustness. We still had the convergence monitoring to handle such cases, though.

The parameter sets not being absolutely identical for both linear systems, moreover, indicates that individual properties of the linear systems do have some impact on the optimal parameter set and that the evolutionary optimization method does properly reflect this. However, we need to outweigh the differences of parameter sets versus the costs of training phases.

### 13.8.2   Parameter Optimization: Linear Elasticity Problem

In continuum mechanics, linear elasticity formulations describe deformation of solid materials under external stresses. We refer to the literature for an extensive overview about such models [19, 32]. These problems are also described by elliptic PDEs and, thus, algebraic multigrid is a considerable option. However, the complexity of the underlying problems typically require specialized AMG approaches [8, 14]. These construct the multigrid hierarchy based on aggregation [39] and, this way, can ensure to properly handle so-called rigid-body-modes. This is crucial in order to obtain robust convergence. A variety of parameter choices still remains regarding the construction of these aggregates, where our optimization approach based on evolutionary strategies is well-suited to find a good set of parameters. Again, this follows the idea of Informed Machine Learning, where the basic strategy is set based on experience but the fine-tuning therein is subject of the ML-based optimization.

Figure 13.4 shows results of the parameter optimization that has been performed for the more challenging but much smaller case of the thin beam and then has been applied to an either thick or thin steel beam. The smaller problem size in the thin case explains the faster times compared to the thick case.

Compared to default SAMG parameters for elasticity problems, about 13% and 21% of the runtime could have been saved with the optimization of the aggregative setup. It essentially became more aggressive here, leading to coarse level problems with less degrees of freedom.

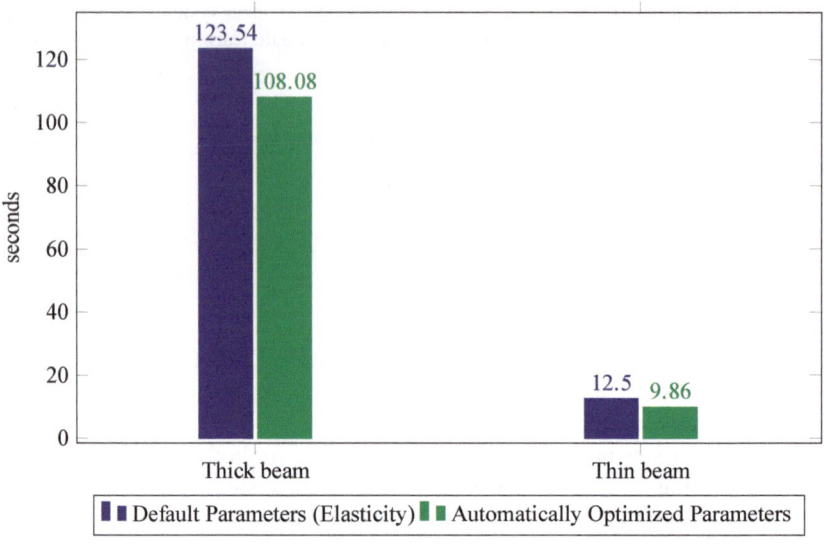

**Fig. 13.4** Results for single linear systems from the simulation of a thick and thin steel beam with default parameters for elasticity problems and optimized settings from an offline training. The problem has about 2.2 million and 140 thousand DOF in the thick and thin case, respectively. As in Fig. 13.3, the training time is excluded

The parameter optimization considered different minimal and maximal aggregate sizes for the aggregative AMG coarsening approach, along with different modes and thresholds for strength-of-connection. This results in a rather large number of almost 45,000 possible combinations. The offline training here took 5,000 evaluations. Here we intended to demonstrate the ability of the optimization methods to identify a good parameter setting. This comprehensive optimization has been applied once in an offline-mode. The result is equally well-applicable to further such simulations with different material thicknesses. Thus, in the end, the training time is irrelevant for the gain in the simulation time. Due to the large amount of possible settings, no manual tuning has been considered here.

Because the simulation requires setups within each step, as the rigid-body-modes evolve with the deformation of the material, these savings carry over to the entire simulation. But an additional setup reusage was hardly applicable here.

### 13.8.3 Setup Reusage: Sequence of Reservoir Simulation Problems

We will demonstrate the potential of reusing linear solver setups with a sequence of five linear systems from one time step of a Black-Oil reservoir simulation with 788,000 discretization cells.

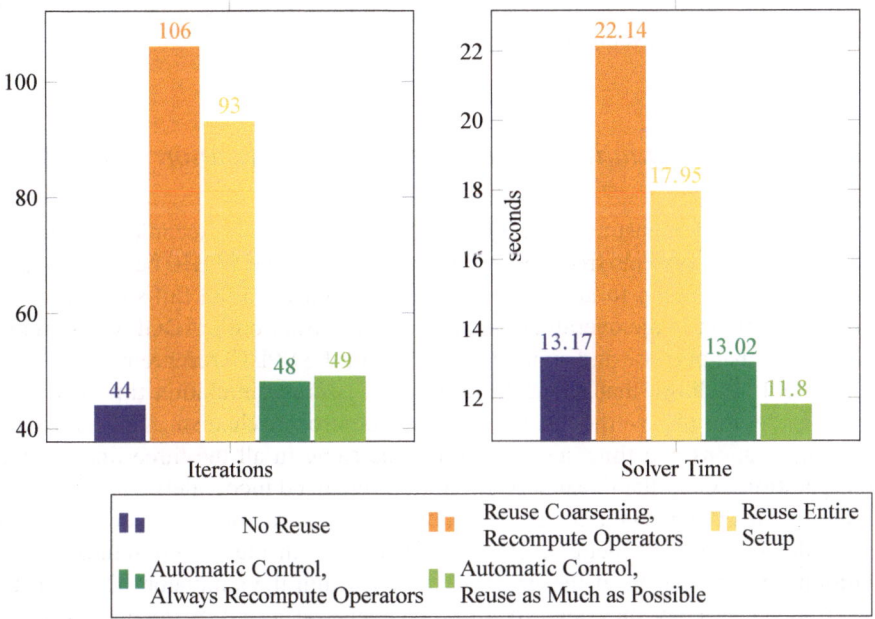

**Fig. 13.5** Iteration counts and solver runtimes for the set of five pressure problems from one time step of an industrial simulation with 788,742 cells. Different setup reusage approaches applied

Clearly, in cases where a rather trivial reusage of setups is possible, our control mechanism can easily do so. However, in this particular case, the control mechanism is actually necessary in order to gain performance benefits from reusing setups without robustness issues. On the one hand, the systems all are from different Newton iterations of the same time step of the simulation and, thus, some setup reusage is expected to be beneficial. On the other hand, trivially reusing full or partial setups within all Newton iterations results in severe robustness issues in this case.

Figure 13.5 illustrates the numerical issues and the performance of different setup reusage approaches. It compares the computation of a fresh setup as a reference (left bar) with trivially reusing[6] or partly reusing setups in all systems with the results of the automatic control mechanism.

The autonomous control properly handles the challenging sequence of linear systems and ensures that a robust solution is provided in all cases. Moreover, by reusing setups as often as possible, it allows for saving about 11% of the runtime.

Clearly, the potential for savings is higher in cases where more parts of the setups can be reused for a larger number of linear systems. The control will still ensure that

---

[6] The lower iteration count for reusing entire compared to partially reusing setups is due to round-off impacts that make SAMG stop early with fully reusing setups here.

setups are recomputed as soon as beneficial for the performance or necessary for the robustness, as in this exemplary case.

### 13.8.4    Full Simulation Result: Reservoir Application (SPE10)

Above we have demonstrated the effects of machine learning techniques with two exemplary linear systems from reservoir simulations of the SPE10 benchmark case [3]. We are now going to apply our control mechanism to the full simulation run and see the effect in the overall application. We use Stanford's ADGPRS simulator [13, 41] for that purpose and simulate 500 days of the SPE10 reservoir.

Figure 13.6 shows that about 29 minutes of overall simulation time had been saved. Including the overhead costs from the control mechanism. The number of Newton iterations and time steps remained the same in all the three linear solver configurations. Only the linear solver iterations changed moderately.

Of the overall saving, about 16 minutes are mere linear solver time. The remainder results from better memory utilization with the reused setups and, in addition, the processing of the linear solutions within the simulator: The adjusted linear solver strategy may lead to different results at round-off level beyond the required solution accuracy. This is inherent to any parameter adjustment with any

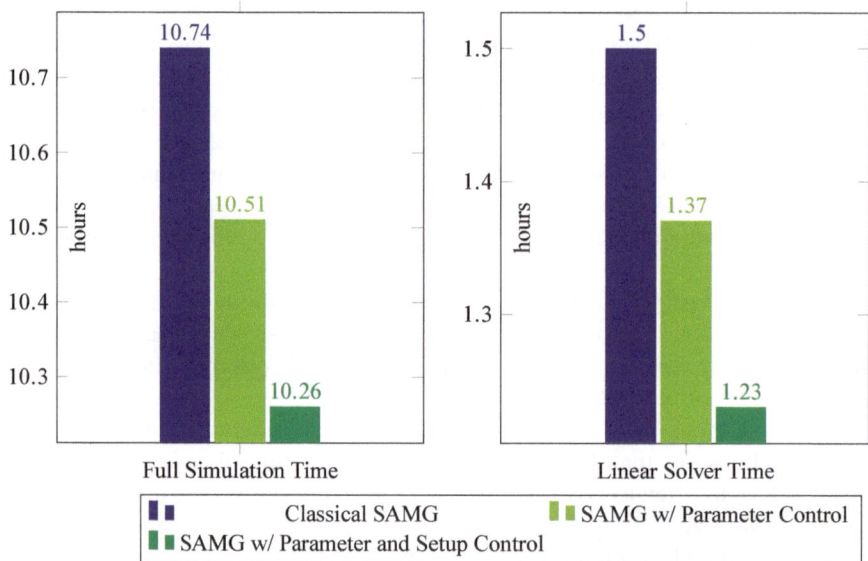

**Fig. 13.6** Full simulation time (left) and linear solver runtimes (right) in the SPE10 simulation with classical SAMG (fixed parameters), SAMG with auto-controlled parameters and SAMG with both auto-controlled parameters and setups. All timings do include the control mechanism

solver method and, thus, has not been investigated further. Of the linear solver savings, 7.75 minutes had been due to parameter optimization and 8.25 minutes due to setup reusage.

### 13.8.5   Full Simulation Result: Groundwater Application

Groundwater simulations are a promising field of application for our autonomous solver control: with small time steps and accordingly rather well-conditioned linear systems, single level incomplete factorization are a sufficiently suited preconditioner in some cases and time steps. Whereas algebraic multigrid is the method of choice in many cases with larger time steps sizes. Typically, simulations do feature both situations and, thus, an automatic control is required to select the appropriate solver setting within different time steps.

We will demonstrate groundwater simulations with Modflow USG 1.8 that is provided by the USGS [23] and consider one single-phase simulation with only water and one two-phase simulation with water and solid particles. In the single-phase case, the linear systems are too ill-conditioned in some time steps for standard-AMG to be applicable robustly. At least without further stabilization techniques applied. In the two-phase case, it is the other way round: the linear systems are well-conditioned enough for ILU to be sufficiently fast in some time steps.

We include $\alpha$SAMG [4] in the comparison in Fig. 13.7, as it can also switch between different solver methods, based on statistical evaluations. Indeed, it performs better than either plain ILU or plain AMG. However, the more versatile machine learning-based solver control provides the most robust approach with the best performance.

### 13.8.6   Full Simulation Result: Computational Fluid Dynamics Application

The OpenFOAM package [40] is widely used for computational fluid dynamics (CFD) applications, also among industrial users. The simulation that we consider here uses the PIMPLE method, an OpenFOAM-included combination of SIMPLE [24] and PISO [15], where SAMG is used as a pressure solver within the solution scheme. The considered problem consists of about 14 million cells and is simulated for twenty thousand time steps. It is run on 240 cores with MPI.

The specialty of the AMG application in OpenFOAM is that only a very few setup calculations are actually necessary. This is already controlled by the AMG integration in the OpenFOAM package. That is, our autonomous solver control will only be used to further optimize the solver parameters, not to further control the

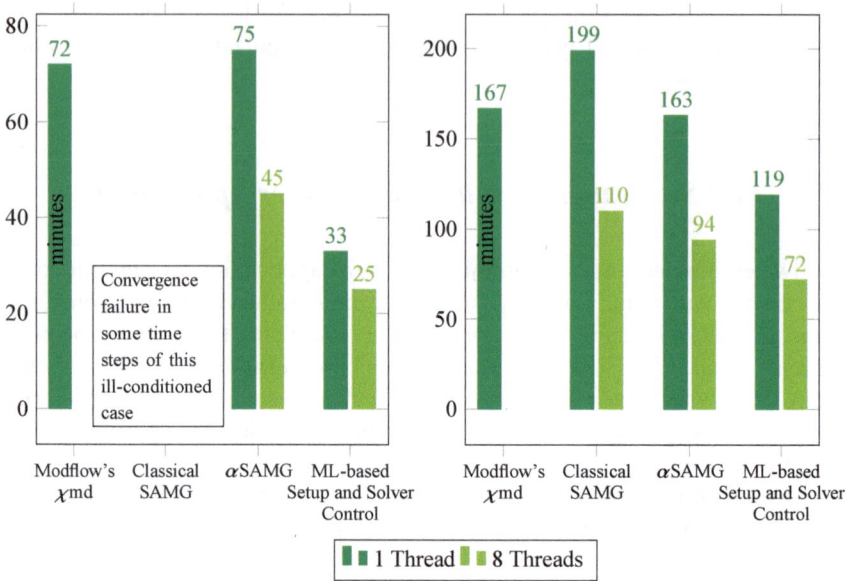

**Fig. 13.7** Full simulation runtime with different linear solvers and control approaches for a Modflow USG 1.8 single-phase simulation with 737,191 cells (left) and a two-phase simulation with 552,600 cells (right). $\chi$md is Modflow's non-parallel single-level solver that serves as reference benchmark here

setup here. Moreover, the parameters have already been tuned manually. Thus, the machine leaning optimization will only be used for some remaining fine-grained tuning: Both again follows the approach of Informed Machine Learning, where all available knowledge already has been applied. It also explains why the potential performance gain is smaller than in cases where less tuned default parameters have been the reference.

The application of our solver control is two-fold. We do first apply an offline training to find a good parameter setting for the setup parameters of AMG. Due to only very few setups being computed within the actual simulation run, an online training would cause a significant overhead, as a fresh setup was calculated with a new setting of parameters within each evaluation. Therefore, we apply the offline training beforehand and not actually as part of the simulation. The outcome of this comprehensive training that took about two hours can be used for all comparable simulations and serves as one-time-optimization. A manual optimization via trial-and-error could hardly be comparably exhaustive. This offline training has emphasized the solution phase of AMG according to the expected amount of setup-reusage, though.

The resulting settings concerned the target size of the coarse-level problem, the strength-of-connection threshold of AMG and the multigrid level where SAMG switches from classical (aggressive) coarsening [35] to sparsified variants [5, 20] that particularly target at distributed memory parallelism, at the expense of a

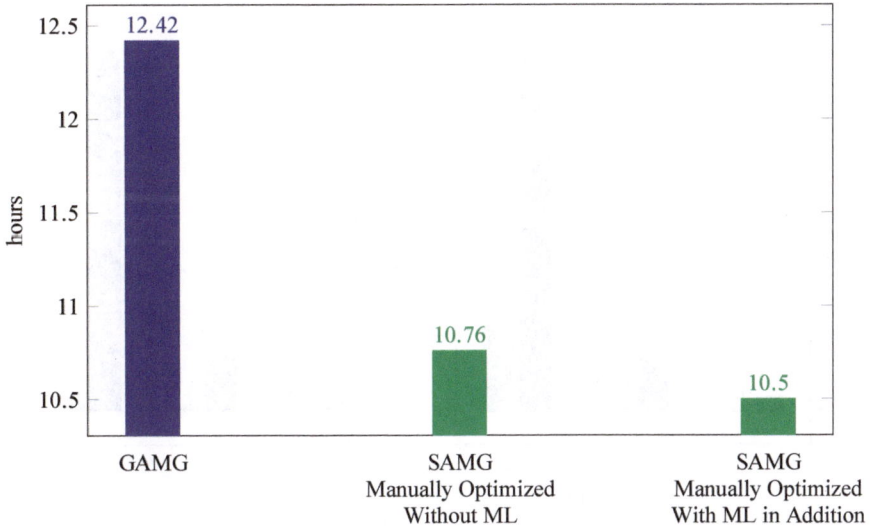

**Fig. 13.8** Full simulation runtime with OpenFOAM with the manually optimized SAMG and SAMG with ML-based control in addition. The OpenFOAM built-in GAMG solver is used as a reference

slightly decreased robustness. In this offline training, the runtime of the solver for a single linear system had been reduced from 1.914 to 0.877 seconds. Most of this performance gain applies to the setup phase and, thus, will not fully carry over in this application. The iteration count was 19 for the original parameters and 21 for the optimized ones.

These parameters had been applied for the full simulation run now. And an online training phase has been used to further optimize the parameters in the solution phase. It decided between the CG and GCR Krylov method and between Gauss-Seidel and Jacobi smoothing. Moreover, it controlled how to evaluate residuals for convergence checks. The evaluation of this rather small number of possible combinations has been well-feasible in an online training, despite the outer setup reusage: the AMG setup is not at all affected by these settings.

As seen in Fig. 13.8, compared to the already optimized settings another 2.5% of runtime, 15 minutes, could have been saved. The relative saving is less than in the offline training, because only eight setups have been made throughout the simulation. In these, the savings had been the highest. Moreover, some efforts had already been applied in the manual tuning of parameters in this application.

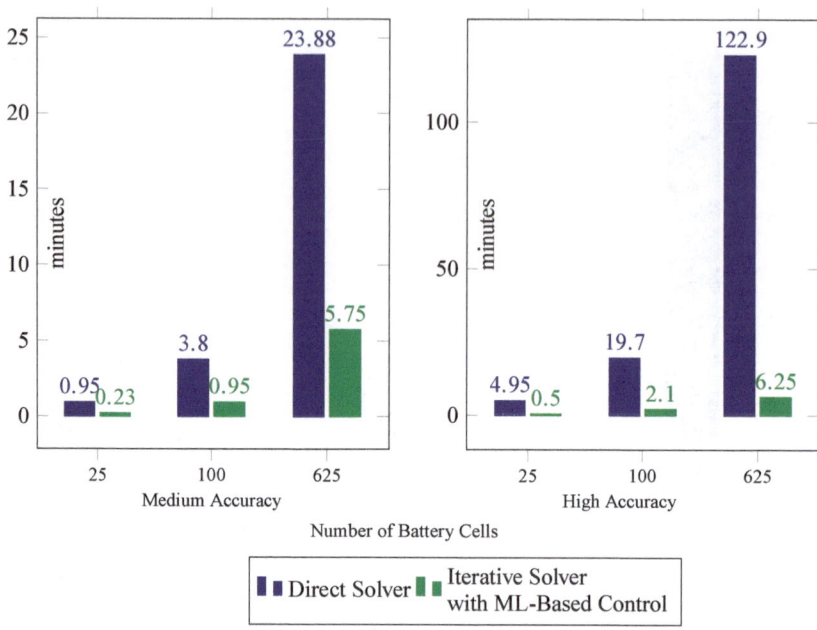

**Fig. 13.9** Runtime to simulate 100 seconds of battery aging at different accuracies. The direct solver serves as robust reference. The ML-based control allows for exploiting iterative schemes

### 13.8.7 Full Simulation Result: Battery Aging Simulation

The simulation of battery aging [25, 26] plays an important role in the efficient usage of batteries in electrical vehicles as well as in energy storage. The challenge for the linear solver is that the properties of the linear systems highly depend on the type of battery cell and the underlying model. Thus, the solver can be confronted with very different types of linear systems within the same simulator. This makes selecting iterative methods in a robust manner rather difficult and direct solvers had been the only considerable option so far. With our autonomous control, however, iterative methods can safely replace the direct solver. The control ensures the appropriate iterative method to be used for an individual type of battery cell. Thus, the better efficiency of iterative methods can be exploited, as seen in Fig. 13.9 (for compatibility reasons, these benchmarks have been performed on an Intel Core i7-4790 and 16 GB RAM). Especially for problems with a higher model resolution and more connected battery cells, with correspondingly larger linear systems.

## 13.9 Conclusions and Future Research

We have used a mechanism based on evolutionary and surrogate machine learning techniques to optimize the usage of linear solvers in different kinds of simulation applications. Internal monitoring mechanisms at the same time ensure robustness of the solver to be maintained in all cases.

The approach follows the idea of Informed Machine Learning: the basic linear solver strategy is defined based on human knowledge for a certain type of simulation. The huge variety of options and aspects to consider with several constraints would make an automated approach for defining the overall solver strategy virtually impossible to realize for production use. The automated approach, however, has turned out to be efficient in further optimizing those parameters where variation is possible within a certain base-strategy of the linear solver. Here it is superior over human experience, as it can exploit also information that are evaluated during a simulation or learned from previous runs. Moreover, the presented approach that exploits tree-based evolutionary learning and surrogate methods requires significantly less evaluations than trivially evaluating all possible parameter combinations.

Significant savings in runtime and, thus, computational resources have been possible for a vast range of simulation applications. We have to keep in mind that this only results from optimizing solver parameters automatically. The underlying solver strategy remains the same.

Future research will focus on further improving the surrogate models both in terms of the surrogate dimensions as well as in comparing different systems based on their surrogate representation.

The idea of Informed Machine Learning can be exploited further by defining different initial sets of parameter search spaces, where the appropriate one can be selected automatically by the method, for instance, based on the surrogate models. This would maintain the exploitation of human knowledge while reducing the need for user interaction. Last but not least, also the consideration of and combination with further optimization methods would be a promising field of further research.

**Acknowledgments** The author gratefully acknowledges that large parts of this development has been funded by the Fraunhofer Cluster of Excellence "Cognitive Internet Technologies". The General Purpose Research Simulator (GPRS) developed by the Reservoir Simulation Research Group (SUPRI-B) at Stanford University was used in this work.

## References

1. Brandt, A.: Algebraic Multigrid Theory: the Symmetric Case. Applied Mathematics and Computation **19**(1–4), 23–56 (1986)
2. Brandt, A., Livne, O.: Multigrid Techniques: 1984 Guide with Applications to Fluid Dynamics. SIAM Classics in Applied Mathematics (2011)

3. Christie, M., Blunt, M.: Tenth SPE Comparative Solution Project: A Comparison of Upscaling Techniques. SPE Reservoir Evaluation and Engineering p. 308–317 (2001)
4. Clees, T., Ganzer, L.: An Efficient Algebraic Multi-Grid Solver Strategy for Adaptive Implicit Methods in Oil Reservoir Simulation (2007)
5. De Sterck, H., Falgout, R., Nolting, J., Yang, U.: Distance-two Interpolation for Parallel Algebraic Multigrid. Numerical Linear Algebra with Applications 15(2–3), 115–139 (2007)
6. Feurer, M., Hutter, F.: Hyperparameter Optimization. In: Automated Machine Learning. Springer (2019)
7. Ghojogh, B., Sharifian, S., Mohammadzade, H.: Tree-based Optimization: A Meta-Algorithm for Metaheuristic Optimization (2018)
8. Griebel, M., Oeltz, D., Schweitzer, M.: An Algebraic Multigrid Method For Linear Elasticity. SIAM Journal on Scientific Computing 25(2) (2002)
9. Gries, S.: System-AMG Approaches for Industrial Fully and Adaptive Implicit Oil Reservoir Simulations. PhD thesis, University of Cologne (2015)
10. Gries, S.: Autonomous Linear Solver Control to Improve Performance of Simulations. In: European Conference on the Mathematics of Oil Recovery (ECMOR) (2022)
11. Grote, M., Huckle, T.: Parallel Preconditioning with Sparse Approximate Inverses. SIAM Journal on Scientific Computing 18(3), 838–853 (1997)
12. Hestenes, M., Stiefel, E.: Methods of Conjugate Gradients for SOlving Linear Systems. Journal of Research of the National Bureau of Standards 49(6) (1952)
13. Hui, C.: Development of Techniques for General Purpose Simulators. PhD thesis, Stanford University (2002)
14. Hülsmann, G., Krechel, A., Plum, H.J., Schweitzer, M., Hu, W., Wu, C., Koishi, M.: Scalable Linear Solvers for Computational Material Design of Filled Rubbers. In: NWC 2019, NAFEMS World Congress. Summary of Proceedings : A world of engineering simulation (2019)
15. Issa, R.: Solution of the Implicitly Discretized Fluid Flow Equations by Operator-Splitting. Journal of Computational Physics 62, 40–65 (1985)
16. Jankowski, D., Jackowski, K.: Evolutionary Algorithm for Decision Tree Induction. In: International Conference on Computer Information Systems and Industrial Management (2014)
17. Jeong, S., Murayama, M., Yamamoto, K.: Efficient Optimization Design Method Using Kriging Model. Journal of Aircraft 42(2), 413–420 (2005)
18. Lindauer, M., Eggensperger, K., Feurer, M., Biedenknapp, A., Deng, D., Benjamins, C., Ruhkopf, T., Sass, R., Hutter, F.: SMAC3: A Versatile Bayesian Optimization Package for Hyperparameter Optimization (2021)
19. Marsden, J., Hughes, T.: Mathematical Foundations of Elasticiy. Dover Publications (1983)
20. Meier-Yang, U.: Parallel Algebraic Multigrid Methods - High Performance Preconditioners. In: Numerical Solutions of Partial Differential Equations on Parallel Computers (2006)
21. Meijerink, J., van der Vorst, H.: An Iterative Solution Method for Linear Systems of which the Coefficient Matrix is a Symmetric M-Matrix. Mathematics of Computation 31(137), 14–162 (1977)
22. Mishev, I., Fedorova, B., Terekhov, S.: Linear Solver Performance Optimization in Reservoir Simulation Studies (2009)
23. Panday, S., Langevin, C., Niswonger, R., Ibaraki, M., Hughes, J.: MODFLOW-USG version 1: An unstructured grid version of MODFLOW for simulating groundwater flow and tightly coupled processes using a control volume finite-difference formulation. U.S. Geological Survey Techniques and Methods, Book 6, Chapter A45 (2013)
24. Patankar, S., Spalding, D.: A Calculation Procedure for Heat Mass and Momentum Transfer in Three Dimensional Parabolic Flows. International Journal on Heat and Mass Transfer 15, 1787 (1972)
25. Puchta, M., Schledde, D.: Virtuelle Batterien in der Entwicklung von Elektrofahrzeugen. Digital Engineering Magazine (2011)

26. Puchta, M., Schwalm, M., Dengler, F.: High-precision, High-dynamic Emulation of Lithium-Ion Cells for the Entire Life Cycle. In: The 30th International Electric Vehicle Symposium and Exhibition (2017)
27. Rainville, F.M., Fortin, F.A., Gardner, M.A., Parizeau, M., Gagné, C.: DEAP: A Python Framework for Evolutionary Algorithms (2012)
28. von Rueden, L., Mayer, S., Beckh, K., Giesselbach, G., Heese, R., Kirsch, B., Pfrommer, J., Pick, A., Ramamurthy, R., Walczak, G., Bauckhage, C., Schuecker, J.: Informed Machine Learning - A Taxonomy and Survey of Integrated Prior Knowledge into Learning Systems (2021)
29. Ruge, J., Stüben, K.: Algebraic Multigrid (AMG). In: Multigrid Methods, SIAM Frontiers in Applied Mathematics, vol. 5 (1986)
30. Saad, Y.: Iterative Methods for Sparse Linear Systems. PWS Publishing Company, Boston (1996)
31. Saad, Y., Schultz, M.: GMRES: A Generalized Minimal Residual Algorithm for Solving Nonsymmetric Linear Systems. SIAM Journal on Scientific and Statistical Computing 7, 856–869 (1986)
32. Slaughter, W.: The Linearized Theory of Elasticity. Birkhäuser (2002)
33. Sobol, I.: Global Sensitivity Indices for Nonlinear Mathematical Models and their Monte Carlo Estimates. Mathematics and Computers in Simulation 55, 1–3, 271–280 (2001)
34. Souza, A., Nardi, L., Oliveira, L., Olukotun, K., Lindauer, M., Hutter, F.: Prior-Guided Bayesian Optimization (2020)
35. Stüben, K.: A Review of Algebraic Multigrid. Journal of Computational and Applied Mathematics (2001)
36. Stüben, K., Ruge, J., Clees, T., Gries, S., Plum, H.J.: Algebraic Multigrid - From Academia to Industry. In: Scientific Computing and Algorithms in Industrial Simulations. Springer (2017)
37. Thum, P., Stüben, K.: Advanced Algebraic Multigrid Application for the Acceleration of Groundwater Simulations. In: XIX International Conference on Water Resources (2012)
38. Trottenberg, U., Oosterlee, C., Schüller, A.: Multigrid. Elsevier Academic Press (2001)
39. Vanek, P., Mandel, J., Brezina, M.: Algebraic Multigrid by Smoothed Aggregation for Second and Fourth Order Elliptic Problems. Computing 56, 179–196 (1996)
40. Weller, H., Tabor, G., Jasak, H., Fureby, C.: A Tensorial Approach to Computational Continuum Mechanics using Object-Oriented Techniques. Computers in Physics 12(6) (1998)
41. Zhou, Y.: Parallel General-Purpose Reservoir Simulation with Coupled Reservoir Models and Multisegment Wells. PhD thesis, Stanford University (2012)

# Chapter 14
# Anomaly Detection in Multivariate Time Series Using Uncertainty Estimation

**Moritz Müller, Gunar Ernis, and Michael Mock**

**Abstract** Today's industrial machines are equipped with several sensors that detect environmental changes and generate time series. One challenging task is the detection of anomalies in multivariate time series to proactively schedule machine maintenance and prevent failures during cost intense processes. Recent deep learning-based anomaly detectors demonstrate remarkable results as they can process large datasets of raw data. A common unsupervised method is to measure the discrepancy or anomaly score between observation and the expected behaviour approximated by a neural network. No approach incorporates multivariate uncertainties quantified by a Bayesian neural network and expert knowledge in the form of probabilistic relations into the anomaly score to enhance anomaly detection. We propose a Bayesian neural network that estimates uncertainty and multivariate time series forecasts. In this chapter, we introduce an anomaly score function based on Hotelling's $T^2$ statistic and the quantile function to estimate appropriate thresholds for the anomaly scores. Our experimental results show that anomaly scores are specifically separable into normal and anomalous regions when the discrepancies exploit probabilistic relations between multivariate features. Moreover, we compare the anomaly score separability and the anomaly detection accuracy against recent state-of-the-art methods. The evaluation shows that uncertainty-driven anomaly scores are competitive in both terms.

## 14.1 Introduction

As more data [29] becomes available through ongoing efforts to optimise and automate traditional manufacturing, machine learning (ML) algorithms [27] begin to play a more significant role in analysing the data to enhance industrial processes and product quality. A crucial challenge in this field is detecting anomalous events

M. Müller · G. Ernis (✉) · M. Mock
Fraunhofer IAIS, Sankt Augustin, Germany
e-mail: moritz.mueller@iais.fraunhofer.de; gunar.ernis@iais.fraunhofer.de;
michael.mock@iais.fraunhofer.de

© The Author(s) 2025
D. Schulz, C. Bauckhage (eds.), *Informed Machine Learning*,
Cognitive Technologies, https://doi.org/10.1007/978-3-031-83097-6_14

and behaviours in observed multivariate time series. A multivariate time series consists of multiple variables—for example, sensor measurements—varying over time. Anomaly detection (AD) describes the challenge of finding patterns in data such as time series that do not conform to the expected behaviour [5]. Such algorithms encounter sensitive irregularities in early production stages to proactively schedule machine maintenance and prevent failures during cost-intensive processes [14]. The research field addressing AD in time series shows various strategies. Many approaches are limited to univariate data streams and cannot identify anomalies across different feature domains [3, 4]. A straightforward approach for addressing multivariate AD is to apply a unique univariate anomaly detector for every variable in a multivariate time series. Nevertheless, it yields many disadvantages:

First, it is inefficient and requires several costly anomaly detectors to maintain and synchronise in practical scenarios. In contrast, a multivariate anomaly detector consists of one algorithm that requires no synchronisation and the maintenance of only one model.

Second, this procedure assumes that the variables are not correlated and analyses each variable separately. These methods fail, particularly in detecting the anomalous relationship between variables that do not conform with relationships observed on normal time segments.

Third, expert knowledge in the form of correlations or known behaviour of different variables is not considered. Hence, there are several multivariate AD approaches [12, 18, 22, 26] based on deep neural networks (DNNs) to detect anomalous sequences in multivariate time series.

The last point can be addressed by including knowledge sources in the algorithm, thus transforming it into an informed machine learning algorithm [24]. In the approach presented in this chapter, we make use of several assumptions about certain distributions and correlation matrices based on statistical relations attributable to scientific knowledge sources (as described in [24]). We represent these sources as probabilistic relations through assumptions about our prior distributions and correlation matrices. Finally, we integrate our knowledge by choosing how the discrepancy between data points is calculated, which corresponds to narrowing down the given hypothesis set.

One can observe that a considerable amount of AD proposals rely heavily on the prediction itself without using any uncertainty estimation. The uncertainty estimation casts light on the trustworthiness of the prediction in DNN and impacts the AD as well. Bayesian neural networks (BNNs) belong to the group of DNNs, enabling us to quantify a given prediction's uncertainties. A simple approach to estimating such a network's uncertainties is to measure the variability of predictions for the same input sequence.

A recent work of [30] proposed a BNN that exploits the model's uncertainty for AD in univariate time series. In particular, the authors approximate the model's uncertainty estimation using Monte Carlo dropout (MC dropout) as suggested by [9]. Furthermore, upon the uncertainty estimation, a predictive region is constructed and used to classify each observation: Every observation outside the region is classified as an anomaly. Their experiments demonstrate that their network considerably enhances the accuracy for univariate AD when high uncertainties occur.

Inspired by these works, this chapter aims to build a BNN that can detect anomalies in multivariate time series based on uncertainty estimation. In particular, we use a BNN that forecasts the time steps of a multivariate time series. Then, we compute an anomaly score that measures the discrepancy between predictions and observations based on the Hotelling's $T^2$ statistic. We then employ the quantile function of the cumulative distribution of the anomaly score distribution to derive an appropriate threshold to determine if observations are normal/anomalous. The first contribution of this chapter is that it examines how uncertainty estimates can be incorporated into the anomaly score. Our experimental results demonstrate that the uncertainty estimates enhance specifically the AD accuracy when the correlation within the uncertainty estimates is considered.

This chapter's second contribution compares the proposed approach against state-of-the-art methods [1, 26, 31] for multivariate AD. In our experiments, we observe that uncertainty-driven methods are indeed competitive with recent works. This is the first work addressing multivariate AD based on the uncertainty estimations across multiple variables.

## 14.2 Background and Related Work

This section describes the background and related work of AD in time series. First, we describe the general problem of detecting anomalies in time series. Then we explain the term unsupervised AD and the concept of BNNs. Finally, we provide an overview of state-of-the-art methods that detect anomalies in time series using an unsupervised learning approach.

### 14.2.1 Problem Formulation and Anomaly Categorization

A time series is a collection of data points taken at successive (equally) spaced points in time. Moreover, it can be decomposed in the following terms:

The trend describes the long-term average movement of a time series. Seasonality describes cyclical fluctuations that occur at regular intervals such as hourly, daily, and so forth in a time series. Finally, a crucial component of time series is inherent noise that describes random fluctuations that are not predictable. Univariate time series consists of one variable that varies over time. In contrast, a multivariate time series comprises a sequence $\mathbf{X} = (\mathbf{x}_1, \mathbf{x}_2, \ldots, \mathbf{x}_t) \in \mathbb{R}^{N \times T}$ with $T$ time steps and $N$ continuous features. A feature vector $\mathbf{x}_i \in \mathbb{R}^N$ of timestamp $i$ is considered to be an anomaly when its corresponding anomaly score $h_{\hat{y}}(\mathbf{x}_i)$ exceeds a specified threshold $\epsilon \in \mathbb{R}$. The anomaly score function $h_{\hat{y}}(\mathbf{x}_i)$ is a distance measurement that describes the deviation between the expected characteristic and the observation. Univariate

AD is already a challenging task as it requires the analysis of the seasonable pattern for one variable. The detection of multivariate-based anomalies is even more complex as we consider several variables of distinctive feature domains.

In general, anomalies can be interpreted as instances of data collections that deviate from the expected values. Hence, a prerequisite for identifying anomalies is to analyse how critical components such as trend, seasonality, and noise appear in regular time series. In the following, we categorise the term anomaly into point, context and collective anomalies as defined by [5]. This anomaly categorisation is crucial for choosing the AD algorithm as they rely on different data characteristics.

### 14.2.1.1   Point Anomalies

The simplest form of anomalies are point anomalies that deviate significantly from most samples and occur without requiring any further context. Since point anomalies are independent of other contextual attributes, such as time, the analysis of point anomalies is usually less extensive than detecting other time series anomalies. For example, consider a time series representing the hourly utilisation of public transport in a city. A sudden drop or rise in utilisation at a single time stamp would be considered a point anomaly.

### 14.2.1.2   Context Anomalies

In [25], context anomalies are identified by analysing behavioural and contextual attributes of instances. Behavioural attributes describe non-contextual characteristics of an instance, whereas contextual attributes such as time or location put the sample features into a context. The traffic at any time stamp describes the behavioural characteristics where the corresponding time stamp contextualises this observation. Seasonal morning or late afternoon traffic peaks are not necessarily considered contextual anomalies. High public transport utilisation is common at rush-hour traffic times. In contrast, increased occupancy of public transportation during the typically low-traffic hours would represent a typical example of a context anomaly as the behavioural characteristic does not fit in the context.

### 14.2.1.3   Collective Anomalies

Collective anomalies appear as a sequence of instances that deviate from the usual sequence pattern. Any other attributes are unnecessary for the detection of this class. Each example of a collective anomaly group is not necessarily a point anomaly. However, their occurrence together as a group is anomalous. A collective anomaly example could be when the traffic utilisation within a specific time window has constantly high values and deviates significantly from the load. High occupancy in public transport frequently occurs during the day. Nevertheless, the demand for such services is changing dynamically. The constant and high occupancy within the context of this window is a clear indicator of a collective anomaly.

### 14.2.2   Unsupervised Anomaly Detection

As the availability of data and computing resources increases, deep learning (DL) approaches outperform traditional ML algorithms in many areas, such as AD [4]. For instance, support vector machines (SVMs) require manual feature selection and extraction techniques. Manual feature selection techniques increase the expenditure of time for data preprocessing as the number of available features grow. In contrast, DL can deal directly with raw input data as the DL model learns which features are relevant to solve the given prediction problem. Further expert knowledge in the form of probabilistic relations can be incorporated to enhance the AD.

In the typical AD-scenario, it is assumed that most data instances are normal and each instance's class is unknown, corresponding to an unsupervised learning task. The goal in AD is to capture the data's main characteristics, which can be carried out in several ways. We follow a common method in DL: First, we learn a synthetical reconstruction of the original input followed by forecasting future values of the given training data. The fundamental assumption is that the optimised DL model maps instances that resemble the majority of samples from the training set to similar regions. Hence, a popular method to reveal anomalies is to measure the discrepancy between the predicted and observed outcomes. This discrepancy measurement is usually referred to as the anomaly score.

### 14.2.3   Bayesian Neural Networks

In the following subsection, we describe the general concept of BNNs and the related types of uncertainties. Traditional neural networks (NNs) are a popular predictive tool nowadays because they can learn non-linear abstractions from the input. However, NNs are often miscalibrated because its corresponding prediction tends to be overconfident [28]. BNNs are a framework to quantify the underlying model uncertainty by encoding the degrees of beliefs along with the prediction. In contrast to a typical NN that assigns point estimates to weights, the weights $w$ in a BNN represent a prior distribution. The prior captures the belief of the weight representation without having any data sample observed before. It is commonly assumed to follow a Gaussian distribution $p(w) \sim N(0, 1)$.

The posterior distribution over the weights $w$

$$p(w|X, Y) = \frac{p(X, Y|w)p(w)}{p(X, Y)} \qquad (14.1)$$

are determined using the Bayes theorem. It expresses the epistemic uncertainty for the given data samples $X$ and $Y$. The core idea is to quantify the uncertainty of the NN predictions to understand how trustworthy the predictions are. Uncertainty estimation is crucial in the sense as it measures the level of confidence for a prediction. For example, an autonomous vehicle can decide to pass the control to a human driver when the detected objects are associated with high uncertainties. In [8], the term uncertainty is split into epistemic, aleatoric and predictive uncertainty:

### 14.2.3.1 Epistemic Uncertainty

Epistemic uncertainty is usually associated with knowledge uncertainty [8], and the lack of available training data causes it. Hence, epistemic uncertainty is reducible as the number of available observations in $X, Y$ increase. Moreover, the likelihood $p(X, Y|w)$ tells us how well the given weight parameter $w$ describes the observed data $X, Y$. One can apply marginalisation to retrieve the unknown evidence

$$p(X, Y) = \int_w p(X, Y, w)dw = \int_w p(X, Y|w)p(w)dw. \qquad (14.2)$$

One key concept of Bayesian inference is integrating unnecessary variables out [2]. The exact calculation of the posterior distribution $p(w|X, Y)$ involves the marginalisation for the whole parameter space. In practice, the complete posterior distribution is intractable, especially with state-of-the-art networks that contain millions of weight parameters.

### 14.2.3.2 Aleatoric Uncertainty

Aleatoric uncertainty describes the inherent noise of the data. In particular, the data distribution characterises the variability of the underlying variables. Even though this uncertainty is irreducible, the range of possible outcomes can be estimated.

### 14.2.3.3 Predictive Uncertainty

The predictive uncertainty

$$p(y^*|x^*) = \int_w \underbrace{p(y^*|x^*, w)}_{\text{likelihood}} \underbrace{p(w|X, Y)}_{\text{posterior}} dw \qquad (14.3)$$

represents the network uncertainty about its output $y^*$ given a new input $x^*$. It consists of the product between the posterior distribution $p(w|X, Y)$ and the likelihood $p(y^*|x^*, w)$ given the model weight $w$ and the new input $x^*$. However, due to the aforementioned posterior intractability problem, the predictive distribution has to be approximated. An efficient method to perform the approximation is Monte Carlo dropout [9]

$$q(y^*|x^*)(y^*) = \underbrace{\int_w p(y^*|x^*, w)q(w|X, Y)\,dw}_{\approx p(y^*|x^*)}, \qquad (14.4)$$

where $w = \{W_i\}_{i=1}^L$ is a set of random variables for a model with $L$ layers. Moreover, the expectation value can be estimated to be [9]

$$\mathbb{E}_{q(y^*|x^*)} \approx \frac{1}{T}\sum_{t=1}^T p(y^*|x^*, w_t), \text{s.t. } w_t \sim q(w|X, Y) \qquad (14.5)$$

by averaging $T$ stochastic forward passes through the network while simultaneously applying dropout. The main idea is that each dropout operation corresponds to a sample $w_t \sim q(w|X, Y)$ from an approximate posterior distribution. A Monte Carlo integration with a sufficiently large sample size $T$ provides an approximate predictive uncertainty.

### 14.2.4 Related Work

Traditional multivariate AD approaches mostly rely on autoencoders (AEs) [20, 23, 31] variational autoencoders (VAEs) [22, 26] or generative adversarial networks (GANs) [1, 18, 21] and show remarkable performances. However, the research field addressing AD in time series in uncertainty estimation has hardly been explored. BNNs can express uncertainties about their reconstruction or prediction for a given input. The BNN proposed by [30] benefits from uncertainty estimates as the valid prediction interval increases proportionally to the model uncertainty observed in univariate time series predictions. Thus this framework reduces the false alarm rate of anomalies as model uncertainty increases at high-variance events. However, the proposed framework exhibits some limitations:

It lacks empirical evidence of its eligibility for AD by using publicly available univariate time series benchmarks. Moreover, the forecast model focuses primarily on predicting univariate time series.

Furthermore, uncertainty estimation and its corresponding predictive interval ignore possible relations [24] (like correlations) between variables.

## 14.3 Detecting Anomalies in Time Series Using Uncertainty Estimation

To explain our proposed method, we introduce the central processing units of the AD pipeline using Fig. 14.1. For each unit, we provide a detailed description in the following sections. The first unit represents the data source that contains a time series $\mathbf{X}_{source} \in \mathbb{R}^{N \times T}$ including $T$ feature vectors of size $N$.

### 14.3.1 Window Processing and Forecast Modelling

Since the available computing resources restrict deep learning models, we process the data source using a small sliding window $\mathbf{X}_{in} = (\mathbf{x}_j, \mathbf{x}_{j+1}, \ldots, \mathbf{x}_{j+T_{in}})$ of length $T_{in} < T$ as highlighted in Fig. 14.2.

Moreover, the observation $\mathbf{Y} = (\mathbf{x}_{j+T_{in}+1}, \ldots, \mathbf{x}_{j+T_{int}+T_{out}})$ represents the corresponding feature sequence of length $T_{out} < T$ we are aiming to forecast using the desired model

$$g_\theta : \mathbb{R}^{N \times T_{in}} \mapsto \mathbb{R}^{N \times B \times T_{out}}, \tag{14.6}$$

$$g_\theta(\mathbf{X}_{in}) \mapsto \hat{\mathbf{Y}}. \tag{14.7}$$

The model parameters $\theta$ are optimised on a training set that we assume contains mainly normal instances.

During inference, we generate $B$ predictions $\hat{Y}_{b,j}$ while applying MC or variational dropout [10] with probability $p \in [0, 1]$ to all $L$ layers of the network $g_\theta$. Each sampled prediction $\hat{\mathbf{Y}}_b$ corresponds to the sampled model configuration $\theta_b = \{\Theta_i\}_{i=1}^L$. In our BNN architecture, we apply variational dropout on as long short-term memory (LSTM)-layers and Monte Carlo dropout on non-recurrent layer structures such as multilayer perceptrons (MLPs). Here, we make the same

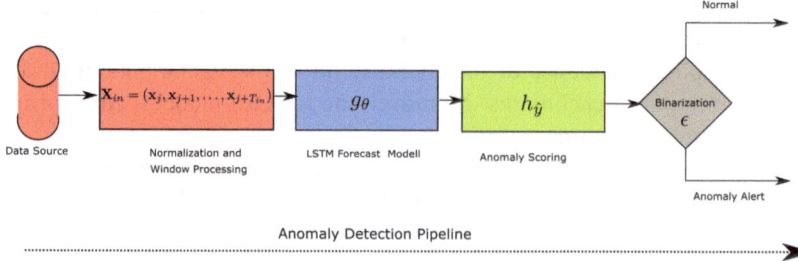

**Fig. 14.1** The multivariate anomaly detection pipeline consists of a data source, data processing unit, a multivariate forecasting model, anomaly score function $h_{\hat{y}}$ and a binarisation operation using a threshold $\epsilon$

## Window Processing

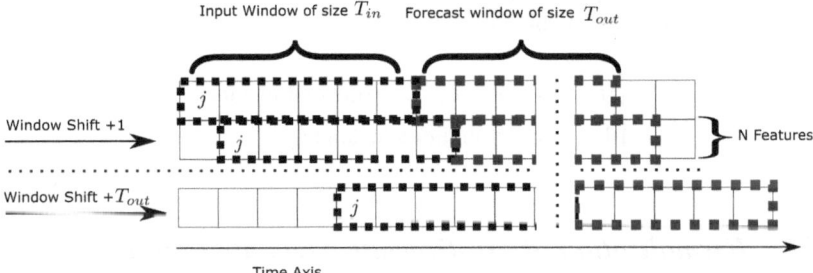

**Fig. 14.2** Time series processing using a sliding window consisting of the input $T_{in}$ and output sequence length $T_{out}$

assumption as made in [30] that parameters follow an isotropic Gaussian prior $\theta \sim \mathcal{N}(0, 1)$. Considering the given prediction ensemble $\hat{\mathbf{Y}}_i \in \mathbb{R}^{N \times B}$ at time step $i$, where $\hat{\mathbf{y}}_i^{(j)}$ denotes the vector of feature $j$. We are assuming that the model output follows a Gaussian distribution

$$\mathbf{Y}_i | \theta \sim \mathcal{N}(\boldsymbol{\mu}_i, \boldsymbol{\Sigma}_i) \tag{14.8}$$

with mean $\boldsymbol{\mu}_i \in \mathbb{R}^N$ and a covariance matrix $\boldsymbol{\Sigma}_i \in \mathbb{R}^{N \times N}$. We approximate the expected value of time $i$ and feature $j$ as

$$E(\mathbf{y}_i^{(j)} | \theta, x_i^{(j)}) \approx \hat{\mu}_i^{(j)} = \frac{1}{B} \sum_{b=1}^{B} \hat{y}_{ib}^{(j)}, \tag{14.9}$$

where $\hat{\boldsymbol{\mu}}_i = (\hat{\mu}_i^{(1)}, \dots, \hat{\mu}_i^{(N)})$ represents the prediction mean vector of all features.

The expected covariance of two given feature predictions $\hat{\mathbf{y}}_i^{(j)}$ and $\hat{\mathbf{y}}_i^{(k)}$ is approximated by the sample covariance matrix $\hat{\boldsymbol{\Sigma}}_i \in \mathbb{R}^{N \times N}$ with entries

$$\hat{\Sigma}_{i,jk} = \frac{1}{B-1} \sum_{b=1}^{B} \left( \hat{y}_{ib}^{(j)} - \hat{\mu}_i^{(j)} \right) \left( \hat{y}_{ib}^{(k)} - \hat{\mu}_i^{(k)} \right) \tag{14.10}$$

and $B - 1$ degrees of freedom. Furthermore, we utilise the scoring function

$$h_{\hat{y}} : \mathbb{R}^{N \times B \times T_{out}} \times \mathbb{R}^{N \times T_{out}} \mapsto \mathbb{R}, \tag{14.11}$$

$$h_{\hat{y}}(\hat{\mathbf{Y}}, \mathbf{Y}) \mapsto a_{j+T_{in}+T_{out}} \tag{14.12}$$

that measures the dissimilarity between the prediction $\hat{\mathbf{Y}}$ and observation $\mathbf{Y}$. The larger the anomaly score $h_{\hat{y}}$ is, the higher the probability that time step $j + T_{in} + T_{out}$

is anomalous. One fundamental assumption in our approach is that each computed anomaly score $a_{j+T_{in}+T_{out}}$ of time step $j + T_{in} + T_{out}$ can be separated by one predefined scalar threshold value $\epsilon \in \mathbb{R}$:

$$D_\epsilon(a_i) = \begin{cases} 0 & \text{if } a_{j+T_{in}+T_{out}} < \epsilon \text{ (normal instance)} \\ 1 & \text{else (anomalous instance).} \end{cases} \qquad (14.13)$$

### 14.3.2  Formalization of Multivariate Anomaly Detection

We can formulate the problem of multivariate AD as finding a configuration $(g_\theta, h_{\hat{y}}, \epsilon)$ that distinguishes normal from abnormal instances as accurately as possible. To differentiate between normal and abnormal patterns in time series in an unsupervised fashion, we require a model $g_\theta$ that needs to capture the standard representation of multivariate time series. Solely reconstruction-based approaches such as VAEs [22, 26] require strong knowledge about the structure of the latent space and model optimisation for different AD problems. Furthermore, these approaches focus heavily on the task of AD itself and are inapplicable to other applications such as forecasting. In contrast, BNN forecast models are more versatile as they learn a compressed latent space, forecast future values, quantify the uncertainty along with the prediction and detect anomalies. We propose to use multi-step predictions with a window size $T_{out} > 1$ to capture point and collective anomalies.

The proposed architecture is highlighted in Fig. 14.3, which uses a combination of AEs and LSTMs. The purpose of the AE is to learn a compact embedding $\mathbf{z}$ of the original input $\mathbf{X}$. In particular, the decoder $g_{\theta_{dec}}$ forces the encoder $g_{\theta_{enc}}$ to learn only those embeddings that the decoder can correctly reconstruct. Since our AD solely relies on the forecasted prediction ensemble from the LSTMs, the decoder $g_{\theta_{dec}}$ is deactivated during inference.

### 14.3.3  Anomaly Scoring

We investigate anomaly score functions $h_{\hat{y}}$ based on epistemic uncertainty. Epistemic uncertainty usually arises from examples that have never or rarely been observed during model optimisation. However, the uncertainty is reduced as the number of training examples increases. Hence, we incorporate epistemic uncertainties into the discrepancy measurement between forecasted and observed instances. The core idea is that low quantified uncertainties penalise discrepancies by a more significant weight, while large uncertainties suppress discrepancies. The intuition of the anomaly score function $h_{\hat{y}}$ is to measure the differences between $\mathbf{Y}_i$ and $\hat{\mathbf{Y}}_i$ that reveals whether the time step $i$ is likely to be an anomaly or not. We present two

**Fig. 14.3** Multivariate forecasting model based on LSTM-based auto-encoder architecture

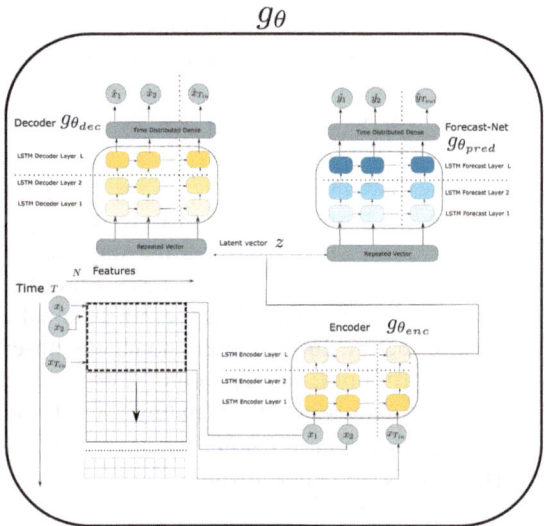

ways to measure these discrepancies: In the first case, where we suggest the $\chi^2$-distance $\mathrm{D}_{\chi^2}$ is used as anomaly score, no correlations between the input variables are taken into account and in the second case, where the Hotelling's $T^2$-distance $\mathrm{D}_{T^2}$ is used as anomaly score, the full correlation matrix $\hat{\boldsymbol{\Sigma}}$ is considered. The larger the anomaly score $h_{\hat{y}}$ is, the higher the probability that time step $i$ is anomalous. This is why we employ hypothesis tests [7, 17] to construct appropriate anomaly score functions, where we define our null hypothesis as

$$H_0 : \hat{\boldsymbol{\mu}}_i = \mathbf{Y}_i, \tag{14.14}$$

where the prediction ensemble mean $\hat{\boldsymbol{\mu}}_i$ is likely to resemble the observation $\mathbf{Y}_i$ if the time step is normal. Vice versa, the alternative hypothesis

$$H_1 : \hat{\boldsymbol{\mu}}_i \neq \mathbf{Y}_i \tag{14.15}$$

states that the prediction ensemble $\hat{\boldsymbol{\mu}}_i$ distinguishes significantly from the observation $\mathbf{Y}_i$ and is therefore likely to be an anomaly. A naïve construction is to assume that each of the $N$ features is independently Gaussian distributed random variables with known means $\boldsymbol{\mu}_i \in \mathbb{R}^N$ and variances $\boldsymbol{\sigma}_i \in \mathbb{R}^N$. In this scenario, the distance $\mathrm{D}_{\chi^2}(\boldsymbol{\mu}_i, \mathbf{Y}_i)^2$

$$\mathrm{D}_{\chi^2}(\boldsymbol{\mu}_i, \mathbf{Y}_i)^2 = \sum_{j=1}^{N} \left( \frac{\mu_i^{(j)} - y_i^{(j)}}{\sigma_i^{(j)}} \right)^2 \tag{14.16}$$

follows a $\chi^2$-distribution [7] with $N$ degrees of freedom and we reject the null hypothesis $H_0$ when the computed statistic

$$\text{Reject } H_0 \text{ if } \chi^2_{N,1-\alpha} < D_{\chi^2}(\mu_i, Y_i)^2 \text{ (anomaly)} \qquad (14.17)$$

exceeds the quantile function $\chi^2_{N,1-\alpha}$ with $N$ degrees of freedom. The Chi-Squared $D_{\chi^2}$ distance is particularly sensitive to strongly mean-shifted statistics. This test is likely to fail in a situation where the variables of the observation $Y_i$ resemble the one represented by the predicted mean. Still, the features differ in the relationship of the variables established in the off-diagonal entries $\Sigma_{i,jk}$ within ($j \neq k$) the covariance matrix. In this case, anomalous observations will likely fall below the $\chi^2_{N,1-\alpha}$ control limit, and $Y_i$ would be considered normal observation erroneously. This can be addressed by extending the Chi-Squared $D_{\chi^2}$ distance by the non-diagonal entries of the covariance matrix. This extension of the $\chi^2$-statistic is called the Hotelling's $T^2$ [11] statistic

$$T_i^2 = (\hat{\mu}_i - Y_i)^\top \hat{\Sigma}^{-1} (\hat{\mu}_i - Y_i), \qquad (14.18)$$

where the scaled

$$D_{T^2}(\mu_i, \hat{\Sigma}, Y_i)^2 = T_{i,N,B-N}^2 = \frac{(B-N)}{N(B-1)} T_i^2 \sim F_{N,B-N} \qquad (14.19)$$

statistic (anomaly score) follows the $F_{N,B-N}$ distribution with $B-N$ degrees of freedom. We reject the null hypothesis $H_0$ when $D_{T^2}$ exceeds the corresponding tabulated threshold $F_{N,B-N,1-\alpha}$ of the significance level $\alpha$:

$$\text{Reject } H_0 \text{ if } F_{N,B-N,1-\alpha} < T_{B,B-N}^2 \text{ (anomaly)}. \qquad (14.20)$$

Hence, the Hotelling's $T^2$ test addresses this problem by being sensitive to mean-shift and counter-relationship anomalies. Vice versa, the Chi-Squared $D_{\chi^2}$ distance is likely to perform better than the Hotelling's $T^2$ when the observation $Y_i$ includes solely mean-shift anomalies.

### 14.3.4  Anomaly Threshold Fitting

After transforming uncertainty estimates into an anomaly score, the final binarisation function $D_\epsilon$ depends on the threshold $\epsilon$ as well and needs to be specified along with the configuration parameters $g_\theta$ and $h_{\hat{y}}$. Figure 14.4 is a rough orientation example that visualises how the score collection is distributed. Our central assumption is that the output of the anomaly score function $h_{\hat{y}}$ can be distinguished by a threshold $\epsilon$ in two regions: The left region (low anomaly scores) corresponds

**Fig. 14.4** Anomaly score distribution generated on an independent validation set $\mathcal{D}_{(val)} = (\mathcal{X}_{val}, \mathcal{Y}_{val})$: The quantile function $\epsilon = Q(1 - \alpha)$ selects the $1 - \alpha$ quantile of the collected anomaly scores $A_{val}$. The vertical dashed line $\epsilon$ corresponds to the $1 - \alpha = 0.9$ quantile

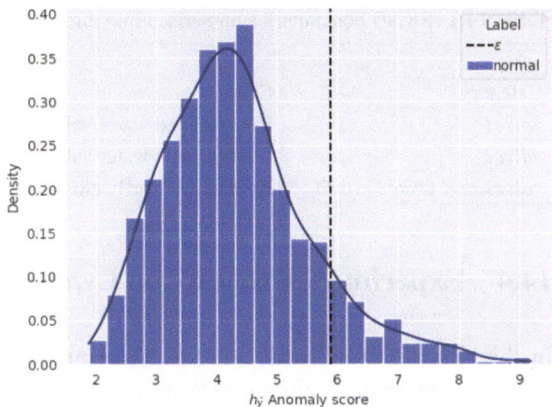

to normal instances and the right section (high anomaly scores) to anomalous cases. Determining an appropriate threshold $\epsilon$ in an unsupervised setting with unlabelled data is a non-trivial task due to the unknown anomaly score distribution. An effective strategy to specify the threshold is to collect all the anomaly scores $A_{val}$ using an independent validation set $\mathcal{D}_{(val)} = (\mathcal{X}_{val}, \mathcal{Y}_{val})$ where we assume that it consists mainly of normal examples. In the following, we consider a cumulative distribution of normal anomaly scores $A_{val}$:

$$CDF_{A_{val}} : \mathbb{R} \mapsto [0, 1], \tag{14.21}$$

$$CDF_{A_{val}}(\epsilon) = P(A_{val} \leq \epsilon) = 1 - \alpha \tag{14.22}$$

The quantile function

$$Q : [0, 1] \mapsto \mathbb{R}, \tag{14.23}$$

$$Q(1 - \alpha) = \inf \left\{ \epsilon \in \mathbb{R} : 1 - \alpha \leq CDF_{A_{val}}(\epsilon) \right\} \tag{14.24}$$

returns a threshold $\epsilon$ where the probability that random draws of normal scores fall below the control limit is $1 - \alpha$. Here, $\alpha$ serves as well as a hyperparameter and represents the probability that a normal instance is considered as anomaly $(D_\epsilon(a_i)) = 1$, given the instance, is normal. A naïve approach for choosing a threshold candidate $\epsilon$ is to set $\epsilon = \max(A_{val})$ as the transition between the scoring area of anomalous and normal instances are usually overlapping. In practice, we suspect that the discrepancies between the observation $\mathbf{Y}_i$ and prediction mean $\hat{\boldsymbol{\mu}}_i$ are relatively large when dealing with high inherent noise. The calculated $\chi^2$ and Hotelling's $T^2$ statistics associated with normal instances are likely to exceed the upper quantiles, respectively. Hence, we suggest computing an upper control limit $\epsilon$ using the quantile function in (14.24) for both statistics.

**Table 14.1** SKAB benchmark summarised into three categories

| Category | Cause | Number of instances | | |
| --- | --- | --- | --- | --- |
| | | Normal | Anomaly | Total |
| *valve1* | Partly closed water valve1 | 11,853 | 6309 | 18,162 |
| *valve2* | Partly closed water valve2 | 2795 | 1517 | 4312 |
| *valve1 and valve2 mixed* | Both valves partly closed | 24,398 | 7826 | 22,474 |

## 14.4 Experimental Setup and Evaluation

In this section, we describe the experimental setup that we use to compare our proposed uncertainty driven method against the state-of-the-art methods such as USAD [1], OmniAnomaly [26] and DAGMM [31].

### 14.4.1 Skoltech Anomaly Benchmark Data Set

We perform our experiments on the publicly available benchmark[1] Skoltech Anomaly Benchmark (SKAB) ver.0.9 data set provided by [15, 16]. It is specifically designed for evaluating multivariate AD algorithms in the field of faults and failures of technical systems. The test platform is a simple water circulation platform that captures eight sensor time series measurements. We summarised the benchmark into three main experiment categories as shown in Table 14.1 and trained a new model for each category.

Figure 14.5a illustrates a normalised time series of the training set describing the healthy state of the water platform, where the test set shown in Fig. 14.5b contains measurements that are associated with collective anomalies or normal instances. The root cause of the anomaly is that specific controls, such as water valves, are modified for a short duration, and it simulates unhealthy states during sensor recordings. For any further technical details, we refer to the website of this project.[1] We split the data of each experiment category into a train, validation, and test set. Based on the SKAB publisher's suggestions, we utilise the first $\sim$33% of data points for the model training and anomaly calibration. Moreover, we further split the optimisation set into a training set $\mathcal{D} = (\mathcal{X}_{train}, \mathcal{Y}_{train})$ and a validation set $\mathcal{D} = (\mathcal{X}_{val}, \mathcal{Y}_{val})$. The training set contains 85% and the validation set 15% consecutive samples of the optimisation set. Each experiment is normalised separately by min-max feature scaling. Afterwards, we determine the parameters $\mu_{train}$ and $\sigma_{train}$ to standardise all train, validation and test sets.

---

[1] https://github.com/waico/SkAB.

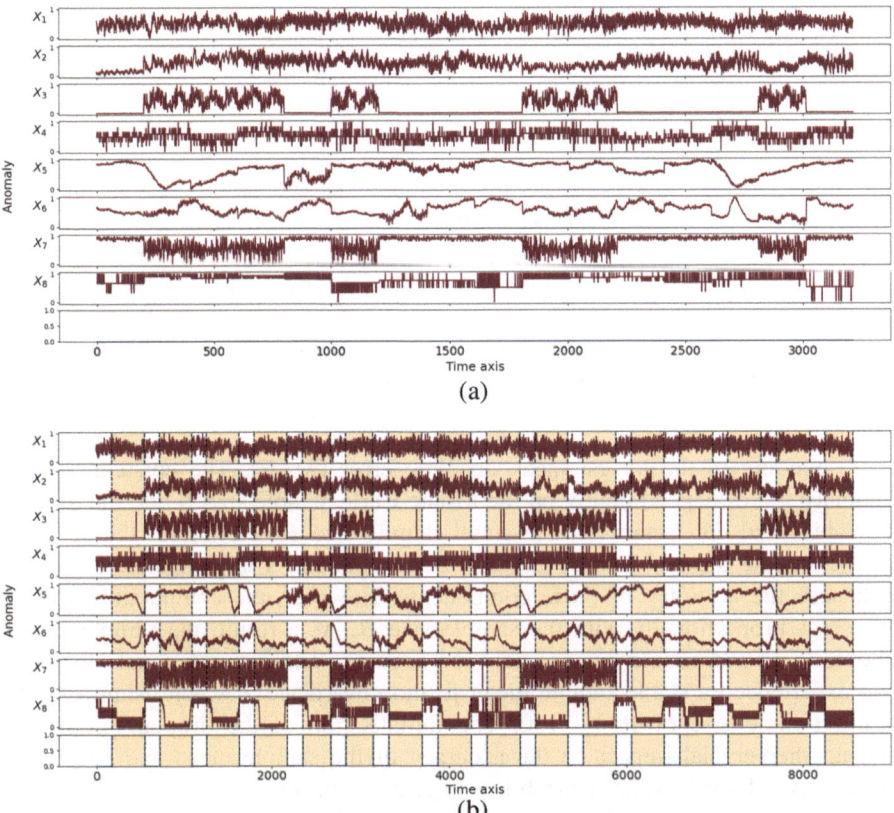

**Fig. 14.5** Multivariate time series of the SKAB anomaly detection benchmark: The time series illustrates the healthy characteristics of the water circulation platform shown in (**a**), taking into account that the water valve *valve1* is open. This time series consists solely of sensor measurements (red), and the control states, such as the water valves that cause anomalies, are not represented. In contrast, (**b**) illustrates time intervals with anomalous sensor measurements caused by partially closed water valves. The intervals are indicated by the orange bars and the anomaly label in the last row. (**a**) Train set: Describes the healthy state of the water circulation platform. The BNN $g_\theta$ uses sliding input windows $\mathbf{X}_{in} \in \mathbb{R}^{8 \times T_{in}}$ to learn to forecast normal feature windows $\hat{\mathbf{Y}} \in \mathbb{R}^{8 \times T_{out}}$. (**b**) Test set: Consists of healthy and unhealthy feature characteristics. We solely utilise this subset in our experiments to assess the AD performances

## 14.4.2 Experimental Hyperparameters

The following experimental results are based on the configuration $(g_\theta, h_{\hat{y}}, \epsilon)$ of hyperparameters shown in Table 14.2. The first two segments represent the hyperparameter of the model $g_\theta$ that impacts the forecasting accuracy, the third segment impacts the anomaly score function $h_{\hat{y}}$, and the last line affects the threshold parameter $\epsilon$. For the model $g_\theta$, we select the best hyperparameter combination

**Table 14.2** Overview of utilised model hyperparameters during evaluation

| Component | Hyperparameters | Symbols | Value |
|---|---|---|---|
| $g_\theta$ | Batch size | $b$ | 40 |
| | Learning rate | $\eta$ | 0.001 |
| | Decay rate | $\gamma$ | 0.96 |
| | Loss function | $\mathcal{L}$ | MSE |
| $g_{\theta_{enc}}$ | Number of encoder hidden units | $H_{enc}$ | 128, 64 |
| $g_{\theta_{dec}}$ | Number of decoder hidden units | $H_{dec}$ | 64, 128 |
| $g_{\theta_{pred}}$ | Number of forecast-net hidden units | $H_{pred}$ | 64, 128 |
| $h_{\hat{y}}$ | Dropout probability | $p$ | 0.2 |
| | Monte Carlo sample size | $B$ | 100 |
| $\epsilon$ | Significance level | $\alpha$ | 0.05 |

corresponding to the lowest mean absolute error (MAE). The epistemic-driven anomaly scores depend highly on the Monte Carlo sample size $B$ and the dropout probability $p$. For instance, a too-small sample size $B$ deteriorates the accuracy of the sample mean $\hat{\mu}$ and covariance $\hat{\Sigma}$. In contrast, a large sample size of $B$ enhances the accuracy of those components. However, the computational effort increases proportionally with the sample size $B$. As a result, the sample size $B$ should yield a desirable balance between accuracy and computational efficiency. We experienced a stable sample mean $\hat{\mu}$ and covariance $\hat{\Sigma}$ when using a Monte Carlo sample size of $B \geq 100$. The initialisation of the dropout probability has another crucial impact on the accuracy of the sample mean and covariance matrix. A relatively small probability (e.g., $p < 0.01$) corresponds to a sample covariance matrix where the diagonal entries are close to zero. This is undesirable as the covariance matrix is prone to have not the full rank and is therefore not invertible. On the contrary, the variance increases and the prediction sample mean is vulnerable to becoming inaccurate the larger we set the dropout probability $p$. We experienced that the dropout probability in the range between $p \in \{0.1, 0.3\}$ provides an accurate prediction and a stable covariance matrix in most experiments. As a result, we utilise the median ($p = 0.2$) of that range. The significance level $\alpha$ substantially impacts AD as the selected threshold is based on the quantile function $\epsilon = Q(1 - \alpha)$. This function determines the $1 - \alpha$ quantile of the collected anomaly scores from the independent dataset $\mathcal{D}_{val} = (\mathcal{X}_{val}, \mathcal{Y}_{val})$. The inverse probability $1 - \alpha$ denotes our confidence that a true normal instance will fall below the threshold. A confidence level of 100% in the context of AD means that we assume that there is no intersection between the anomaly scores of normal and anomalous instances. Nevertheless, time series often occur continuously, and the transition from normal to anomalous observation is usually seamless. Hence, an intersection of the anomaly scores between those classes is unavoidable. For the following experiments, we utilise a 95% confidence level ($\alpha = 0.05$) as proposed by [30].

### 14.4.3   Evaluation Metrics

We evaluate the performance of our AD-approaches regarding various metrics, each covering different aspects. Further, we utilise binary classification metrics such as the false alarm rate (FAR)

$$FAR = \frac{FP}{FP + TN} \ \text{(lower is better)} \ (\downarrow) \tag{14.25}$$

which is also known as the false positive rate (FPR) and represents the expectancy of the false positive (FP) ratio. The missing alarm rate (MAR)

$$MAR = \underbrace{\frac{FN}{TP + FN}}_{1-\text{recall}} \ \text{(lower is better)} \ (\downarrow) \tag{14.26}$$

describes the inverse of the recall as the expectancy of the false negative (FN) ratio. In addition, we report the precision, recall and the $F_1$-score that are crucial to assess the AD algorithm performances. As the final AD heavily relies on the unsupervised selected threshold, we additionally access the separability of the anomaly score by analysing the area under the curve (AUC) of the receiver operating characteristics (ROC). A good balance between relatively low FAR and MAR is desirable as AD algorithms are used as an early warning system in industrial applications. End-users distrust the warning systems and ignore reported alerts when the FAR is relatively large. Vice versa, an AD detector with low sensitivity is unlikely to provide early warnings of unhealthy or atypical events that could lead to a disaster.

### 14.4.4   Discussion of Utilized Anomaly Detection Metrics

The AD performance depends heavily on the BNN $g_\theta$, the discriminator function $h_{\hat{y}}$ and the threshold $\epsilon$ determined in an unsupervised manner. The selected evaluation metrics are mainly inspired by works that address multivariate AD. Su et al. [26] argue that the AD is already correct if an alert for anomalies is triggered within any subset of a ground truth anomaly segment. In their evaluation, they propose a so-called point adjustment of their anomaly detection. It overwrites the predictions of an entire sequence with the ground truth anomaly information when an arbitrary sample and the corresponding ground truth label are associated with an anomaly alert. Based on the prediction adjustment, they determine the precision, recall, and $F_1$ as AD metrics. The key problem with this evaluation technique is that the ground truth information for unseen samples is usually unknown in practical applications.

Furthermore, their modifications distort the evaluation results as the ground truth information overwrites many samples. Hence, in our entire evaluation, we turned off the point adjustment for their VAE called OmniAnomaly.

## 14.4.5 Experimental Results and Analysis

This section shows experimental results on the forecasting and AD performances on multivariate time series. We assess the AD performance of the BNN and the proposed anomaly score $h_{\hat{y}}$ on the SKAB dataset and contrast our AD performance against recent state-of-the art methods such as DAGMM[2] (AE + Gaussian Mixture model (GMM)) [31], USAD[3] (GAN inspired AE network) [1] and Omni-Anomaly[4] (VAE) [26]. The anomaly score distribution of DAGMM represents the density estimation under the framework of GMM, and considerable energies correspond to anomalies while USAD outputs an anomaly score. It is based on the reconstruction errors of a GAN inspired network that incorporates two AE. OmniAnomaly outputs a reconstruction probability that indicates how likely a given input can be decoded. Therefore, high reconstruction probabilities correspond to normal instances and low probabilities to anomalies. Figure 14.6 illustrates the anomaly score distribution of the mixed categories *valve1* and *valve2*, respectively. Here we utilise each experiment's first ~33% time steps for both categories as a training set, and the rest ~ 66% as an evaluation set described in Sect. 14.4.1. We observe that epistemic-driven uncertainty methods such as Hotelling's $T^2$ (Fig. 14.6a) that incorporate the full covariance matrix and its correlation leads to a desirable separability. In particular, it shows the highest separability between the normal and anomalous bell curves with an AUC score of 0.899, followed by the Chi-Squared $D_{\chi^2}$ (Fig. 14.6b) yielding an AUC score of 0.759. We can explain this result by the fact that the Hotelling's $T^2$ considers the relationship between variables, whereas, in contrast, the Chi-Squared $D_{\chi^2}$ approach assumes that the variables are independently distributed.

In the following, we analyse the anomaly scores and the separability of DAGMM (Fig. 14.7a), USAD (Fig. 14.7b) and OmniAnomaly (Fig. 14.7c). OmniAnomaly shows poor separability by yielding an AUC of 0.341 that is worse compared to the AUC = 0.5 of a random classifier. The anomaly and the normal bell curves mostly overlap; hence, an accurate separation between those classes is impossible. On the other hand, USAD anomaly scores turn out to be robust as its reconstruction errors are separable by showing a large AUC of 0.878. Nevertheless, the BNN, in combination with the Hotelling's $T^2$ method, outperforms all evaluated AD algorithms in terms of separability.

---

[2] https://github.com/danieltan07/dagmm.

[3] https://github.com/manigalati/usad.

[4] https://github.com/NetManAIOps/OmniAnomaly.

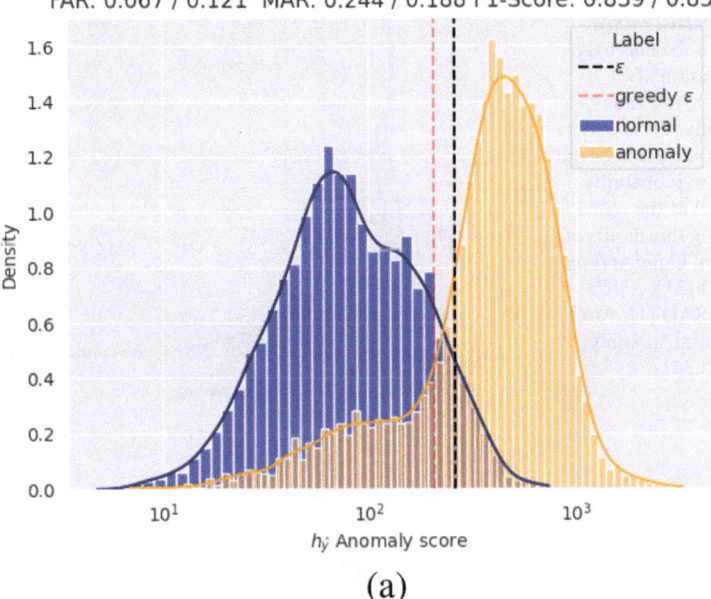

(a)

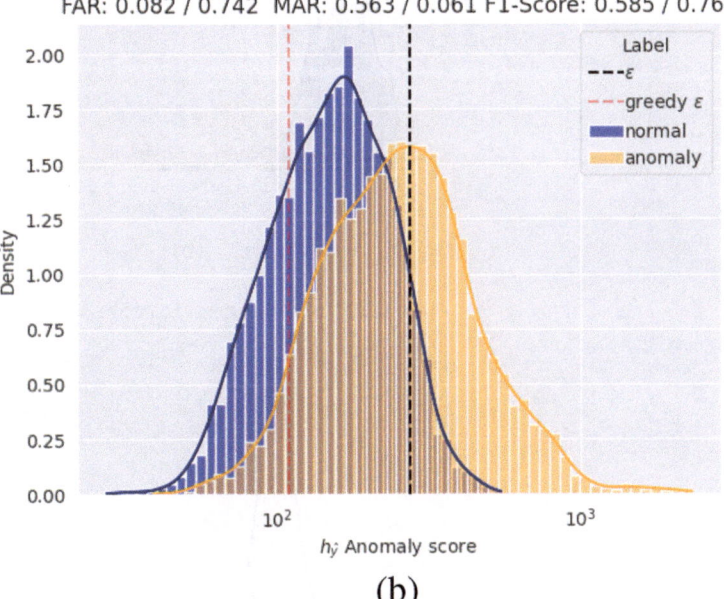

(b)

**Fig. 14.6  SKAB category *valve1 and valve2 mixed*:** This visualisation shows the anomaly score distributions of the SKAB time series considering *valve1 and valve2 mixed*. The first score in the headings ($\epsilon$/greedy$\epsilon$) corresponds to the fitted threshold $\epsilon$ with significance level $\alpha = 0.05$ represented by the black dashed line. The second score is associated with the threshold (red dashed line) that yields the highest possible $F_1$ considering all threshold candidates. (**a**) Hotelling's $T^2$, AUC $= \mathbf{0.899}$ . (**b**) Chi-Squared $D_{\chi^2}$, AUC $= 0.759$

**Fig. 14.7 SKAB category**
***valve1 and valve2 mixed***:
Illustration of the anomaly
score distribution of
DAGMM, USAD, and
OmniAnomaly: For a better
comparability, we mirror the
reconstruction probability
scores as well as the
corresponding thresholds of
the VAE from OmniAnomaly.
(**a**) DAGMM [31], AUC:
0.652. (**b**) USAD [1], AUC:
0.878. (**c**) OmniAnomaly
[26], AUC: 0.341

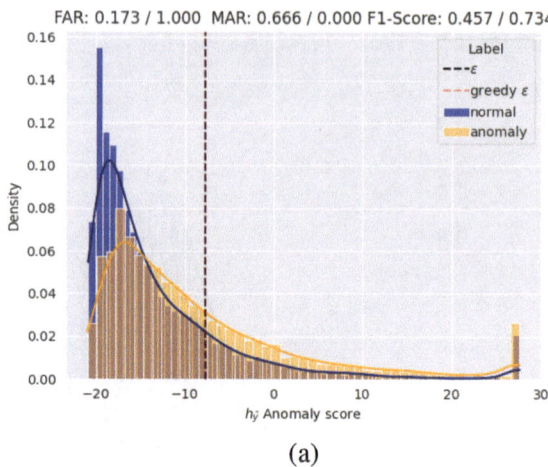

(a)

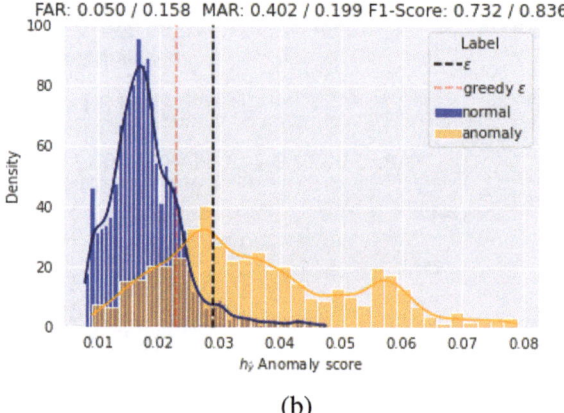

(b)

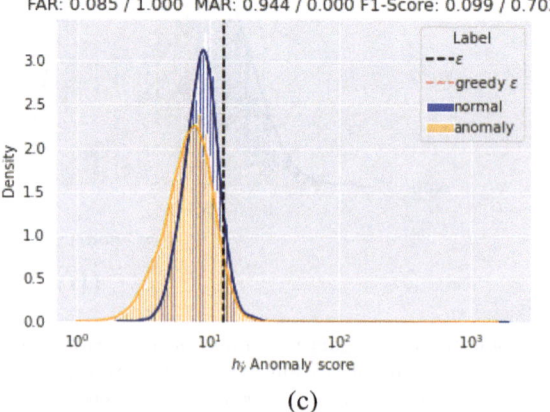

(c)

**Table 14.3** Multivariate AD results based on the SKAB categories *valve1 and valve2 mixed*

| Metric | Prec. (↑) | Rec. (↑) | FAR (↓) | MAR (↓) | F₁ (↑) | AUC (↑) |
|---|---|---|---|---|---|---|
| Hotelling's $T^2$ + Q [13] | **0.999** | 0.468 | **0.000** | 0.532 | 0.637 | 0.882 |
| Isolation Forest [19] | 0.793 | 0.006 | 0.002 | 0.994 | 0.012 | \ |
| AutoEncoder [6] | 0.995 | 0.214 | 0.001 | 0.786 | 0.353 | 0.155 |
| Omni-Anomaly [26] | 0.430 | 0.084 | 0.132 | 0.916 | 0.141 | 0.341 |
| USAD [1] | 0.942 | 0.598 | 0.050 | 0.402 | 0.732 | 0.878 |
| DAGMM [31] | 0.726 | 0.334 | 0.173 | 0.666 | 0.457 | 0.652 |
| BNN + Hotelling's $T^2$ | 0.942 | **0.756** | 0.067 | **0.244** | **0.839** | **0.899** |
| BNN + Chi-Squared $D_{\chi^2}$ | 0.884 | 0.437 | 0.082 | 0.563 | 0.585 | 0.759 |

The bold values should highlight the relevant best values in the respective category

### 14.4.5.1  Anomaly Detection Analysis

Table 14.3 summarizes the AD results evaluated on the combined SKAB *valve1* and *valve2* dataset. In addition to the mentioned AD algorithms, we evaluate AD implementations like AE [6], Isolation Forest [19] and Hotelling's $T^2$ + Q [13] proposed for the SKAB benchmark. Moreover, the other results correspond to DAGMM, USAD, OmniAnomaly and the BNN with its anomaly scoring functions. The Hotelling's $T^2$ + Q method demonstrates the lowest FAR and largest precision. Nevertheless, it turns out that it ignores more than half of all anomalous time steps. The reason for the imbalanced precision and recall values is that the algorithm selects too large thresholds and therefore ignores a high proportion of anomalies.

As OmniAnomaly and DAGMM show low separability for their anomaly scores, their corresponding AD results in undesirably low F₁ scores. Furthermore, the anomaly alerts of USAD are relatively reliable due to its low FAR. However, it ignores ∼40% of all anomalies. On the contrary, the BNN in combination with the Hotelling's $T^2$ outperforms all other methods concerning the recall, MAR, FAR, F₁ and AUC metric. Another significant observation is that AUC of the BNN driven methods increase proportionally to the F₁ score. In particular, the visualised anomaly score distribution in Fig. 14.6 illustrates that the unsupervised selected threshold is especially close to the best possible threshold when the distribution yields a high separability. Hence, further enhancements in the separability of the anomaly score functions are likely to correspond to an improvement in AD.

## 14.5  Discussion of Experimental Results

The following sections discuss the proposed anomaly scores and threshold approaches. We start by reasoning why the proposed quantile-based threshold method is more appropriate than the tabulated control limits from the $\chi_N^2$ and $F_{N,B-N}$ distribution, the AD performance achieved on the SKAB dataset in greater detail. Compared to the state-of-the-art, we contextualise the observed AD performance of the BNN.

### 14.5.1 Quantile Based Threshold Versus Tabulated Control Limits

The threshold selection method significantly impacts the AD accuracy as it separates anomaly scores into two classes. In our concept, we hypothesised that thresholds which solely depend on the tabulated control limits are likely to correspond to unfavourable AD accuracy. For instance, the control limits generated from the $F_{N,B-N}$ distribution depend on the degree of freedom $B - N$, the number of features $N$, and the significance level $\alpha$. One problem with this threshold selection method is the assumption that the difference between $\hat{\mu}_i - Y_i$ is approximately zero for each normal time step. We argue that time series include components with irreducible inherent noise that also occurs in normal data; hence, the residuals can be large even for non-anomalous instances. However, the generated control limits are independent of such factors. Consequently, we suggest collecting the uncertainty-driven anomaly scores on an independent validation set where we assume it contains mostly normal instances. Afterwards, we determine the threshold $\epsilon = Q(1 - \alpha)$ corresponding to the $1 - \alpha$ quantile. To validate our described concerns, we contrast in Fig. 14.8 the $F_{8,92}$ distribution against the scaled Hotelling's $T^2_{8,92}$ distribution generated on the SKAB validations set ($N = 8$, $B = 100$, $\alpha = 0.05$).

We recognise that the shape of both bell curves illustrated in Fig. 14.8a and b are similar. Nevertheless, both distributions do not share a common scale. Consequently, most Hotelling's $T^2_{8,92}$ anomaly scores exceed the tabulated control limit $F_{8,92,0.95} \approx 2$ of the $F_{8,92}$ distribution. It confirms our concerns that the tabulated control limit $F_{8,92,0.95}$ is unsuitable for being used as a threshold in AD. In the following, we consider the $\chi^2_{8,92}$ probability density function and the Chi-Squared $D_{\chi^2}$ anomaly score distribution as illustrated in Fig. 14.9. Notably, neither the shape of the bell curves nor the scale of both distributions resembles. In this example, most Chi-Squared $D_{\chi^2}$ anomaly scores corresponding to normal instances exceed the control limit $\chi^2_{8,0.95} \approx 15$. Similar to the control limits generated from the $F_{8,92,0.95}$ distribution, we can conclude that the threshold $\chi^2_{8,0.95}$ is not suitable for separating Chi-Squared $D_{\chi^2}$ anomaly scores into normal and anomalous regions accurately. In contrast, the threshold selection method that is based on the quantile function $\epsilon = Q(0.95)$ is close to the best possible threshold and corresponds to accurate AD results as shown in Fig. 14.6.

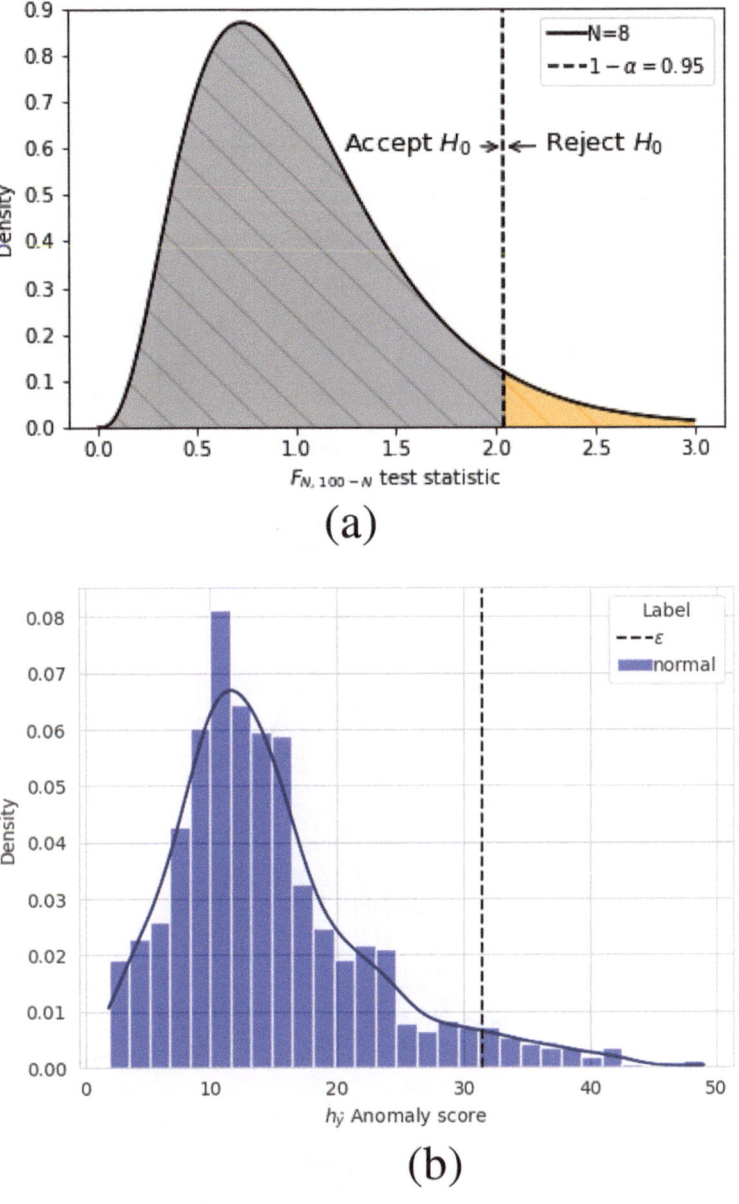

**Fig. 14.8** Comparison between the probability density function $F_{8,92}$ (left) and the Hotelling's $T^2_{8,92}$ distribution (right): (**a**) illustrates the control limit $F_{8,92,0.95}$ and (**b**) the quantile based threshold $\epsilon = Q(0.95)$ as a vertical dashed line. Both values correspond to the $1 - \alpha = 0.95$ confidence level. (**a**) Probability density function of $F_{8,92}$. (**b**) Hotelling's $T^2_{8,92}$ distribution

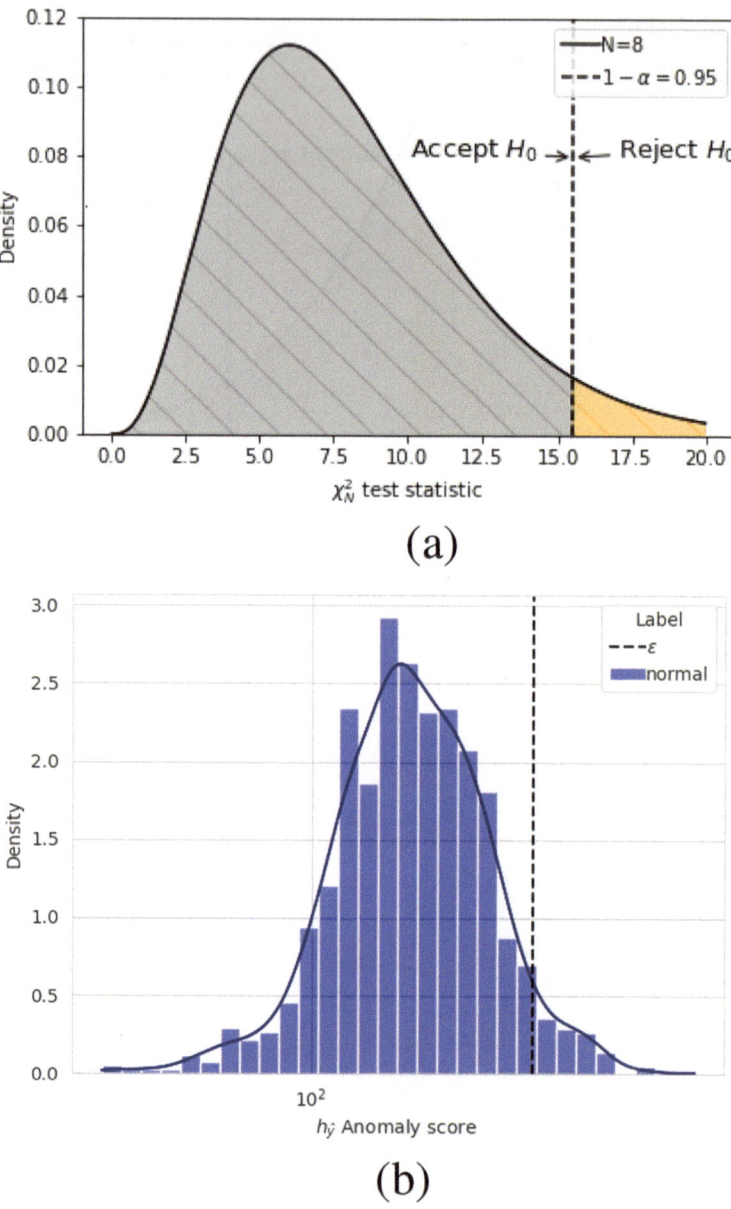

**Fig. 14.9** Comparison between the probability density function $\chi_8^2$ (left) and the Chi-Squared $D_{\chi^2}$ anomaly score distribution (right): (**a**) shows the control limit $\chi_{8,0.95}^2$ and (**b**) the quantile based threshold $\epsilon = Q(0.95)$. Both values correspond to the $1 - \alpha = 0.95$ confidence level. (**a**) Probability density function of $\chi_8^2$. (**b**) Chi-Squared $D_{\chi^2}$ distribution

### 14.5.2 Competitiveness to Recent Work

The estimated densities of DAGMM demonstrate high separability on the SKAB dataset. A possible explanation is that DAGMM ignores the temporal relationship between the inputs. With OmniAnomaly, we observe that the resulting reconstruction probabilities are not sufficiently distinguishable and are consequently associated with a low AD accuracy. We suspect that OmniAnomaly achieves its superior AD accuracy mainly when the point-adjustment method is applied as described in Sect. 14.4.4. Moreover, the GAN-inspired AE architecture USAD outputs a reconstruction error that demonstrates high separability on the SKAB dataset. Furthermore, USAD outperforms DAGMM and OmniAnomaly at AD in our experiments by yielding larger $F_1$ scores. The BNN in combination with the epistemic driven Hotelling's $T^2$ score surpasses DAGMM and USAD in terms of separability of the anomaly scores and AD. After this evaluation, we confirm that uncertainty-driven methods are competitive to existing AEs and VAEs to detect anomalies in multivariate time series.

## 14.6 Conclusion

This is the first study that addresses anomaly detection in multivariate time series and incorporates time series prediction and uncertainty estimation. This chapter examines how epistemic uncertainties, quantified using a Bayesian neural network, can be exploited to detect multivariate anomalies when combined with informed machine learning techniques, like incorporating knowledge sources. We demonstrate a Chi-Squared $D_{\chi^2}$ and Hotelling's $T^2$ approach that translates those epistemic uncertainties into an anomaly score. We propose a quantile-based threshold selection to separate anomaly scores into normal and anomalous regions. Our evaluation results show that the pairwise probabilistic relationship between variables in the Hotelling's $T^2$ corresponds to a better separability of anomaly scores and detection accuracy. Furthermore, the evaluation outlines the competitiveness of our method on multivariate sensor recordings captured from a water platform by comparing it against recent state-of-the-art approaches.

**Acknowledgments** This contribution was supported by the Fraunhofer Cluster of Excellence "Cognitive Internet Technologies".

## References

1. Audibert, J., Michiardi, P., Guyard, F., Marti, S., Zuluaga, M.A.: USAD: Unsupervised anomaly detection on multivariate time series. In: Proceedings of the 26th ACM SIGKDD International Conference on Knowledge Discovery & Data Mining, pp. 3395–3404 (2020)

2. Bishop, C.M.: Bayesian neural networks. Journal of the Brazilian Computer Society **4**(1) (1997)
3. Braei, M., Wagner, S.: Anomaly detection in univariate time-series: A survey on the state-of-the-art. arXiv preprint arXiv:2004.00433 (2020)
4. Chalapathy, R., Chawla, S.: Deep learning for anomaly detection: A survey. arXiv preprint arXiv:1901.03407 (2019)
5. Chandola, V., Banerjee, A., Kumar, V.: Anomaly detection: A survey. ACM computing surveys (CSUR) **41**(3), 1–58 (2009)
6. Chen, J., Sathe, S., Aggarwal, C., Turaga, D.: Outlier detection with autoencoder ensembles. In: Proceedings of the 2017 SIAM international conference on data mining, pp. 90–98. SIAM (2017)
7. Cowan, G.: Statistical data analysis. Oxford University Press (1998)
8. Gal, Y.: Uncertainty in deep learning. University of Cambridge **1**(3), 4 (2016)
9. Gal, Y., Ghahramani, Z.: Dropout as a bayesian approximation: Representing model uncertainty in deep learning. In: international conference on machine learning, pp. 1050–1059 (2016)
10. Gal, Y., Ghahramani, Z.: A theoretically grounded application of dropout in recurrent neural networks. Advances in neural information processing systems **29**, 1019–1027 (2016)
11. Hotelling, H.: The generalization of student's ratio. In: Breakthroughs in statistics, pp. 54–65. Springer (1992)
12. Hundman, K., Constantinou, V., Laporte, C., Colwell, I., Soderstrom, T.: Detecting spacecraft anomalies using LSTMs and nonparametric dynamic thresholding. In: Proceedings of the 24th ACM SIGKDD international conference on knowledge discovery & data mining, pp. 387–395 (2018)
13. Joe Qin, S.: Statistical process monitoring: basics and beyond. Journal of Chemometrics: A Journal of the Chemometrics Society **17**(8–9), 480–502 (2003)
14. Kamat, P., Sugandhi, R.: Anomaly detection for predictive maintenance in industry 4.0-a survey. In: E3S Web of Conferences, vol. 170, p. 02007. EDP Sciences (2020)
15. Katser, I., Kozitsin, V., Lobachev, V., Maksimov, I.: Unsupervised offline changepoint detection ensembles. Applied Sciences **11**(9), 4280 (2021)
16. Katser, I.D., Kozitsin, V.O.: Skoltech anomaly benchmark (SKAB). https://www.kaggle.com/dsv/1693952 (2020). https://doi.org/10.34740/KAGGLE/DSV/1693952. [Online; accessed 14-April-2021]
17. Lehmann, E., Romano, J.: Testing Statistical Hypotheses. Springer Texts in Statistics. Springer New York (2006). URL https://books.google.de/books?id=K6t5qn-SEp8C
18. Li, D., Chen, D., Jin, B., Shi, L., Goh, J., Ng, S.K.: MAD-GAN: Multivariate anomaly detection for time series data with generative adversarial networks. In: International Conference on Artificial Neural Networks, pp. 703–716. Springer (2019)
19. Liu, F.T., Ting, K.M., Zhou, Z.H.: Isolation forest. In: 2008 eighth IEEE international conference on data mining, pp. 413–422. IEEE (2008)
20. Malhotra, P., Ramakrishnan, A., Anand, G., Vig, L., Agarwal, P., Shroff, G.: LSTM-based encoder-decoder for multi-sensor anomaly detection. arXiv preprint arXiv:1607.00148 (2016)
21. Niu, Z., Yu, K., Wu, X.: LSTM-based VAE-GAN for time-series anomaly detection. Sensors **20**(13), 3738 (2020)
22. Park, D., Hoshi, Y., Kemp, C.C.: A multimodal anomaly detector for robot-assisted feeding using an LSTM-based variational autoencoder. IEEE Robotics and Automation Letters **3**(3), 1544–1551 (2018)
23. Provotar, O.I., Linder, Y.M., Veres, M.M.: Unsupervised anomaly detection in time series using LSTM-based autoencoders. In: 2019 IEEE International Conference on Advanced Trends in Information Theory (ATIT), pp. 513–517. IEEE (2019)
24. von Rueden, L., Mayer, S., Beckh, K., Georgiev, B., Giesselbach, S., Heese, R., Kirsch, B., Walczak, M., Pfrommer, J., Pick, A., Ramamurthy, R., Garcke, J., Bauckhage, C., Schuecker, J.: Informed machine learning - a taxonomy and survey of integrating prior knowledge into learning systems. IEEE Transactions on Knowledge and Data Engineering pp. 1–1 (2021). https://doi.org/10.1109/tkde.2021.3079836

25. Song, X., Wu, M., Jermaine, C., Ranka, S.: Conditional anomaly detection. IEEE Transactions on knowledge and Data Engineering **19**(5), 631–645 (2007)
26. Su, Y., Zhao, Y., Niu, C., Liu, R., Sun, W., Pei, D.: Robust anomaly detection for multivariate time series through stochastic recurrent neural network. In: Proceedings of the 25th ACM SIGKDD International Conference on Knowledge Discovery & Data Mining, pp. 2828–2837 (2019)
27. Weichert, D., Link, P., Stoll, A., Rüping, S., Ihlenfeldt, S., Wrobel, S.: A review of machine learning for the optimization of production processes. The International Journal of Advanced Manufacturing Technology **104**(5), 1889–1902 (2019). https://doi.org/10.1007/s00170-019-03988-5
28. Wilson, A.G., Izmailov, P.: Bayesian deep learning and a probabilistic perspective of generalization. arXiv preprint arXiv:2002.08791 (2020)
29. Wuest, T., Weimer, D., Irgens, C., Thoben, K.D.: Machine learning in manufacturing: advantages, challenges, and applications. Production & Manufacturing Research **4**(1), 23–45 (2016)
30. Zhu, L., Laptev, N.: Deep and confident prediction for time series at uber. In: 2017 IEEE International Conference on Data Mining Workshops (ICDMW), pp. 103–110. IEEE (2017)
31. Zong, B., Song, Q., Min, M.R., Cheng, W., Lumezanu, C., Cho, D., Chen, H.: Deep autoencoding gaussian mixture model for unsupervised anomaly detection. In: International conference on learning representations (2018)